AF179799

Solar Sailing for Deep-Space Explorations

Chen Gao · Wei Wang

Solar Sailing for Deep-Space Explorations

Orbit Design, Station Keeping, and Formation Flying

Springer

Chen Gao
School of Mechanics and
Transportation Engineering
Northwestern Polytechnical University
Xi'an, China

Wei Wang
School of Aeronautics and Astronautics
Shanghai Jiao Tong University
Shanghai, China

ISBN 978-981-95-0108-3 ISBN 978-981-95-0109-0 (eBook)
https://doi.org/10.1007/978-981-95-0109-0

© The Editor(s) (if applicable) and The Author(s), under exclusive license to Springer Nature Singapore Pte Ltd. 2026

This work is subject to copyright. All rights are solely and exclusively licensed by the Publisher, whether the whole or part of the material is concerned, specifically the rights of translation, reprinting, reuse of illustrations, recitation, broadcasting, reproduction on microfilms or in any other physical way, and transmission or information storage and retrieval, electronic adaptation, computer software, or by similar or dissimilar methodology now known or hereafter developed.
The use of general descriptive names, registered names, trademarks, service marks, etc. in this publication does not imply, even in the absence of a specific statement, that such names are exempt from the relevant protective laws and regulations and therefore free for general use.
The publisher, the authors and the editors are safe to assume that the advice and information in this book are believed to be true and accurate at the date of publication. Neither the publisher nor the authors or the editors give a warranty, expressed or implied, with respect to the material contained herein or for any errors or omissions that may have been made. The publisher remains neutral with regard to jurisdictional claims in published maps and institutional affiliations.

This Springer imprint is published by the registered company Springer Nature Singapore Pte Ltd.
The registered company address is: 152 Beach Road, #21-01/04 Gateway East, Singapore 189721, Singapore

If disposing of this product, please recycle the paper.

Preface

Solar sailing is a groundbreaking propulsion technology that uses the solar radiation pressure to generate low thrust, eliminating the need for traditional chemical propellants. By deploying large and reflective sails to capture and reflect solar radiation, solar sails provide continuous acceleration, making them ideal for long-duration space missions. Solar sails have diverse applications in deep space explorations. In the Earth-Moon system, they enable cost-effective lunar missions, supporting artificial orbits and station keeping for cis-lunar observation or communication satellites. Beyond the Moon, solar sails offer promise for interplanetary missions, such as monitoring solar activity from unique vantage points in the Sun-Earth system or exploring planets and asteroids via displaced heliocentric orbits. Their capability for formation flying in deep space also opens new possibilities for distributed observatories and multi-satellite missions, where synchronized operations are essential. This book explores the intricacies of solar sail orbit design, station keeping, and formation flying, offering a comprehensive framework for understanding the dynamics and control of solar sailing in diverse and complex gravitational environments. By covering both theoretical foundations and practical applications, it serves as a definitive guide for researchers, engineers, and mission planners aiming to harness the full potential of solar sailing in the next era of deep space explorations.

This book summarizes the authors' recent work on solar sailing, with a particular emphasis on its applications in deep space explorations. Chapter 1 provides a comprehensive introduction to solar sailing, underscoring its importance in deep space missions and offering a detailed exploration of orbit dynamics and control. Chapters 2–5 examine the motion of solar sails in the cis-lunar space, investigating various Earth-Moon dynamic models. These chapters focus specifically on the applications of solar sailing for the design and station-keeping of artificial orbits, highlighting the potential of solar sails for observation, navigation, and communication in cis-lunar missions. In Chaps. 6–9, the research shifts to the topic of solar sail formation flying in the Earth-Moon, Sun-Earth/Moon, and heliocentric systems. These chapters explore solar sail formation from the perspectives of coupled orbit-attitude optimal control, distributed control architectures, and consensus strategies under different communication typologies while also addressing the bounds of relative motion in such formations. This broadens the scope of solar sailing's potential for complex, multi-satellite missions and collaborative space operations in various gravitational environments.

The authors have received valuable comments from colleagues and friends, particularly from our supervisors: Prof. Jianping Yuan (Northwestern Polytechnical University), Prof. Josep Masdemont (Universitat Politècnica de Catalunya), Prof. Gerard Gómez (Universidad de Barcelona), Prof. Giovanni Mengali and Prof. Alessandro Antonio Quarta (Università di Pisa), and Prof. Hexi Baoyin (Tsinghua University). In addition, we express our gratitude for the financial support from the China Scholarship Council (Grant Nos. 201506290133 and 201806290088), the Fundamental Research Funds for the Central Universities (Grant No. D5000240142), and the National Natural Science Foundation of China (Grant Nos. 11902175 and 12272223).

Xi'an, China — Chen Gao
Shanghai, China — Wei Wang
May 2025

Contents

Chapter 1
Introduction

1.1 Solar Sailing

The Russian scientist Konstantin Tsiolkovsky (1857–1935) predicted that *"Mankind will not always stay on the Earth; the pursuit of light and space will lead him to penetrate the bounds of the atmosphere, timidly at first, but in the end to conquer the whole of solar space"*.[1] Since the successful launch of the first satellite, Sputnik-1, in 1957, humanity has achieved remarkable milestones in artificial satellite missions and human spaceflight. Driven by rapid advancements in science and technology, and motivated by the desire to maintain a competitive edge in the space race, space agencies have increasingly focused on deep space exploration activities. In the 21st century, numerous deep space missions and projects have been proposed and executed. For example, National Aeronautics and Space Administration (NASA) has launched Artemis program for the Earth-Moon system, the New Frontiers program for the Solar System, and the James Webb Space Telescope for the study of the Universe. Additionally, the China National Space Administration (CNSA) has carried on the Chang'e project for lunar exploration and the Tianwen-1 mission to Mars. The European Space Agency (ESA) has conducted the Rosetta Mission for comet exploration and the Solar Orbiter mission to study the Sun, while the Japan Aerospace Exploration Agency (JAXA) has undertaken the Hayabusa and Hayabusa-2 missions targeting near-Earth asteroids. Deep space exploration missions inherently require spacecraft to travel vast distances from Earth, often spanning decades. One of the most critical challenges in these missions is the propulsion system, particularly given that traditional chemical-based propulsion technologies rely on finite fuel supplies, thereby limiting mission duration and range. To address these limitations, several new propulsion technologies have been developed, among which solar sailing emerges as a particularly promising option.

Solar sailing is a propulsion technique that utilizes the interaction of solar photons with a reflective membrane to achieve thrust. Although individual photons have no

[1] K. Tsiolkovsky. Beyond the Planet Earth, 1920.

© The Author(s), under exclusive license to Springer Nature Singapore Pte Ltd. 2026

C. Gao and W. Wang, *Solar Sailing for Deep-Space Explorations*,
https://doi.org/10.1007/978-981-95-0109-0_1

mass, they carry momentum. When they strike a reflective surface, such as a sail, they impart a tiny force, i.e. solar radiation pressure (SRP) force, that can be harnessed for propulsion. Based on this principle, a solar-sail spacecraft typically comprises three key components: payloads, a deployment mechanism, and a sail. The sail is constructed from a highly reflective film designed with a large surface-to-mass ratio, which significantly enhances its efficiency in harnessing solar radiation for propulsion. Therefore, compared to traditional propulsion systems, the key advantages of solar sailing include continuous thrust, infinite specific impulse, and independence from fuel consumption. Unlike chemical rockets, which provide a large thrust but quickly exhaust their fuel, solar sailing offer continuous, albeit small, thrust. Over time, this sustained acceleration enables spacecraft to reach high velocities, whichi is crucial for deep space explorations. In addition, rocket engines is limited by the amount of fuel carried onboard. Solar sailing, however, do not require fuel, resulting in an effectively infinite specific impulse. As long as the spacecraft is within the influence of sunlight and the sail film maintains its integrity, it can continue to accelerate. Moreover, since solar sailing relies on sunlight, it eliminates the need to carry heavy fuel loads, thereby reducing the spacecraft's mass and increasing its potential range and flexibility for deep space missions.

In addition, from a dynamical perspective, deep space activities encounter more complex dynamical environments compared to those in the near-Earth environment. This complexity is particularly evident when solar radiation pressure acceleration is incorporated into the dynamical models that govern the equations of motion of the spacecraft. For near-Earth spacecraft, the reference mathematical model used to study the dynamics is the Two Body Problem or its low-perturbation variant due the fact that the gravitational force from the central body significantly outweighs any perturbations. However, for deep space spacecraft, a suitable mathematical model is the high-perturbation Two-Body Problem, which may include variations such as the SRP-perturbed Two Body Problem, the Restricted Three Body Problem (RTBP), the Restricted Four Body Problem (RFBP), or even ephemeris models. Therefore, deep space exploration missions present significantly greater challenges in terms of trajectory design, orbit and attitude control, mission analysis, and other related aspects.

This book explores the unique advantages of solar sailing technology and discusses its applications in deep space exploration, particularly within the Earth-Moon, Sun-Earth, and Sun-planets systems. It delves into the complexities of dynamics and control of solar-sail spacecraft in these dynamical environments, with a special focus on orbit design, station keeping, and formation flying. Through detailed analysis, this work aims to enhance the understanding of solar sail applications for deep space missions, offering insights into the potential for advanced missions that would be otherwise difficult or impossible to achieve with conventional propulsion systems.

1.2 The State of Art

Although solar sailing has only recently been regarded as a practical continuous low-thrust propulsion system, its origins trace back to the early 20th century with Konstantín Tsiolkovsky and his collaborator Friedrich Tsander. In 1924, Tsander was the first to publish a practical paper on solar sailing, inspired by Tsiolkovsky's proposition of accelerating spacecraft using sunlight in 1921. Despite this early work, the concept of solar sailing did not attract much attention over the next three decades. It was not until 1950 that Carl Wiley and Richard Garwin reintroduced and published the idea of solar sailing [1]. In recent decades, solar sailing has experienced a resurgence, driven by its potential applications and successful demonstration missions. Notable missions include JAXA's Interplanetary Kite-craft Accelerated by Radiation of the Sun (IKAROS) [2], NASA's NanoSail-D2 [3] and the Planetary Society's LightSail-1 and LightSail-2 [4, 5], and SIASAIL-I [6], all illustrated in Fig. 1.1. Some details of these four solar sailing missions are in the following:

1. **IKAROS**: IKAROS is launched by JAXA in 2010 and is the first solar-sail spacecraft to successfully complete orbit missions. The spacecraft has a total mass of 307 kg, with a deployable sail area of approximately 196 m^2. The sails were deployed and tensioned through four tip points, each with a mass of 0.5 kg. Thin-film solar cells are integrated into the sails to provide energy for the spacecraft. IKAROS is successfully deployed on June 10, 2010, and it verifies the reflectivity control devices on orbit and attitude control by adjusting the reflectivity of 80 electrochromic material panels located on the outer edge of the sails. Additionally, IKAROS demonstrates a successful flyby of Venus, further proving the viability of solar sailing technology.
2. **NanoSail-D2**: NanoSail-D2 is a small satellite built by NASA's Marshall Space Flight Center and Ames Research Center to study the deployment of a solar sail in space. It is a three-unit CubeSat, measuring 30 cm $\times$ 10 cm $\times$ 10 cm with a mass of 4 kg. Launched in 2010, the mission aimed to showcase the deployment and operation of a solar sail in low-Earth orbit. The spacecraft features a 10 m^2 sail made from ultra-thin reflective material. After a delayed deployment, NanoSail-D2 successfully unfolds its sail, marking it one of the first successful solar sail missions in Earth's orbit. The mission demonstrated solar sailing's potential for deorbiting space debris and passive propulsion, paving the way for future advancements in space propulsion technology.
3. **LightSail-1**: LightSail-1, launched by the Planetary Society in May 2015, is a groundbreaking solar sail mission designed to demonstrate solar sailing technology in low Earth orbit. Weighing approximately 4.5 kg, the spacecraft featured a sail with a total area of about 32 m^2, made from Mylar, a lightweight and reflective material. Although the primary goal is to deploy and unfurl the sail, the mission faces technical challenges that prevent full sail deployment. Nevertheless, LightSail-1 successfully validates the technologies necessary for future solar sail missions and laid the groundwork for subsequent missions.

Fig. 1.1 Demonstration missions for solar sailing: IKAROS (Top-left) [2], NanoSail-D2 (Top-right) [3], LightSail-1 (Middle-left) [4], and LightSail-2 (Middle-right) [5], SIASAIL-I (Bottom) [6]

4. **LightSail-2**: Following LightSail-1, LightSail-2 is launched on June 25, 2019. It is notable for being the first spacecraft to be driven entirely by SRP in Earth's orbit. On July 31, 2019, the Planetary Society officially announced that the orbital altitude of LightSail-2 had increased measurably, demonstrating that solar sailing is a viable propulsion technology for CubeSats. As of now, LightSail-2 remains in orbit, with its altitude having gradually reduced from 730 km to 630 km.
5. **SIASAIL-I**: SIASAIL-I is based on a 6U CubeSat, designed and produced by the Shenyang Institute of Automation, Chinese Academy of Sciences. The deployment mechanism of SIASAIL-I consists of a two-stage deployment system, which enables directional transformation, stretching, and membrane deployment within the limited space of the CubeSat. The system occupies only 0.5U (48 mm $\times$ 90 mm $\times$ 95 mm) and has a total mass of less than 1 kg. It was successfully

launched from the Xiaoxiang-1 07 satellite on August 31, 2019, and successfully deployed a 0.78 m $\times$ 0.78 m sail in Low Earth Orbit in December.

1.2.1 Orbit Design

Celestial mechanics traditionally focuses on the study of how gravitational forces influence the orbital motion between celestial bodies. A fundamental principle in this field is that gravitational force follows an inverse-square law, whereby the strength of the force is proportionally to the square of the distance between the objects. Of particular interest is the special case in which one celestial body, such as the Earth, possesses an infinitesimal mass with respect to a significantly more massive body, such as the Sun. This simplification enables the motion analysis of a smaller body under the gravitational influence of other larger ones, without accounting for the gravitational effects of the smaller body on the larger masses. These systems exhibit a variety of *natural* orbits, such as Keplerian orbits in Kepler's problem and libration point orbits in the Restricted Three-Body Problem (RTBP). Unlike natural celestial bodies, the artificial spacecraft is equipped with propulsion systems, allowing its motion to be influenced not only by gravitational forces from celestial bodies, but also by thrust generated by the spacecraft itself. Consequently, the spacecraft orbit can be designed and controlled artificially, offering *artificial* orbits advantages that are unparalleled compared to natural ones.

The peculiar propulsion mechanism of a solar sail enables the fulfilment of a class of high-energy orbits [7] as, for example, rectilinear trajectories [8–10], or closed orbits whose orbital plane does not contain the primary body [11]. Those non-Keplerian trajectories may be quasi-periodic orbits around artificial Lagrange points [12], useful for monitoring solar plasma storms [13], or heliocentric/planetocentric displaced orbits, capable of observing the special region of a celestial body [14]. After the first publication of solar sail-based (circular) displaced orbits by McInnes [15, 16], extensive studies have been carried out to investigate the dynamics and control problems for such orbits [17–19]. Based on the two-body assumption, the families of circular displaced orbits have been thoroughly discussed [20]. It has been shown that the inherently unstable families of displaced orbits can be rendered linearly stable by some simple control laws [21, 22]. In particular, for solar sail coated with electrochromic material panels [23], the elliptic displaced orbits can be achieved since the propulsive acceleration modulus can be modulated within a suitable range with reflectivity control devices [24, 25]. Moreover, the high-performance solar sail also facilitates the families of angular momentum reversal orbits [26–30].

Within the framework of the RTBP, such as the dynamics of spacecraft in the Sun-Earth/Moon and Earth-Moon systems, solar sailing can be employed to generate and sustain artificial libration points and nearby (quasi-) periodic orbits. These are of particular interest for mission applications over extended time intervals. We begin by considering solar sailing in the Sun-Earth/Moon system. McInnes [31, 32] first discussed the existence and stability of artificial libration points in the Sun-Earth/Moon

RTBP, for both ideal and non-ideal solar sails. Baoyin et al. [33] studied the effects of SRP on the distribution of artificial libration points taking the eccentricity of the Sun-Earth/Moon system into consideration. According to the stability analysis, it is expected that (quasi-) periodic orbits exist around the artificial libration points [34–36]. Baoyin et al. [37] studied solar-sail halo orbits at artificial Sun-Earth/Moon L_1 point, and provided third-order approximations using the Lindstedt-Poincaré method. Later, Waters et al. [35] computed more families of orbits for a solar sail around an arbitrary artificial libration point high above the ecliptic plane. Besides, solar-sail periodic orbits were designed by Biggs el al. [38] and Gong et al. [39], in the elliptic Sun-Earth/Moon system, for Earth's pole observation. For the transfers between artificial LPOs, Soldini et al. [40] and Duan [34] proposed a solar-sail maneuver method to perform heteroclinic connections between Lissajous orbits. Waters et al. [41, 42] discussed the invariant manifolds associated to the artificial LPOs and their applications for homoclinic/heteroclinic transfers. From a dynamical point of view, and using dynamical systems tools, the Sun-Earth/Moon-sail system was investigated by Farrés et al. [43–45]. Starting from the 2D family of artificial libration points parameterized by the sail orientation, Farrés found a high order approximation of the motion on the center manifold performing the reduction to center manifold, and computed the families of periodic orbits by means of a continuation method, including planar and vertical periodic orbits and halo-type ones.

While extensive research has been conducted on solar sailing in the Sun-Earth/Moon system, studies within the Earth-Moon system remain relatively scarce. Compared to the Sun-Earth/Moon-sail system, analyzing the motion of a sail in the Earth-Moon system is more complex due to the non-autonomous nature of the problem. This complexity arises from the fact that the Sun rotates around the Earth-Moon barycenter, introducing time-dependent dynamics into the model. In 1993, McInnes [46] was the first to discuss the motion of a solar-sail spacecraft in the vicinity of the Earth-Moon L_2 point, computing displaced periodic orbits whose displacement is determined by the sail attitude angles, which can be used for lunar far-side communications. This work was completed and extended by Simo et al. [47]. The idea was to analytically find particular solutions of the linear equations of motion, including the SRP acceleration, around the L_2 point. Gong et al. [48] further introduced the inclination of the Moon's orbital plane with respect to the ecliptic plane into the linearized dynamical model and generated the quasi-periodic orbits. In particular, when the inclination is set to zero, the quasi-periodic orbits degenerate into the periodic ones discussed by McInnes [46] and Simo et al. [47]. In relation with the motion around the Earth-Moon triangular points, Qian et al. [49] summarized the one-third subharmonic resonance period orbits following the aforementioned works. By adjusting the frequency of the sail's attitude angles to match the natural frequency of the natural system, Chujo [50] proposed a quasi-periodic orbit, which can be achieved with small lightness number, above the L_2 point. When the motion regime is no longer limited to the vicinity of the libration points, the full nonlinear equations have to be exploited within the framework of the Earth-Moon RTBP model, where semi-analytical and numerical implementations are usually necessary. Ozimek et al. [51] investigated the possibility of solar sailing for lunar south pole coverage in a SRP perturbed RTBP model and some of the most interesting solar-sail orbits were refined in a model that considered numerical ephemerides for the location of the two primaries. Furthermore, this work was also completed and extended by Wawrzyniak et al. [52]. More recently, Heiligers et al. [53, 54] has taken a further step in the understanding of the natural dynamics of a solar sail in the Earth-Moon Circular RTBP, computing some families of periodic orbits departing from resonant periodic Lyapunov and halo orbits from a dynamical point of view. In this work, the families were computed solving a two-point boundary value problem and using a continuation procedure, where the lightness number and the pitch angle of the sail were the continuation parameters. Concerning the same problem, Jorba-Cuscó et al. [55–57] take a further step by considering two more realistic dynamical models that incorporate the gravitational effect of the Sun: the Bicircular Problem (BCP) and Quasi-Bicircular Problem (QBCP) models, both of which also include the SRP acceleration. In fact, including the Sun's gravity into the Earth-Moon-sail system does not increase the dynamical complexity of the model, since the SRP depends periodically on time with the same period as the Sun around the Earth-Moon system. Using a continuation method, Jorba-Cuscó et al. studied the evolution of the families of resonant periodic orbits, including the shape and linear stability, considering the SRP acceleration.

The way of thrust generation of solar sailing dictates that the magnitude of the SRP acceleration is quite small, and its direction must be opposite to that of the Sun, which limits its self-control capacity. Compared to solar sailing, a SEP system is another low-thrust technology which possesses a high specific impulse but relies on a finite onboard propellant source. This mechanism makes SEP system useful for some high-energy missions but unsuitable for long-term ones. Based on these concerns, Leipold and Götz [58] proposed a hybrid sail propulsion that combines a solar sail and a SEP system on the same spacecraft. In this way, the two separate systems complement each other: the SEP system can provide thrust large enough and in almost any direction, in particular towards the Sun where the solar sail is unable to generate thrust, meanwhile, the solar sail can provide part of the acceleration in order to save propellant and lower the demand on the SEP system. We should note that for a hybrid-sail spacecraft part of the sail area is covered by flexible thin-film solar cells (TFSC) which act as a power source for the SEP system and other subsystems [58]. Such hybrid systems have already been suggested for many applications, including, for instance, supporting displaced geostationary orbits [59], optimal trajectories for Earth pole sitters [60], generation of the artificial libration points in the Sun-Earth/Moon system [61], and orbital control of the displaced orbits in the Earth-Moon system [62].

1.2.2 Station Keeping

Station keeping refers to the process of maintaining a spacecraft's position or orbit within a specified region over time. In traditional propulsion systems, this is accomplished through fuel-based methods, such as thrusters. In contrast, solar sailing achieves station keeping by utilizing solar radiation pressure (SRP) acceleration. This approach offers distinct advantages and challenges compared to conventional propulsion systems, as it eliminates the need for propellant and enables continuous operation over extended durations. From a methodological perspective, station-keeping strategies can be categorized into two main types: the first involves the application of control theory techniques to address specific station-keeping challenges, while the second leverages spacecraft orbit dynamics and employs dynamical system tools to stabilize the motion [63].

First we present some work on the first station-keeping strategy. Bookless et al. [64] considered using solar sail propulsion to provide station-keeping at quasi-periodic orbits around Sun-Earth L_1 and L_2 points. The optimal control laws were developed to minimize the control requirements and possible control techniques, including solar sail area variation and solar sail pitch and yaw angle variation, were investigated. Aiming at the SRP-assisted station-keeping of Lissajous and halo orbits around the Sun-Earth/Moon L_1 point, Lou et al. [65] designed an active disturbance rejection controller. Specifically, the sail is equipped with an RCD, and therefore, its reflectivity rate is considered as a control variable, in addition to the two sail angles. Shahid et al. [66] investigated the use of SRP for station-keeping at the Sun-Earth/Moon L_2 point. They developed a nonlinear higher-order feedback controller that accommodates nonaffine SRP inputs. Both underactuated (only sail pitch and roll angle variations) and fully actuated (area variation and two angle variations) control cases were tested, showing that underactuated control can maintain bounded tracking errors, while fully actuated control achieves tracking errors on the order of a few meters. Focusing on solar-sail station-keeping in the L_4 region of binary asteroids, Guzzetti et al. [67] first constructed reference solutions within a higher-fidelity modified Elliptic Restricted Three-Body Problem (ERTBP). They then implemented a turn-and-hold station-keeping strategy, a variant of the prediction and correction method, to offset orbit determination errors. Considering the degradation of the sail's optical properties, Huang et al. [68] proposed an integrated guidance and control strategy that updates the reference orbit based on in-situ estimation, coupled with an active disturbance rejection controller. In this strategy, attitude angles and reflective rate serve as the control variables. Moore et al. [69] addressed the problem of controlling a solar sail in a stable hover/orbit after the deployment of a series of small probes or a single large lander. The problem is characterized by an instantaneous change in the sail's acceleration. The linear quadratic regulator and time-delayed feedback control are designed to accommodate gradual small changes and significant instantaneous changes in the sail's characteristic acceleration, respectively.

Now, we turn to the second station-keeping strategy. Although research in this area is relatively scarce compared to the first strategy, we have still gathered and analyzed several relevant studies. A structure-preserving controller was developed by Xu et al. [70] to stabilize a hyperbolic solar-sail Hamiltonian system. Within the framework of this controller, the invariant (stable, unstable, and center) manifolds of the equilibrium are used to stabilize the Hamiltonian system. Using this controller, a stable Lissajous orbit is generated for the planar solar-sail system, ensuring its bounded motion. Then, the Hamiltonian structure-preserving control was extended by Soldini et al. [71] using the SRP acceleration for the stabilization of unstable periodic orbits in the Sun-Earth Circular Restricted Three Body Problem (CRTBP). High-amplitude orbits with complex and conjugate eigenvalues were considered, and the most efficient control parameters were found to be the reflective area and the in-plane angle. Farrés et al. [72] first parameterized the 2D family of equilibria using the two sail angles in the Sun-Earth CRTBP-SRP model and designed a control strategy that utilizes knowledge of the position of the invariant manifolds and how they vary with changes in sail orientation. They then extended their work to the ERTBP-SRP model and demonstrated that similar station-keeping strategies are also effective [73]. More extensions on this station keeping method can be found in [74, 75]. Duan et al. [76] recently proposed a station-keeping strategy using solar sail maneuvers, achieved by changing the sail parameters, which can be understood as 'jumps' in position rather than in velocity within the phase space. Based on a systematic analysis of the impact of a maneuver (an instantaneous sail reorientation) on the motion of spacecraft, station-keeping can be achieved by periodically performing maneuvers to prevent the spacecraft from escaping.

1.2.3 Formation Flying

Althoug the use of solar sail has been widely discussed in space missions, some scientific tasks require an extremely large reflector area that cannot be realized in pracitcal operations, rendering it difficult to compromise between technical requirements and current manufacturing level. A feasible solution to the problem is to disaggregate the payload among multiple spacecraft, with each functional module conveying the indispensable mass only. As thus, the characteristic acceleration of a solar sail would be significantly increased with a reduction of the payload mass. In addition, the concept of solar sail formation flying provides a means for increasing the observation capability, by arranging multiple sails in a formation to construct a virtual sensor to achieve greater resolution than a single one.

The application of solar sail formation for in-situ observations of geomagnetic tail in magnetosphere missions entails the elaborated selection of the reference orbit, and the nominal Sun-synchronous orbit was suggested as a favorable option for formation deployment [77–79]. In fact, the employment of invariant relative motion conditions were proved to be effective for designing the semi-natural (loose) solar sail formations [80–82]. To enable the prescribed (strict) relative motion geometry, active formation control laws, using the electrochromic devices for reflectivity modulation of the sail, are indispensable [83]. The concept of solar sail formation around displaced orbits [84–86] and artificial Lagrange point orbits [87–89] was also discussed, it was shown that the relative motion between two sails (referred to as chief and deputy, or leader and follower) can be stabilized with simple control laws, and the associated stable regions were numerically found for semi-natural formations. However, inherent weakness exists in such a chief-deputy topology, because the unique chief sail, which represents the only information source about the reference state for the deputy, is a single point of the failure. In this case, the whole system is not able to effectively cope with unexpected emergencies. Another limitation associated with the chief-deputy strategy is the absence of a mutual feedback information flow throughout the formation structure. As a result, unfavorable situations would arise if a fault takes place in the communication links. A possible improvement consists in including the information exchange among the deputies into the feedback control. To this end, the formation architecture was extended to consists of multiple (i.e. more than two) solar sails, and a series of distributed steering control laws were developed via consensus [90, 91].

1.3 Organization of the Book

Thus far, we have provided a brief introduction to the orbit dynamics and control of solar sailing from the perspectives of orbit design, station-keeping, and formation flying. For a more comprehensive overview of solar sailing, the reader is referred to Refs. [7, 92–94], wherein attitude and structural motion of solar sail, as well as mission application aspects are included. The content of this book is organized into ten chapters, covering topics such as orbit design, station-keeping, and formation flying in the Earth-Moon, Sun-Earth, and heliocentric systems, described as follows.

- Chapter 2 investigates the motion of a hybrid sail around the cis-lunar libration points within the Earth-Moon ERTBP framework, focusing on both the computation and station-keeping of families of displaced orbits. The equations of motion for the hybrid sail are derived and analyzed, considering the eccentricities of the Earth-Moon and Sun-Earth/Moon systems, as well as the inclination angle of the Earth-Moon orbital plane relative to the ecliptic. Given that the eccentricities and inclination are small, quasi-periodic displaced orbits are generated by solving the first-order approximations of the equations of motion. The displacement of these quasi-periodic orbits is determined by the reflectivity rate and two attitude angles of the sail. The model errors arising from linearization are quantitatively analyzed. Considering both the model errors and the inherent instability of motion around the cis-lunar libration points, a backstepping controller is proposed for orbital maintenance. The sail's reflectivity rate and orientation are optimized to minimize the overall propellant consumption of the SEP system.
- In Chap. 3, we analyze the motion of a hybrid sail in the Earth-Moon CRTBP, focusing on the evolution of families of resonant periodic orbits and the development of an orbit-attitude station-keeping controller. The force and torque models

for a hybrid sail equipped with reflectivity control devices (RCD) are established, with the RCD serving as the orbit-attitude control actuator. The orbital and attitude equations of motion are derived within the CRTBP framework. By incorporating the SRP acceleration into the orbital equations of motion, the evolution of resonant halo and Lyapunov orbits is analyzed as the reflectivity rate and attitude angles are varied. A coupled orbit-attitude control strategy is proposed, consisting of three components: an optimal periodic orbital controller, solar electric propulsion acceleration optimization, and a robust backstepping attitude controller.

- Chapter 4 focuses on the study of the influence of SRP on the dynamics around the dynamical substitutes of the L_1 and L_2 points in the Earth-Moon QBCP. The equations governing the QBCP, including the SRP acceleration, are first established. Known results regarding the low-order resonant periodic orbits of the QBCP around the L_1 and L_2 points in the RTBP are then summarized. Starting from the dynamical substitutes of the two cis-lunar libration points and the low-order resonant periodic orbits, the evolution of families of resonant periodic orbits is studied as the sail parameters, which define its orientation and efficiency, are varied. In particular, the characteristic curves, maximal Floquet exponent, and linear normal behavior of the periodic orbits are analyzed. Furthermore, the interconnections between the families and the influence of the SRP on the stability of the orbits are explored.
- Most of the solar sail libration point orbits obtained in the QBCP-SRP model are unstable, and therefore, Chap. 5 focuses on the station keeping. Specifically, dynamical reshaping approaches are explored, with the goal of reshaping the dynamical structure around the nominal orbit and canceling the unstable Floquet modes using SRP acceleration. The lightness number and two orientation angles (the pitch and yaw angles) of the sail are chosen as control variables. Within the algorithms, two different nominal points, i.e. the isochronous point and the minimum distance point, are considered to examine the role of nominal points in the stability and convergence of the controller. Additionally, the Jet Transport technique is employed in the implementation of the procedure to derive the control law explicitly, represented as high-order Taylor polynomials in terms of the deviation between the spacecraft's state and the associated nominal point.
- Chapter 6 addresses the formation flying for spinning solar sails in the Earth-Moon CRTBP, where the follower is controlled to track a predefined trajectory around the leader while maintaining their attitudes alignment. This requires focusing on the coupled dynamics of orbit and attitude motion. Specifically, the RCD serves as the joint orbit and attitude control actuator, necessitating consideration of the saturation and coupling of orbit-attitude control variables. Additionally, constraints on the relative attitude angles, their rates of change, and the optimal switching of RCDs are considered. As a result, the problem becomes a constrained optimal coupled orbit-attitude control problem, which is then tackled using model predictive control by solving a convex quadratic programming problem with linear constraints.
- Chapter 7 develops a cooperative control paradigm for solar sail formation flying in the Sun-[Earth+Moon] CRTBP. The formation is deployed around an L_2-type

artificial equilibrium point, where a truncated Fourier series is used to parameterize the nominal artificial halo orbit. According to whether the topological structure is leaderless or accommodates a virtual leader, two typical cases are studied in detail, and the group consensus is guaranteed by utilizing protocols that are formulated on a general directed communication topology. In particular, the virtual leader case is further investigated depending on whether the state information of the virtual leader is accessible to all of the sails or to a part of the formation only. The coordinated architecture facilitates each sail to explore the mutual information couplings with its available adjacent neighbors, so that a synchronized formation tracking can be achieved while increasing the system flexibility.

- Chapter 8 establishes a distributed (chief-deputy) solar sail formation framework where the chief sail covers a heliocentric displaced orbit, whereas each deputy sail manages to track a prescribed relative trajectory by adjusting the sail attitude and reflectivity modulation ratio. For chief sail following a planet-following displaced orbit, the feasible reagions as a function of orbital parameters have been determined with a semi-analytical approach. In particular, distributed formation architectures have been designed for both the full state feedback case and the relative velocity-free case, the latter being accomplished with the incorporation of a passive filter that eliminates the relative velocity requirement. The underlying consensus protocols allow each deputy sail to update their relative states in a cooperative way, such that an asymptotic formation coordination can be assured even relying on local data interactions only. The presented approach circumvents an undesirable situation in which a malfunction of the chief sail would result in a failure of the whole formation system.
- Chapter 9 presents a methodology for modeling the relative motion between heliocentric circular displaced orbits. Similar to classical Keplerian orbital elements, a new set of (circular) displaced orbital elements has been proposed to characterize the sail-based spacecraft (e.g. solar sail and electric sail) motion along displaced orbits. The newly defined displaced orbital elements are then utilized to derive the closed-form solution to the relative motion problem between circular displaced orbits, without the need of linearizing or solving nonlinear equations. The obtained relative motion solution provides a straightforward parametric interpretation of invariant manifold in the configuration space, and the relative motion bounds, namely, extreme values of relative position components can be calculated from an analytical viewpoint, both for quasi-periodic orbits in the incommensurable case and periodic orbits in the 1 : 1 commensurable case. Moreover, in some degenerate cases, the extreme values of relative distance bounds can also be solved analytically. The closed-form expressions of quantitative relative distance bounds are of great significance for safe mission progress because the knowledge of maximum distance is crucial information for communication design, whereas the minimum distance is essential for collision-avoidance purposes.
- Paralleling the approach in Chaps. 9 and 10 proposes a semi-analytical approach for evaluating the extreme values of relative motion between two sail-based spacecraft that move along elliptic displaced orbits. For the incommensurable case, the radial, along-track and cross-track relative motion bounds correspond to the roots of a sort

of algebraic equations. Whereas for the 1 : 1 commensurable case, a Fourier series expansion is adopted to derive the time-explicit expression of spacecraft relative motion. In this case, the extreme values for the radial and along-track motion are the real roots of a sextic equation, whereas for the cross-track motion, they satisfy a quartic equation. The presented approach is able to predicte the inner and outer bounds of relative motion between elliptic displaced orbits. Compared to the pure numerical integration that is time consuming, this method is more efficient in computation.

References

1. Garwin R L. Solar sailing: A practical method of propulsion within the solar system [J]. Jet Propulsion, 1958, 28 (3): 188–190.
2. Tsuda Y, Mori O, Funase R, et al. Achievement of IKAROS-Japanese deep space solar sail demonstration mission [J]. Acta Astronautica, 2013, 82 (2): 183–188.
3. Johnson L, Young R, Montgomery E, et al. Status of solar sail technology within NASA [J]. Advances in Space Research, 2011, 48 (11): 1687–1694.
4. Betts B, Nye B, Vaughn J, et al. Lightsail 1 mission results and public outreach strategies [C]. 4th International Symposium on Solar Sailing, Kyoto, Japan, 2017.
5. Spencer D A, Betts B, Bellardo J M, et al. The LightSail 2 solar sailing technology demonstration [J]. Advances in Space Research, 2021, 67 (9): 2878–2889.
6. Liu J, Zhao P, Wu C, et al. SIASAIL-I solar sail: From system design to on-orbit demonstration mission [J]. Acta Astronautica, 2022, 192: 133–142.
7. McKay R J, Macdonald M, Biggs J. Survey of highly non-Keplerian orbits with low-thrust propulsion [J]. Journal of Guidance, Control, and Dynamics, 2011, 34 (3): 645–666.
8. Mengali G, Quarta A A. Optimal heliostationary missions of high-performance sailcraft [J]. Acta Astronautica, 2007, 60 (8–9): 676–683.
9. Quarta A A, Mengali G. Solar sail capabilities to reach elliptic rectilinear orbits [J]. Journal of Guidance, Control, and Dynamics, 2011, 34 (3): 923–926.
10. Quarta A A, Mengali G. Optimal solar sail transfer to linear trajectories [J]. Acta Astronautica, 2013, 82 (2): 189–196.
11. McInnes C R. Passive control of displaced solar sail orbits [J]. Journal of Guidance, Control, and Dynamics, 1998, 21 (6): 975–982.
12. Baoyin H, McInnes C R. Solar sail halo orbits at the Sun-Earth artificial L_1 point [J]. Celestial Mechanics and Dynamical Astronomy, 2006, 94 (2): 155–171.
13. Salazar F J T, McInnes C R, Winter O C. Intervening in Earth's climate system through space-based solar feflectors [J]. Advances in Space Research, 2016, 58 (1): 17–29.
14. Forward R L. Statite–A spacecraft that does not orbit [J]. Journal of Spacecraft and Rockets, 1991, 28 (5): 606–611.
15. McInnes C R, L. Simmons J F L. Solar sail halo orbits I: Heliocentric case [J]. Journal of Spacecraft and Rockets, 1992, 29 (4): 466–471.
16. McInnes C R, L. Simmons J F L. Solar sail halo orbits II: Geocentric Case [J]. Journal of Spacecraft and Rockets, 1992, 29 (4): 472–479.
17. McInnes C R. Dynamics, stability, and control of displaced non-Keplerian orbits [J]. Journal of Guidance, Control, and Dynamics, 1998, 21 (5): 799–805.
18. Bookless J, McInnes C R. Dynamics and control of displaced periodic orbits using solar-sail propulsion [J]. Journal of Guidance, Control, and Dynamics, 2006, 29 (3): 527–537.
19. Xu M, Xu S. Nonlinear dynamical analysis for displaced orbits above a planet [J]. Celestial Mechanics and Dynamical Astronomy, 2008, 102 (4): 327–353.

20. McInnes C R. The existence and stability of families of displaced two-body orbits [J]. Celestial Mechanics and Dynamical Astronomy, 1997, 67 (2): 167–180.
21. McInnes C R. Passive control of displaced solar sail orbits [J]. Journal of Guidance, Control, and Dynamics, 1998, 21 (6): 975–982.
22. Gong S, Li J, Baoyin H. Analysis of displaced solar sail orbits with passive control [J]. Journal of Guidance, Control, and Dynamics, 2008, 31 (3): 782–785.
23. Mengali G, Quarta A A. Heliocentric trajectory analysis of Sun-pointing smart dust with electrochromic control [J]. Advances in Space Research, 2016, 57 (4): 991–1001.
24. Gong S, Li J. Solar sail heliocentric elliptic displaced orbits [J]. Journal of Guidance, Control, and Dynamics, 2014, 37 (6): 2021–2026.
25. Tamakoshi D, Kojima H. Solar sail orbital control using reflectivity variations near the Earth-Moon L_2 Point [J]. Journal of Guidance, Control, and Dynamics, 2017, 41 (2): 1764–1776.
26. Vulpetti G. Sailcraft at high speed by orbital angular momentum reversal [J]. Acta Astronautica, 1997, 40 (10): 733–758.
27. Vulpetti G. General 3D H-reversal trajectories for high-speed sailcraft [J]. Acta Astronautica, 1999, 44 (1): 67–73.
28. Mengali G, Quarta A A, Romagnoli D, et al. H^2-reversal trajectory: A new mission application for high-performance solar sails [J]. Advances in Space Research, 2011, 48 (11): 1763–1777.
29. Zeng X, Baoyin H, Li J, et al. Three-dimensional time optimal double angular momentum reversal trajectory using solar sails [J]. Celestial Mechanics and Dynamical Astronomy, 2011, 111 (4): 415–430.
30. Zeng X, Alfriend K T, Vadali S R. Solar sail planar multireversal periodic orbits [J]. Journal of Guidance, Control, and Dynamics, 2014, 37 (2): 674–681.
31. McInnes C R, McDonald A J, Simmons J F. et al. Solar sail parking in restricted three-body systems [J]. Journal of Guidance, Control, and Dynamics, 1994, 17 (2): 399–406.
32. McInnes C R. Artificial Lagrange points for a partially reflecting flat solar sail [J]. Journal of Guidance, Control, and Dynamics, 1994, 22 (1): 185–187.
33. Baoyin H, McInnes C R. Solar sail equilibria in the elliptical restricted three-body problem [J]. Journal of Guidance, Control, and Dynamics, 2006, 29 (3): 538–543.
34. Duan X, Gómez G, Masdemont J J, et al. A picture of solar-sail heteroclinic enhanced connections between Lissajous libration point orbits [J]. Communications in Nonlinear Science and Numerical Simulation, 2020, 85, 105252.
35. Waters T J, McInnes C R. Periodic orbits above the ecliptic in the solar-sail restricted three-body problem [J]. Journal of Guidance, Control, and Dynamics, 2007, 30 (3), 687–693.
36. Verrier P, Waters T, Sieber J. Evolution of the L_1 halo family in the radial solar sail circular restricted three-body problem [J]. Celestial Mechanics and Dynamical Astronomy, 2014, 120 (4): 373–400.
37. Baoyin H, McInnes C R. Solar sail halo orbits at the Sun-Earth artificial L_1 point[J]. Celestial Mechanics and Dynamical Astronomy, 2006, 94, 155–171.
38. Biggs J D, McInnes C R, Waters T. Control of solar sail periodic orbits in the elliptic three-body problem [J]. Journal of Guidance, Control, and Dynamics, 2009, 32 (1): 318–320.
39. Gong S, Li J. Solar sail periodic orbits in the elliptic restricted three-body problem [J]. Celestial Mechanics and Dynamical Astronomy, 2015, 121, 121–137.
40. Soldini S, Masdemont J J, Gómez G. Dynamics of solar radiation pressure-assisted maneuvers between Lissajous orbits [J]. Journal of Guidance, Control, and Dynamics, 2019, 42 (4): 769–793.
41. Waters T J, McInnes C R. Invariant manifolds and orbit control in the solar sail three-body problem [J]. Journal of Guidance, Control, and Dynamics, 2008, 31 (3): 554–562.
42. Waters T J, McInnes C R. Solar sail dynamics in the three-body problem: Homoclinic paths of points and orbits [J]. International Journal of Non-Linear Mechanics, 2008, 43(6) 490–496.
43. Farrés A. Contributions to the dynamics of a solar sail in the Earth-Sun system [D]. Ph.D Dissertation, University of Barcelona, 2009.
44. Farrés A, Jorba À. Dynamics of a solar sail near a halo orbit [J]. Acta Astronautica, 2010, 67 (7–8): 979–990.

45. Farrés A, Jorba Á. Periodic and quasi-periodic motions of a solar sail close to SL_1 in the Earth-Sun system [J]. Celestial Mechanics and Dynamical Astronomy, 2010, 107, 233–253.
46. McInnes C R. Solar sail trajectories at the lunar L_2 Lagrange point [J]. Journal of Spacecraft and Rockets, 1993, 30(6): 782–784.
47. Simo J, McInnes C R. Solar sail orbits at the Earth-Moon libration points [J]. Communications in Nonlinear Science and Numerical Simulation, 2009, 14 (12): 4191–4196.
48. Gong S, Li J, Simo J. Orbital motions of a solar sail around the L_2 Earth-Moon libration point [J]. Journal of Guidance, Control, and Dynamics, 2014, 37 (4): 1349–1356.
49. Qian Y J, Liu Z X, Yang X D. Novel subharmonic resonance periodic orbits of a solar sail in Earth-Moon system [J]. Journal of Guidance, Control, and Dynamics, 2019, 42 (11): 2532–2540.
50. Chujo T. Quasi-periodic orbits of small solar sails with time-varying attitude around Earth-Moon libration points [J]. Astrodynamics, 2024, 8 (1): 161–174.
51. Ozimek M T, Grebow D J, Howell K C. Design of solar sail trajectories with applications to lunar south pole coverage [J]. Journal of Guidance, Control, and Dynamics, 2009, 32(6): 1884–1897.
52. Wawrzyniak G G, Howell K C. Numerical techniques for generating and refining solar sail trajectories [J]. Advances in Space research, 2011, 48 (11): 1848–1857.
53. Heiligers J, Hiddink S, Noomen R, et al. Solar sail Lyapunov and Halo orbits in the Earth-Moon three-body problem [J]. Acta Astronautica, 2015, 116: 25–35.
54. Heiligers J, Macdonald M, Parker J S. Extension of Earth-Moon libration point orbits with solar sail propulsion [J]. Astrophysics and Space Science, 2016, 361: 1–20.
55. Jorba-Cuscó M, Farrés A, Jorba Á. Periodic and quasi-periodic motion for a solar sail in the Earth-Moon system [C]. Proceedings of the 67th International Astronautical Congress, Guadalajara, México, 2016.
56. Jorba-Cuscó M, Farrés A, Jorba Á. Solar sail resonant periodic orbits in the augmented Earth-Moon Quasi-Bicircular Problem [C]. Proceedings of the 67th International Astronautical Congress, Bremen, Germany, 2018.
57. Jorba-Cuscó M, Farrés A, Jorba Á. On the stabilizing effect of solar radiation pressure in the Earth-Moon system [J]. Advances in Space Research, 2021, 67 (9): 2812–2822.
58. Leipold M, G'otz M, Hybrid photonic/electric propulsion[C]. Kayser-Threde, GmBH Report SOL4-TR-KTH-001, Munich, Germany, 2002.
59. Heiligers J, Ceriotti M, McInnes C R, et al. Displaced geostationary orbit design using hybrid sail propulsion [J]. Journal of Guidance, Control, and Dynamics, 2011, 34 (6): 1852–1866.
60. Ceriotti M, McInnes C R. Generation of optimal trajectories for Earth hybrid pole sitters [J]. Journal of Guidance, Control, and Dynamics, 2011, 34 (3): 847–859.
61. Baig S, McInnes C R. Artificial three-body equilibria for hybrid low-thrust propulsion [J]. Journal of Guidance, Control, and Dynamics, 2008, 31 (6): 1644–1655.
62. Tamakoshi D, Kojima H. Solar sail orbital control using reflectivity variations near the Earth-Moon L_2 point [J]. Journal of Guidance, Control, and Dynamics, 2018, 41 (2): 417–430.
63. Shirobokov M, Trofimov S, Ovchinnikov M. Survey of station-keeping techniques for libration point orbits [J]. Journal of Guidance, Control, and Dynamics, 2017, 40 (5): 1085–1105.
64. Bookless J, McInnes C. Control of Lagrange point orbits using solar sail propulsion [J]. Acta Astronautica, 2008, 62 (2–3): 159–176.
65. Lou Z, Zhang K, Wang Y, et al. Active disturbance rejection station-keeping control for solar-sail libration-point orbits [J]. Journal of Guidance, Control, and Dynamics, 2016, 39 (8): 1913–1917.
66. Shahid K, Kumar K D. Nonlinear station-keeping control in the vicinity of the Sun-Earth L_2 point using solar radiation pressure [J]. Journal of Aerospace Engineering, 2016, 29 (3): 04015073.
67. Guzzetti D, Sood R, Chappaz L, et al. Stationkeeping analysis for solar sailing the L_4 region of binary asteroid systems [J]. Journal of Guidance, Control, and Dynamics, 2019, 42 (6): 1306–1318.

68. Huang J, Biggs J D, Bai Y, et al. Integrated guidance and control for solar sail station-keeping with optical degradation [J]. Advances in Space Research, 2021, 67 (9): 2823–2833.
69. Moore I, Ceriotti M, McInnes C R. Station-keeping for a solar sail during lander/probe deployment using feedback control [J]. Acta Astronautica, 2022, 201: 182–197.
70. Xu M, Xu S. Structure-preserving stabilization for Hamiltonian system and its applications in solar sail [J]. Journal of Guidance, Control, and Dynamics, 2009, 32 (3): 997–1004.
71. Soldini S, Colombo C, Walker S J. Solar radiation pressure Hamiltonian feedback control for unstable libration-point orbits [J]. Journal of Guidance, Control, and Dynamics, 2017, 40 (6): 1374–1389.
72. Farrés A, Jorba Á. A dynamical system approach for the station keeping of a solar sail [J]. The Journal of the Astronautical Sciences, 2008, 56 (2): 199–230.
73. Farrés A, Jorba Á. On the station keeping of a solar sail in the elliptic Sun-Earth system [J]. Advances in Space Research, 2011, 48 (11): 1785–1796.
74. Farrés A, Jorba Á. Station keeping of a solar sail around a halo orbit [J]. Acta Astronautica, 2014, 94 (1): 527–539.
75. Farrés A, Jorba Á. Station keeping strategies for a solar sail in the solar system [J]. Recent Advances in Celestial and Space Mechanics, 2016, 23: 83–115.
76. Duan X, Gómez G, Masdemont J J, et al. Propellant-free station-keeping design of a solar sail around the Sun-Earth collinear equilibrium points [C]. Proceedings of the 73rd International Astronautical Congress, Paris, France, 2022.
77. Gong S, Gao Y, Li J. Solar sail formation flying on an inclined Earth orbit [J]. Acta Astronautica, 2011, 68 (1–2): 226–239.
78. Mu J, Gong S, Li J. Reflectivity-controlled solar sail formation flying for magnetosphere mission [J]. Aerospace Science and Technology, 2013, 30 (1): 339–348.
79. Parsay K, Schaub H. Designing solar sail formations in sun-synchronous orbits for geomagnetic tail exploration [J]. Acta Astronautica, 2015, 107: 218–233.
80. Parsay K, Schaub H. Drift-free solar sail formations in elliptical Sun-synchronous orbits [J]. Acta Astronautica, 2017, 139: 201–212.
81. Parsay K, Schaub H, Schiff C, et al. Improving magnetosphere in situ observations using solar sails [J]. Advances in Space Research, 2018, 61 (1): 74–88.
82. Parsay K, Schaub H. Preliminary results on optimal establishment of solar sail formations [J]. The Journal of the Astronautical Sciences, 2019, 66 (1): 32–45.
83. Mu J, Gong S, Li J. Coupled control of reflectivity modulated solar sail for GeoSail formation fFlying [J]. Journal of Guidance, Control, and Dynamics, 2015, 38 (4): 740–721.
84. Gong S, Baoyin H, Li J. Solar sail formation flying around displaced solar orbits [J]. Journal of Guidance, Control, and Dynamics, 2007, 30 (4): 1147–1151.
85. Gong S, Baoyin H, Li J. Relative orbit design and control of formation around displaced solar orbits [J]. Aerospace Science and Technology, 2008, 12 (2): 195–201.
86. Gong S, Li J, Baoyin H. Formation flying solar-sail gravity tractors in displaced orbit for towing near-Earth asteroids [J]. Celestial Mechanics and Dynamical Astronomy, 2009, 105 (1–3): 159–177.
87. Biggs J D, McInnes C R. Solar sail formation flying for deep-space remote sensing [J]. Journal of Spacecraft and Rockets, 2009, 46 (3): 670–678.
88. Gong S, Li J, Baoyin H. Relative motion around artificial Lagrange points [J]. Transactions of the Japan Society for Aeronautical and Space Sciences, 2009, 51 (174): 220–226.
89. Shahid K, Kumar K D. Formation control at the Sun-Earth L_2 libration point using solar radiation pressure [J]. Journal of Spacecraft and Rockets, 2010, 47 (4): 614–626.
90. Wang W, Mengali G, Quarta A A, et al. Multiple solar sail formation flying around heliocentric displaced orbit via consensus [J]. Acta Astronautica, 2019, 154: 256–267.
91. Wang W, Baoyin H, Mengali G, et al. Solar sail cooperative formation flying around L_2-type artificial equilibrium points [J]. Acta Astronautica, 2020, 169: 224–239.

92. Fu B, Sperber E, Eke F. Solar sail technology-A state of the art review [J]. Progress in Aerospace Sciences, 2016, 86: 1–19.
93. Zhao P, Wu C, Li Y. Design and application of solar sailing: A review on key technologies [J]. Chinese Journal of Aeronautics, 2023, 36 (5): 125–144.
94. Berthet M, Schalkwyk J, Celik O, et al. Space sails for achieving major space exploration goals: Historical review and future outlook [J]. Progress in Aerospace Sciences, 2024, 150: 101047.

Chapter 2
Motion of a Hybrid Sail Around the Cis-Lunar Libration Points Within the Earth-Moon Elliptic Restricted Three Body Problem

Displaced orbits around the collinear libration points in cis-lunar space offer significant advantages for observation and communication missions. This chapter analyzes the motion of a hybrid sail, which integrates a solar electric propulsion system with a solar sail, within the Earth-Moon Elliptic Restricted Three Body Problem. The conditions for hybrid sail displaced orbits are derived analytically, under the assumption that both the Earth-Moon system and the Sun-Earth/Moon system exhibit small orbital eccentricities. Due to the inherent instability of the quasi-periodic displaced orbits, a backstepping controller is designed to ensure long-term orbit maintenance while minimizing fuel consumption of the solar electric thruster. Illustrative examples demonstrate the effectiveness of the proposed approach.

2.1 Introduction

Solar sailing enables a class of high-energy displaced orbits that are difficult, if not impossible, to achieve with conventional propulsion systems [1]. For instance, heliocentric displaced orbits are valuable for observing the polar regions of celestial bodies [2], while orbits around artificial equilibrium points can be used for monitoring solar plasma storms [3] or relaying interplanetary communications [4]. In particular, displaced orbits around collinear libration points in the Earth-Moon system have garnered significant attention due to their unique advantages. For example, displaced orbits above the L_1 point, with appropriate displacement, can be used to monitor future human outposts [5], while those near the L_2 point can provide continuous telecommunication between Earth and the far side of the Moon [6, 7].

Thus far, considerable effort has been devoted to the study of displaced orbits in the circular Earth-Moon system using a solar sail [7–9], or a hybrid propulsion system that combines a solar sail with a solar electric propulsion (SEP) system [10, 11]. However, the frameworks used in these studies are based on the Circular Restricted

© The Author(s), under exclusive license to Springer Nature Singapore Pte Ltd. 2026
C. Gao and W. Wang, *Solar Sailing for Deep-Space Explorations*,
https://doi.org/10.1007/978-981-95-0109-0_2

Three Body Problem (CRTBP), neglecting the contributions of orbital eccentricities in the Earth-Moon and Sun-Earth/Moon systems. This oversight introduces secular errors, which become particularly significant in realistic scenarios. When the eccentricities are considered, the analysis becomes more complex, evolving into the Elliptic Restricted Three Body Problem (ERTBP). In this case, the dynamical system becomes non-autonomous, and the location of the collinear libration points oscillates.

The aim of this chapter is to investigate the motion of a solar sail around the cis-lunar libration points in the Earth-Moon ERTBP. Additionally, the orbital eccentricity of the Sun-Earth/Moon system and the inclination of the Moon's orbital plane relative to the ecliptic plane are incorporated into the dynamical model. Due to model errors, the inherent instability of the motion, and initial orbit injection errors, an effective station-keeping controller is proposed for long-term orbit maintenance. Considering the limited orbital controllability of the solar sail, an auxiliary SEP system is included to ensure orbital stabilization.

The chapter is organized as follows. Section 2.2 provides a brief introduction to the commonly used models of solar radiation pressure (SRP) force. Section 2.3 presents the equations of motion for a hybrid sail in the Earth-Moon ERTBP. In Sect. 2.4, quasi-periodic orbits around the cis-lunar L_1 and L_2 points are constructed using a first-order approximation, followed by an analysis of the linearization errors. To keep the spacecraft within a predefined neighborhood of the nominal orbit, a station-keeping controller based on the backstepping technique is proposed in Sect. 2.5. Specifically, the sail's reflectivity and orientation are optimized to minimize the overall propellant consumption of the SEP system. The proposed strategy is validated through numerical simulations in Sect. 2.6. Finally, Sect. 2.7 provides conclusions and remarks.

2.2 Solar Radiation Pressure Force

The SRP force is the force exerted by the impact of incident photons upon a body. For a flat and Lambertian sail surface with area S, the exerted SRP force can be written as

$$\boldsymbol{F} = PS(\hat{\boldsymbol{n}} \cdot \hat{\boldsymbol{n}}_S)\left[(c_a + a_d)\hat{\boldsymbol{n}}_S + \left(\frac{2}{3}c_d + 2c_s(\hat{\boldsymbol{n}} \cdot \hat{\boldsymbol{n}}_S)\right)\hat{\boldsymbol{n}}\right] \tag{2.1}$$

where c_a, c_d and c_s are the coefficients of absorption, diffusion, and specular refection, respectively, and $c_a + c_d + c_s = 1$; $\hat{\boldsymbol{n}}$ and $\hat{\boldsymbol{n}}_S$ are the unitary vectors normal to the sail and along the Sun-sail direction, respectively; P is the SRP exerted on the sail surface which can be expressed as

$$P = \frac{W_E}{c}(\frac{R_\oplus}{r_S})^2, \qquad W_E = \frac{L_S}{4\pi R_\oplus^2} \tag{2.2}$$

where $R_\oplus = 1$ au is the Sun-Earth distance, r_S is the Sun-sail distance, L_S is the solar luminosity, and c is the speed of light.

For an ideal sail surface, which can reflect the incident photons perfectly, $c_a = 0$, $c_d = 0$ and $c_s = 1$. Hence, the SRP force exerted on an ideal sail is

$$\boldsymbol{F} = 2PS(\hat{\boldsymbol{n}} \cdot \hat{\boldsymbol{n}}_S)^2 \hat{\boldsymbol{n}} \tag{2.3}$$

Accordingly, the acceleration $\boldsymbol{a}_{SRP}$ due to the SRP force for an ideal sail can now be written as

$$\boldsymbol{a}_{SRP} = \frac{\boldsymbol{F}}{m} = \frac{2PS}{m}(\hat{\boldsymbol{n}} \cdot \hat{\boldsymbol{n}}_S)^2 \hat{\boldsymbol{n}} = \frac{SL_s}{2\pi mcr_S^2}(\hat{\boldsymbol{n}} \cdot \hat{\boldsymbol{n}}_S)^2 \hat{\boldsymbol{n}} \tag{2.4}$$

where m is the mass of the solar sail.

According to Eq. (2.4), there is an inverse square law between the SRP acceleration and Sun-sail distance which is same with the acceleration due to Sun's gravity. Let M_S be the mass of the Sun, the lightness number β of a sail can be defined as

$$\beta = \frac{SL_S}{2\pi GM_S mc} = \frac{\sigma^*}{\sigma} \tag{2.5}$$

where $\sigma = m/S$ is the surface density of the sail, and $\sigma^* = L_S/2\pi GM_S c \approx 1.53\,\mathrm{g/m^2}$.

In this way, the SRP acceleration can be written in terms of Sun's gravitational acceleration as

$$\boldsymbol{a}_{SRP} = \beta\frac{GM_S}{r_S^2}(\hat{\boldsymbol{n}} \cdot \hat{\boldsymbol{n}}_S)^2 \hat{\boldsymbol{n}} \tag{2.6}$$

Note that the lightness number β is an important performance parameter to quantify the efficiency of a solar sail. It is defined as the ratio of the SRP acceleration to the solar gravitational acceleration when the sail is perpendicular to the Sun-sail direction. Some values of this parameter for past missions are $\beta = 0.001$ (IKAROS) [12], $\beta = 0.004$ (NanoSail-D2) [13] and $\beta = 0.011$ (LightSail-1) [14], while for the canceled Sunjammer mission or the future planned Geostorm mission, it increases up to $\beta = 0.036$ [15].

Furthermore, the efficiency of a solar sail can be quantified by its characteristic acceleration κ which is the SRP acceleration when the solar sail is at Earth's distance from the Sun $R_\oplus = 1$ au and the sail is perpendicular to the Sun-sail direction. For an idea solar sail,

$$\kappa = \frac{2PS}{m} \tag{2.7}$$

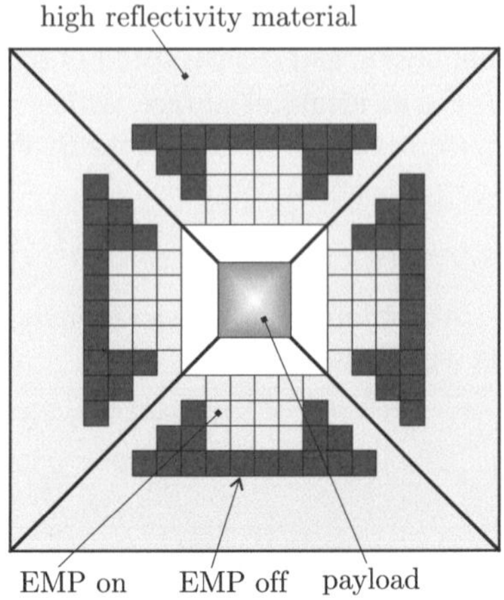

Fig. 2.1 Conceptual scheme of a solar sail with electrochromic material panels

According to Eqs. (2.6) and (2.7), the following relationship holds:

$$\kappa = \beta \frac{GM_S}{R_\oplus^2} \tag{2.8}$$

where $GM_S/R_\oplus^2 \approx 0.00593\,\text{m/s}^2$.

Thus, the SRP acceleration can also be expressed as

$$\boldsymbol{a}_{SRP} = \kappa \frac{R_\oplus^2}{r_S^2} (\hat{\boldsymbol{n}} \cdot \hat{\boldsymbol{n}}_S)^2 \hat{\boldsymbol{n}} \tag{2.9}$$

Reflectivity control devices (RCD) are a thin layer of electrochromic material panels (EMP) mounted on the sail membrane, see Fig. 2.1. The optical property of the EMP can be changed by power on or off. In this way, the reflectivity of the EMP is switched electrically between specular and diffuse modes. $c_a = c_d = 0$ and $c_s = 1$ when the EMP is switch-on whereas $c_s = c_d = 0$ and $c_a = 1$ when the EMP is switch-off. As solar sail attitude control devices, the efficiency of RCD has been successfully verified in the IKAROS mission. Let S denote the total area of sail, S_c the area covered by EMP, S_c^{off} and S_c^{on} the area covered by switch-off and switch-on EMP, then $S_c = S_c^{\text{off}} + S_c^{\text{on}}$. According to Eq. (2.1), the SRP force due to the switched-off EMP is

$$\boldsymbol{F}^{\text{off}} = P S_c^{\text{off}} (\hat{\boldsymbol{n}} \cdot \hat{\boldsymbol{n}}_S) \hat{\boldsymbol{n}}_S \tag{2.10}$$

For an idea sail, the optical properties of the switch-on EMP are same with the ones of normal sail membrane. Thus, the SRP force exerted on the other area of the sail is

$$\boldsymbol{F}^{\text{on}} = 2P(S - S_c^{\text{off}})(\hat{\boldsymbol{n}} \cdot \hat{\boldsymbol{n}}_S)^2 \hat{\boldsymbol{n}} \tag{2.11}$$

Defining $u = S_c^{\text{off}}/S$ as the sail reflectivity rate, the total SRP acceleration for the sail covered by RCD is

$$\boldsymbol{a}_{SRP} = \frac{\boldsymbol{F}^{\text{on}} + \boldsymbol{F}^{\text{off}}}{m} = \frac{\beta}{2} \frac{GM_S}{r_S^2} (\hat{\boldsymbol{n}} \cdot \hat{\boldsymbol{n}}_S)[u\hat{\boldsymbol{n}}_S + 2(1-u)(\hat{\boldsymbol{n}} \cdot \hat{\boldsymbol{n}}_S)\hat{\boldsymbol{n}}] \tag{2.12}$$

or

$$\boldsymbol{a}_{SRP} = \frac{\kappa}{2} (\frac{R_{\oplus}}{r_S})^2 (\hat{\boldsymbol{n}} \cdot \hat{\boldsymbol{n}}_S)[u\hat{\boldsymbol{n}}_S + 2(1-u)(\hat{\boldsymbol{n}} \cdot \hat{\boldsymbol{n}}_S)\hat{\boldsymbol{n}}] \tag{2.13}$$

where $u \in [0, u_{\max}]$ with $u_{\max} = S_c/S$ the maximum sail reflectivity rate.

2.3 Equations of Motion

Consider a hybrid sail in which a SEP system is coupled with a solar sail with electrochromic material panels (EMP). The corresponding acceleration $\boldsymbol{a}$ of the hybrid sail is the sum of the contribution of the SEP system and the solar radiation pressure (SRP) alone, that is

$$\boldsymbol{a} = \boldsymbol{a}_{SEP} + \boldsymbol{a}_{SRP} \tag{2.14}$$

where $\boldsymbol{a}_{SEP}$ is defined by its magnitude a_{SEP} and unit thrust vector $\hat{\boldsymbol{m}}$, given by

$$\boldsymbol{a}_{SEP} = a_{SEP}\, \hat{\boldsymbol{m}} \tag{2.15}$$

To describe the motion in the elliptic Earth-Moon system, Fig. 2.2 presents a non-uniformly rotating reference frame $\mathcal{T}_R(O;\ \hat{\boldsymbol{x}}_r, \hat{\boldsymbol{y}}_r, \hat{\boldsymbol{z}}_r)$: the origin O of the reference frame is centered at the Earth-Moon system barycenter, while the $\hat{\boldsymbol{x}}_r$ axis points to the Moon, the $\hat{\boldsymbol{z}}_r$ axis is positive in the direction of the angular velocity vector, and the $\hat{\boldsymbol{y}}_r$ axis completes the right-handed frame. Let a_{EM}, e_{EM} and f_{EM} denote the semimajor axis, eccentricity, and true anomaly of the Earth-Moon system, the Earth-Moon distance is

$$R_{EM} = \frac{a_{EM}(1 - e_{EM}^2)}{1 + e_{EM} \cos f_{EM}} \tag{2.16}$$

Within the framework of ERTBP, the dimension for acceleration is

$$[a] = \frac{G(M_E + M_M)}{R_{EM}^2} \tag{2.17}$$

where G is the gravitational constant, and M_E and M_M are the masses of the Earth and Moon, respectively.

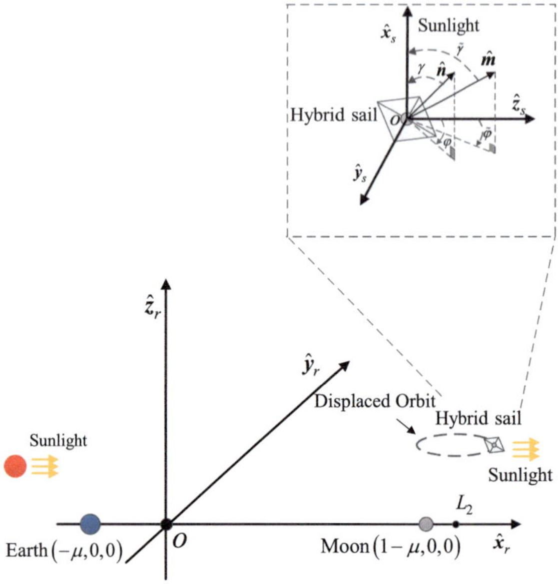

Fig. 2.2 Geometry of a hybrid-sail displaced orbit in the Earth-Moon system

Note that, compared to the Sun-Earth/Moon distance, the displacement of the sail from the Earth-Moon barycenter is a small number, which can be neglected. Therefore, it is reasonable to assume that the distance between the Sun and the barycentre of the Earth-Moon system R_{SEM} is equal to that between the Sun and the sail r_S, that is,

$$r_S \approx R_{SEM} = \frac{a_{SEM}\left(1 - e_{SEM}^2\right)}{1 + e_{SEM}\cos f_{SEM}} \tag{2.18}$$

where a_{SEM}, e_{SEM}, and f_{SEM} are the semi-major axis, eccentricity, and true anomaly of the Sun-Earth/Moon system respectively.

In this chapter, we consider a solar sail covered by the RCD where the SRP acceleration is as in Eq. (2.13). Bearing in mind Eqs. (2.16) and (2.17) and substituting Eq. (2.18) into Eq. (2.13), the dimensionless propulsive acceleration in case of $a_{SEM} = R_{\oplus} = 1$ au becomes

$$\boldsymbol{a} = \frac{\bar{\kappa}}{2}\left(\frac{1 - e_{EM}^2}{1 - e_{SEM}^2}\right)^2 \left(\frac{1 + e_{SEM}\cos f_{SEM}}{1 + e_{EM}\cos f_{EM}}\right)^2 (\hat{\boldsymbol{n}} \cdot \hat{\boldsymbol{n}}_S)\left[u\hat{\boldsymbol{n}}_S + 2(1-u)(\hat{\boldsymbol{n}} \cdot \hat{\boldsymbol{n}}_S)\hat{\boldsymbol{n}}\right] + \boldsymbol{a}_{SEP}\hat{\boldsymbol{m}} \tag{2.19}$$

where $\bar{\kappa} = \kappa / \left[G(M_E + M_M)/a_{EM}^2\right]$ is the dimensionless characteristic acceleration, $\hat{\boldsymbol{n}}$ and $\hat{\boldsymbol{n}}_S$ are the unitary vectors normal to the sail and along the Sun-sail direction, and u is the sail's reflectivity rate.

Then, the equations of motion of the sail in the Earth-Moon ERTBP become

$$
\begin{aligned}
x'' - 2y' &= \frac{1}{1 + e_{EM} \cos f_{EM}} (\Omega_x + a_x) \\
y'' + 2x' &= \frac{1}{1 + e_{EM} \cos f_{EM}} (\Omega_y + a_y) \\
z'' + z &= \frac{1}{1 + e_{EM} \cos f_{EM}} (\Omega_z + a_z)
\end{aligned} \tag{2.20}
$$

where $[\boldsymbol{r}]_{\mathcal{T}_R} = [x, y, z]^{\mathrm{T}}$ is the position vector of the sail, $[\boldsymbol{a}]_{\mathcal{T}_R} = [a_x, a_y, a_z]^{\mathrm{T}}$, and the expression for the potential Ω is

$$
\Omega = \frac{1}{2}(x^2 + y^2 + z^2) + \frac{1-\mu}{r_1} + \frac{\mu}{r_2} \tag{2.21}
$$

where μ is the mass parameter of the Earth-Moon system, and r_1 and r_2 are the Earth-sail distance and Moon-sail distance in the rotating reference frame, respectively, as

$$
r_1 = \sqrt{(x-\mu)^2 + y^2 + z^2}, \qquad r_2 = \sqrt{(x - 1 + \mu)^2 + y^2 + z^2} \tag{2.22}
$$

Let $\boldsymbol{r}_{L_i} = [x_{L_i}, 0, 0]^{\mathrm{T}}$ be the position vector of the L_i point in frame $\mathcal{T}_R$, and denote $\delta \boldsymbol{r}_{L_i} = [\delta x, \delta y, \delta z]^{\mathrm{T}}$ as a small displacement from the L_i point, such that $\boldsymbol{r} = \boldsymbol{r}_{L_i} + \delta \boldsymbol{r}_{L_i}$. From Eq. (2.20), the linearized equations of motion relative to the L_i point can be written as

$$
\begin{aligned}
\delta x'' - 2\,\delta y' &= \frac{1}{1 + e_{EM} \cos f_{EM}} (\Omega_{xx}\, \delta x + a_x) \\
\delta y'' + 2\,\delta x' &= \frac{1}{1 + e_{EM} \cos f_{EM}} \left(\Omega_{yy}\, \delta y + a_y\right) \\
\delta z'' + \delta z &= \frac{1}{1 + e_{EM} \cos f_{EM}} (\Omega_{zz}\, \delta z + a_z)
\end{aligned} \tag{2.23}
$$

where

$$
\begin{aligned}
\Omega_{xx} &= 1 + \frac{2(1-\mu)}{(x_{L_i} + \mu)^3} + \frac{2\mu}{(x_{L_i} - 1 + \mu)^3} \\
\Omega_{yy} &= 1 - \frac{1-\mu}{(x_{L_i} + \mu)^3} - \frac{\mu}{(x_{L_i} - 1 + \mu)^3} \\
\Omega_{zz} &= 1 - \frac{1-\mu}{(x_{L_i} + \mu)^3} - \frac{\mu}{(x_{L_i} - 1 + \mu)^3}
\end{aligned} \tag{2.24}
$$

2.4 Quasi-Periodic Displaced Orbits

2.4.1 Quasi-Periodic Solutions

In principle, the general solution to Eq. (2.23) is difficult to obtain, but only the bounded solution is useful for displaced orbit design. To this end, the problem will be solved by exploiting the fact that the eccentricity $e_{EM} = 0.0549 \ll 1$, which facilitates an approximate form of Eq. (2.23), viz,

$$\begin{aligned} \delta x'' - 2\delta y' &= \Omega_{xx}\delta x + s_x \\ \delta y'' + 2\delta x' &= \Omega_{yy}\delta y + s_y \\ \delta z'' + z &= \Omega_{zz}\delta z + s_z \end{aligned} \tag{2.25}$$

where $[\boldsymbol{s}]_{\mathcal{T}_R} = [\boldsymbol{a}]_{\mathcal{T}_R}/(1 + e_{EM}\cos f_{EM}) = \left[s_x, s_y, s_z\right]^{\mathrm{T}}$.

To describe the SRP exerted on the sail, Fig. 2.2 introduces the sunlight reference frame $\mathcal{T}_S(O;\ \hat{\boldsymbol{x}}_s, \hat{\boldsymbol{y}}_s, \hat{\boldsymbol{z}}_s)$: the origin O is centered at the sail, the $\hat{\boldsymbol{x}}_s$ axis is in the direction of the sunlight, the $\hat{\boldsymbol{z}}_s$ axis is perpendicular to the ecliptic plane, and the $\hat{\boldsymbol{y}}_s$ axis satisfies the right-handed rule. Therefore, the sunlight unit vector $\hat{\boldsymbol{n}}_S$ and sail normal unit vector $\hat{\boldsymbol{n}}$ in frame $\mathcal{T}_S$ can be written as

$$\left[\hat{\boldsymbol{n}}_S\right]_{\mathcal{T}_S} = [1, 0, 0]^{\mathrm{T}} \tag{2.26}$$

$$\left[\hat{\boldsymbol{n}}\right]_{\mathcal{T}_S} = \left[\cos\gamma, \sin\gamma\sin\varphi, \sin\gamma\cos\varphi\right]^{\mathrm{T}} \tag{2.27}$$

where γ and φ are the pitch angle and the clock angle respectively, as shown in Fig. 2.2.

In addition, angles $\tilde{\gamma}$ and $\tilde{\varphi}$ are used to define the direction of $\hat{\boldsymbol{m}}$ in frame $\mathcal{T}_S$ according to Fig. 2.2 as

$$\left[\hat{\boldsymbol{m}}\right]_{\mathcal{T}_S} = \left[\cos\tilde{\gamma}, \sin\tilde{\gamma}\sin\tilde{\varphi}, \sin\tilde{\gamma}\cos\tilde{\varphi}\right]^{\mathrm{T}}$$

Since the problem is formulated in frame $\mathcal{T}_R$, it is useful to express $\boldsymbol{s}$ in that frame. Assume that the Earth-Moon system covers a heliocentric orbit whose perigee coincides with the origin of the frame $\mathcal{T}_R$ at the initial time $t = 0$. As shown in Fig. 2.3, the transformation matrix $\mathbf{T}_{RS}$ from $\mathcal{T}_S$ to $\mathcal{T}_R$ is given by

$$\mathbf{T}_{RS} = \begin{bmatrix} A & B & -\cos f_{EM}\sin i \\ C & D & \sin f_{EM}\sin i \\ \cos f_{SEM}\sin i & -\sin f_{SEM}\sin i & \cos i \end{bmatrix} \tag{2.28}$$

with

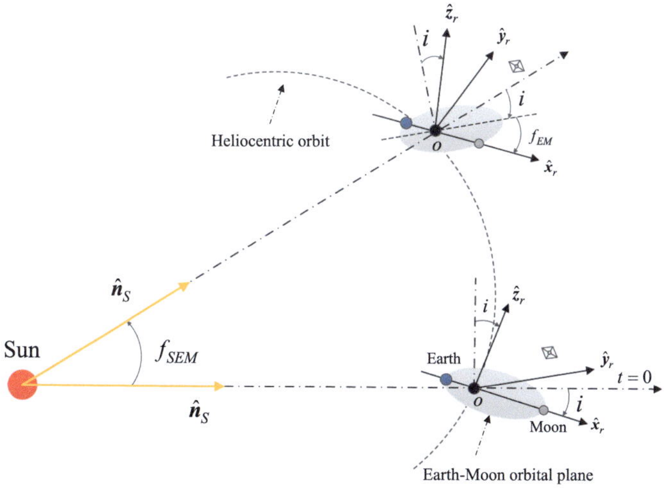

Fig. 2.3 Geometrical relationship of Sun, Earth, and Moon

$$\begin{aligned} A &= \cos f_{EM} \cos f_{SEM} \cos i + \sin f_{EM} \sin f_{SEM} \\ B &= -\cos f_{EM} \sin f_{SEM} \cos i + \sin f_{EM} \cos f_{SEM} \\ C &= -\sin f_{EM} \cos f_{SEM} \cos i + \cos f_{EM} \sin f_{SEM} \\ D &= \sin f_{EM} \sin f_{SEM} \cos i + \cos f_{EM} \cos f_{SEM} \end{aligned} \tag{2.29}$$

where i is the inclination of the Earth-Moon orbital plane with respect to the ecliptic plane. Bearing in mind that f_{SEM} and f_M are time-dependent, it is preferable to relate f_{SEM} to f_M such that the problem can be simplified. To this end, recall the relationship

$$\begin{aligned} f'_{SEM} &= \frac{\mathrm{d} f_{SEM}}{\mathrm{d}t} \cdot \frac{\mathrm{d}t}{\mathrm{d} f_{EM}} \\ &= \frac{\sqrt{G(M_S + M_E + M_M) a_{SEM}(1 - e_{SEM}^2)} (1 + e_{SEM} \cos f_{SEM})^2}{[a_{SEM}(1 - e_{SEM}^2)]^2} \\ &\cdot \frac{[a_{EM}(1 - e_{EM}^2)]^2}{\sqrt{G(M_E + M_M) a_{EM}(1 - e_{EM}^2)} (1 + e_{EM} \cos f_{EM})^2} \\ &= C \sqrt{\frac{(1 - e_{EM}^2)^3}{(1 - e_{SEM}^2)^3}} \frac{(1 + e_{SEM} \cos f_{SEM})^2}{(1 + e_{EM} \cos f_{EM})^2} \end{aligned} \tag{2.30}$$

where M_E, M_M and M_S are the masses of the Earth, Moon and Sun, respectively, and $\mu_S = M_S/(M_E + M_M)$.

Taking the Taylor series expansion in Eq. (2.30) and retaining only the first-order terms of e_{EM} and e_{SEM}, Eq. (2.30) becomes

$$f'_{SEM} \approx C(1 - 2e_{EM} \cos f_{EM} + 2e_{SEM} \cos f_{SEM}) \tag{2.31}$$

The definite integral expression for f'_{SEM} is

$$\begin{aligned}\int_0^{f_{SEM}} \mathrm{d}f_{SEM} &= C\int_0^{f_{EM}} \mathrm{d}f_{EM} - 2Ce_{EM}\int_0^{f_{EM}} \cos f_{EM}\mathrm{d}f_{EM} \\ &+ 2e_{SEM}\int_0^{f_{SEM}} \frac{\cos f_{SEM}}{(1 - 2e_{EM}\cos f_{EM} + 2e_{SEM}\cos f_{SEM})}\mathrm{d}f_{SEM}\end{aligned} \tag{2.32}$$

Retaining only the first-order terms of e_{EM} and e_{SEM}, f_{SEM} can be written as a function of f_{EM} approximately as

$$f_{SEM} \approx Cf_{EM} - 2\,Ce_{EM} \sin f_{EM} + 2e_{SEM} \sin f_{SEM} \tag{2.33}$$

Using Taylor series expansion of trigonometric functions and omitting high-order terms of e_{EM} and e_{SEM}, the following relations hold

$$\begin{aligned}\sin(f_{SEM}) &\approx \sin(Cf_{EM} - 2\,Ce_{EM}\sin f_{EM} + 2e_{SEM}\sin f_{SEM}) \\ &\approx \sin(Cf_{EM}) + \cos(Cf_{EM})(-2\,Ce_{EM}\sin f_{EM} + 2e_{SEM}\sin f_{SEM})\end{aligned} \tag{2.34}$$

Therefore,

$$\begin{aligned}\sin(f_{SEM}) &\approx \frac{\sin(Cf_{EM}) - 2\,Ce_{EM}\cos(Cf_{EM})\sin f_{EM}}{1 - 2e_{SEM}\cos(Cf_{SEM})} \\ &\approx \sin(Cf_{EM}) + 2e_{SEM}\sin(Cf_{EM})\cos(Cf_{EM}) - 2\,Ce_{EM}\sin f_{EM}\cos(Cf_{EM})\end{aligned} \tag{2.35}$$

In the same way, the following relation holds

$$\cos f_{SEM} \approx \cos(Cf_{EM}) - 2e_{SEM}\sin(Cf_{EM})\sin(Cf_{EM}) + 2\,Ce_{EM}\sin f_{EM}\sin(Cf_{EM}) \tag{2.36}$$

Since the mean inclination angle $i = 5.145$ deg is a small number, it is reasonable to use the approximation $\cos i \approx 1$. Thus, the transformation matrix $\mathbf{T}_{RS}$ in Eq. (2.28) becomes

$$\mathbf{T}_{RS} \approx \begin{bmatrix} T_1 & T_2 & -\cos f_{EM}\sin i \\ -T_2 & T_1 & \sin f_{EM}\sin i \\ \cos f_{SEM}\sin i & -\sin f_{SEM}\sin i & 1 \end{bmatrix} \tag{2.37}$$

with

$$\begin{aligned} T_1 = & -e_{SEM} \cos f_{EM} - C e_{EM} \cos(C f_{EM}) + \cos(f_{EM} - C f_{EM}) \\ & + e_{SEM} \cos(f_{EM} - 2C f_{EM}) + C e_{EM} \cos(2 f_{EM} - C f_{EM}) \end{aligned} \tag{2.38}$$

$$\begin{aligned} T_2 = & -e_{SEM} \sin f_{EM} + C e_{EM} \sin(C f_{EM}) + \sin(f_{EM} - C f_{EM}) \\ & + e_{SEM} \sin(f_{EM} - 2C f_{EM}) + C e_{EM} \sin(2 f_{EM} - C f_{EM}) \end{aligned} \tag{2.39}$$

Combining Eqs. (2.19), (2.26), (2.27) and (2.37) and taking the SEP acceleration $\boldsymbol{a}_{SEP} = 0$, the $\boldsymbol{s}$ in frame $\mathcal{T}_R$ can be expressed as

$$[\boldsymbol{s}]_{\mathcal{T}_R} = \frac{\bar{\kappa}(1 - e_{EM}^2)^2}{2(1 - e_{SEM}^2)^2} \frac{(1 + e_{SEM} \cos f_{SEM})^2}{(1 + e_{EM} \cos f_{EM})^3} \cdot \mathbf{T}_{RS} \cdot \begin{bmatrix} u \cos\gamma + 2(1-u)\cos^3\gamma \\ 2(1-u)\cos^2\gamma \sin\gamma \sin\varphi \\ 2(1-u)\cos^2\gamma \sin\gamma \cos\varphi \end{bmatrix}. \tag{2.40}$$

Retaining the first-order terms of e_{EM}, e_{SEM} and $\sin i$, the components of $\boldsymbol{s}$ can be approximated by

$$\begin{aligned} s_x \approx\ & c_{x1} \cos f_{EM} + c_{x2} \sin f_{EM} + c_{x3} \cos(C f_{EM}) + c_{x4} \sin(C f_{EM}) \\ & + c_{x5} \cos(f_{EM} - C f_{EM}) + c_{x6} \sin(f_{EM} - C f_{EM}) + c_{x7} \cos(f_{EM} - 2C f_{EM}) \\ & + c_{x8} \sin(f_{EM} - 2C f_{EM}) + c_{x9} \cos(2 f_{EM} - C f_{EM}) + c_{x10} \sin(2 f_{EM} - C f_{EM}) \\ s_y \approx\ & c_{y1} \cos f_{EM} + c_{y2} \sin f_{EM} + c_{y3} \cos(C f_{EM}) + c_{y4} \sin(C f_{EM}) \\ & + c_{y5} \cos(f_{EM} - C f_{EM}) + c_{y6} \sin(f_{EM} - C f_{EM}) + c_{y7} \cos(f_{EM} - 2C f_{EM}) \\ & + c_{y8} \sin(f_{EM} - 2C f_{EM}) + c_{y9} \cos(2 f_{EM} - C f_{EM}) + c_{y10} \sin(2 f_{EM} - C f_{EM}) \\ s_z \approx\ & c_{z0} + c_{z1} \cos(C f_{EM}) + c_{z2} \sin(C f_{EM}) + c_{z3} \cos f_{EM} \end{aligned} \tag{2.41}$$

where the coefficients $c_{x1}, \ldots, c_{x10}$, $c_{y1}, \ldots, c_{y10}$, and $c_{z0}, \ldots, c_{z3}$ are given in Appendix A.

Note that both the components s_x and s_y contain five time-dependent variables $f_{EM}, C f_{EM}, f_{EM} - C f_{EM}, f_{EM} - 2C f_{EM}$ and $2 f_{EM} - C f_{EM}$ respectively, whereas the component s_z only consists of two time-dependent variables f_{EM} and $C f_{EM}$. Thus, the general solution to Eq. (2.25) can be written as

$$\begin{aligned} \delta x(f_{EM}) =\ & \delta x_1 \cos f_{EM} + \delta x_2 \sin f_{EM} + \delta x_3 \cos(C f_{EM}) + \delta x_4 \sin(C f_{EM}) \\ & + \delta x_5 \cos(f_{EM} - C f_{EM}) + \delta x_6 \sin(f_{EM} - C f_{EM}) + \delta x_7 \cos(f_{EM} - 2C f_{EM}) \\ & + \delta x_8 \sin(f_{EM} - 2C f_{EM}) + \delta x_9 \cos(2 f_{EM} - C f_{EM}) + \delta x_{10} \sin(2 f_{EM} - C f_{EM}) \\ \delta y(f_{EM}) =\ & \delta y_1 \cos f_{EM} + \delta y_2 \sin f_{EM} + \delta y_3 \cos(C f_{EM}) + \delta y_4 \sin(C f_{EM}) \\ & + \delta y_5 \cos(f_{EM} - C f_{EM}) + \delta y_6 \sin(f_{EM} - C f_{EM}) + \delta y_7 \cos(f_{EM} - 2C f_{EM}) \\ & + \delta y_8 \sin(f_{EM} - 2C f_{EM}) + \delta y_9 \cos(2 f_{EM} - C f_{EM}) + \delta y_{10} \sin(2 f_{EM} - C f_{EM}) \\ \delta z(f_{EM}) =\ & \delta z_0 + \delta z_1 \cos(C f_{EM}) + \delta z_2 \sin(C f_{EM}) + \delta z_3 \cos f_{EM} \end{aligned} \tag{2.42}$$

Substituting Eqs. (2.41) and (2.42) into Eq. (2.25), the coefficients of $\delta x(f_{EM})$, $\delta y(f_{EM})$ and $\delta z(f_{EM})$ can be obtained by solving three algebraic equations through equaling the coefficients of trigonometric functions of both sides as

$$\begin{cases} (-1-\Omega_{xx})\delta x_1 - 2\delta y_2 = c_{x1} \\ (-1-\Omega_{xx})\delta x_2 + 2\delta y_1 = c_{x2} \\ (-C^2-\Omega_{xx})\delta x_3 - 2C\delta y_4 = c_{x3} \\ (-C^2-\Omega_{xx})\delta x_4 + 2C\delta y_3 = c_{x4}, \\ [-(1-C)^2-\Omega_{xx}]\delta x_5 - 2(1-C)\delta y_6 = c_{x5} \\ [-(1-C)^2-\Omega_{xx}]\delta x_6 + 2(1-C)\delta y_5 = c_{x6} \\ [-(1-2C)^2-\Omega_{xx}]\delta x_7 - 2(1-2C)\delta y_8 = c_{x7} \\ [-(1-2C)^2-\Omega_{xx}]\delta x_8 + 2(1-2C)\delta y_7 = c_{x8} \\ [-(2-C)^2-\Omega_{xx}]\delta x_9 - 2(2-C)\delta y_{10} = c_{x9} \\ [-(2-C)^2-\Omega_{xx}]\delta x_{10} + 2(2-C)\delta y_9 = c_{x10} \end{cases} \tag{2.43}$$

$$\begin{cases} (-1-\Omega_{yy})\delta y_1 + 2\delta x_2 = c_{y1} \\ (-1-\Omega_{yy})\delta y_2 - 2\delta x_1 = c_{y2} \\ (-C^2-\Omega_{yy})\delta y_3 + 2C\delta x_4 = c_{y3} \\ (-C^2-\Omega_{yy})\delta y_4 - 2C\delta x_3 = c_{y4}, \\ [-(1-C)^2-\Omega_{yy}]\delta y_5 + 2(1-C)\delta x_6 = c_{y5} \\ [-(1-C)^2-\Omega_{yy}]\delta y_6 - 2(1-C)\delta x_5 = c_{y6} \\ [-(1-2C)^2-\Omega_{yy}]\delta y_7 + 2(1-2C)\delta x_8 = c_{y7} \\ [-(1-2C)^2-\Omega_{yy}]\delta y_8 - 2(1-2C)\delta x_7 = c_{y8} \\ [-(2-C)^2-\Omega_{yy}]\delta y_9 + 2(2-C)\delta x_{10} = c_{y9} \\ [-(2-C)^2-\Omega_{yy}]\delta y_{10} - 2(2-C)\delta x_9 = c_{y10} \end{cases} \tag{2.44}$$

$$\begin{cases} (1-\Omega_{zz})\delta z_0 = c_{z0}, \\ (1-\Omega_{zz}-C^2)\delta z_1 = c_{z1} \\ (1-\Omega_{zz}-C^2)\delta z_2 = c_{z2} \\ -\Omega_{zz}\delta z_3 = c_{z3} \end{cases} \tag{2.45}$$

According to Eqs. (2.43), (2.44) and (2.45), the coefficients of motion depend on the dimensionless characteristic acceleration $\bar{\kappa}$, reflectivity rate u and attitude angles γ and φ. For given $\bar{\kappa}$ and u, the coefficients will be determined uniquely by attitude angles. Note from Eq. (2.45) that the coefficients δz_1, δz_2 and δz_3 for out-of-plane motion are determined by c_{z1}, c_{z2} and c_{z3}, which include small numbers e_{EM}, e_{SEM} or $\sin i$ only. Hence δz_1, δz_2 and δz_3 are much smaller than δz_0, and the displacement is dominated by δz_0. Therefore, the value of δz_0 should be maximized so as to obtain a higher displacement. In this sense, the optimal values of the nominal attitude angles are $\gamma_n = 35.264$ deg, $\varphi_n = 0$. Thus, the coefficients of $\delta x(f_{EM})$, $\delta y(f_{EM})$ and $\delta z(f_{EM})$ can be written explicitly as

$$\begin{cases} \delta x_1 = \dfrac{-(1+\Omega_{yy})c_{x1} + 2c_{y2}}{(1+\Omega_{xx})(1+\Omega_{yy}) - 4} \\ \delta x_3 = \dfrac{-(C^2+\Omega_{yy})c_{x3} + 2Cc_{y4}}{(C^2+\Omega_{xx})(C^2+\Omega_{yy}) - 4C^2} \\ \delta x_5 = \dfrac{-[(1-C)^2+\Omega_{yy}]c_{x5} + 2(1-C)c_{y6}}{[(1-C)^2+\Omega_{xx}][(1-C)^2+\Omega_{yy}] - 4(1-C)^2} \\ \delta x_7 = \dfrac{-[((1-2C^2)+\Omega_{yy})]c_{x7} + 2(1-2C)c_{y8}}{[(1-2C^2)+\Omega_{xx}][(1-2C^2)+\Omega_{yy}] - 4(1-2C)^2} \\ \delta x_9 = \dfrac{-[(2-C)^2+\Omega_{yy}]c_{x9} + 2(2-C)c_{y10}}{[(2-C)^2+\Omega_{xx}][(2-C)^2+\Omega_{yy}] - 4(2-C)^2} \\ \delta x_2 = \delta x_4 = \delta x_6 = \delta x_8 = \delta x_{10} = 0 \end{cases} \tag{2.46}$$

$$\begin{cases} \delta y_2 = \dfrac{-(1+\Omega_{xx})c_{y2} + 2c_{x1}}{(1+\Omega_{xx})(1+\Omega_{yy}) - 4} \\ \delta y_4 = \dfrac{-(C^2+\Omega_{xx})c_{y4} + 2Cc_{x3}}{(C^2+\Omega_{xx})(C^2+\Omega_{yy}) - 4C^2} \\ \delta y_6 = \dfrac{-[(1-C)^2+\Omega_{xx}]c_{y6} + 2(1-C)c_{x5}}{[(1-C)^2+\Omega_{xx}][(1-C)^2+\Omega_{yy}] - 4(1-C)^2} \\ \delta y_8 = \dfrac{-[((1-2C^2)+\Omega_{xx})]c_{y8} + 2(1-2C)c_{x7}}{[(1-2C^2)+\Omega_{xx}][(1-2C^2)+\Omega_{yy}] - 4(1-2C)^2} \\ \delta y_{10} = \dfrac{-[(2-C)^2+\Omega_{xx}]c_{xy10} + 2(2-C)c_{x9}}{[(2-C)^2+\Omega_{xx}][(2-C)^2+\Omega_{yy}] - 4(2-C)^2} \\ \delta y_1 = \delta y_3 = \delta y_5 = \delta y_7 = \delta y_9 = 0. \end{cases} \tag{2.47}$$

$$\begin{cases} \delta z_0 = \dfrac{c_{z0}}{1-\Omega_{zz}} \\ \delta z_1 = \dfrac{c_{z1}}{-C^2+1-\Omega_{zz}} \\ \delta z_3 = -\dfrac{c_{z3}}{\Omega_{zz}} \\ \delta z_2 = 0 \end{cases} \tag{2.48}$$

According to Eqs. (2.46) and (2.47), $\delta x(f_{EM})$ only contains cosine functions while $\delta y(f_{EM})$ only contains sine functions in case of $\gamma_n = 35.264$ deg and $\varphi_n = 0$. Table 2.1 lists the parameters of a displaced orbit which indicate that

$\cos(2f_{EM} - Cf_{EM})$ is the dominant component in $\delta x(f_{EM})$, while $\sin(f_{EM} - Cf_{EM})$ and $\sin(2f_{EM} - Cf_{EM})$ are dominant components in $\delta y(f_{EM})$. Other trigonometric functions with small coefficients will result in slight motion fluctuations around that of dominating ones. Besides, $\bar{\kappa}$ will not change the motion proportion because all coefficients are linear with $\bar{\kappa}$, see Appendix A.

Compared with the framework established in the CRTBP [7, 8], more components are found to exist in $\delta x(f_{EM})$, $\delta y(f_{EM})$ and $\delta z(f_{EM})$ in the ERTBP. In contrast to the analysis in Ref. [7], it can be drawn from Eq. (2.42) that the quasi-periodic displaced orbits will not degenerate to periodic orbits, due to the effect of the eccentricities of the Earth-Moon and Sun-Earth/Moon systems.

2.4.2 *Error Analysis*

The model errors induced by the linearization procedure are quantitatively analyzed here, and will be compared with the results obtained in the CRTBP [8]. For a reference quasi-periodic displaced orbit, the required acceleration $\boldsymbol{a}_N$ with the full nonlinear dynamics is given by Eq. (2.20), while the acceleration $\boldsymbol{a}_{SRP}$ is calculated from Eq. (2.19). The absolute and relative errors are defined as [8]

$$\boldsymbol{a}_{AE} = \boldsymbol{a}_{SRP} - \boldsymbol{a}_N \tag{2.49}$$

$$a_{RE} = \frac{\|\boldsymbol{a}_{AE}\|}{\|\boldsymbol{a}_{SRP}\|} \tag{2.50}$$

With the same sail parameters listed in Table 2.1 and setting both e_{EM} and e_{SEM} to 0, the coefficients of the displaced orbit based on the CRTBP can be obtained from Eq. (2.20), as given in Table 2.2. Figure 2.4 shows the absolute and relative errors in the framework of the CRTBP and ERTBP respectively, indicating that e_{EM} and e_{SEM} should be taken into consideration such that the model errors can be significantly reduced. On the other hand, it can be seen from Fig. 2.4b that the value of the relative error in the ERTBP is larger than 1 most of the time, indicating the infeasibility of canceling model errors using SRP only. Hence, the auxiliary SEP system is necessary for long-term orbital maintenance due to the model errors and the inherent instability of the motion around the cis-lunar libration points.

Table 2.1 Parameters of quasi-periodic displaced orbit around the L_1 and L_2 points based on the ERTBP

Point	γ [deg]	φ [deg]	$\bar{\kappa}$	u	$\delta x(f_{EM})$	$\delta y(f_{EM})$	$\delta z(f_{EM})$
L_1	35.264	0	0.002	0.5	$\delta x_1 = -1.82 \times 10^{-5}$	$\delta y_2 = 4.94 \times 10^{-7}$	$\delta z_0 = -2.00 \times 10^{-4}$ $\delta z_1 = -5.08 \times 10^{-5}$ $\delta z_3 = 2.17 \times 10^{-5}$
					$\delta x_3 = -2.75 \times 10^{-5}$	$\delta y_4 = 2.96 \times 10^{-5}$	
					$\delta x_5 = 4.91 \times 10^{-5}$	$\delta y_6 = 1.16 \times 10^{-5}$	
					$\delta x_7 = 1.54 \times 10^{-5}$	$\delta y_8 = 1.20 \times 10^{-6}$	
					$\delta x_9 = -8.60 \times 10^{-5}$	$\delta y_{10} = 3.87 \times 10^{-5}$	
L_2	35.264	0	0.1	0.5	$\delta x_1 = -1.00 \times 10^{-4}$	$\delta y_2 = 1.28 \times 10^{-3}$	$\delta z_0 = 6.03 \times 10^{-3}$ $\delta z_1 = 1.53 \times 10^{-3}$ $\delta z_3 = -1.45 \times 10^{-3}$
					$\delta x_3 = 5.95 \times 10^{-4}$	$\delta y_4 = -1.84 \times 10^{-3}$	
					$\delta x_5 = 1.71 \times 10^{-3}$	$\delta y_6 = -3.33 \times 10^{-2}$	
					$\delta x_7 = 2.41 \times 10^{-5}$	$\delta y_8 = -1.01 \times 10^{-3}$	
					$\delta x_9 = 1.01 \times 10^{-2}$	$\delta y_{10} = -2.81 \times 10^{-2}$	

Table 2.2 Parameters of quasi-periodic displaced orbit around the L_2 point based on the CRTBP

Point	γ [deg]	φ [deg]	$\bar{\kappa}$	u	$\delta x(f_{EM})$	$\delta y(f_{EM})$	$\delta z(f_{EM})$
L_2	35.264	0	0.1	0.5	$\delta x_1 = -1.00 \times 10^{-4}$	$\delta y_2 = 1.28 \times 10^{-3}$	$\delta z_0 = 6.03 \times 10^{-3}$ $\delta z_1 = 1.34 \times 10^{-3}$ $\delta z_3 = 0$
					$\delta x_3 = 0$	$\delta y_6 = -3.33 \times 10^{-2}$	
					$\delta x_5 = 1.71 \times 10^{-3}$	$\delta y_4 = 0$	
					$\delta x_7 = 0$	$\delta y_8 = 0$	
					$\delta x_9 = 0$	$\delta y_{10} = 0$	

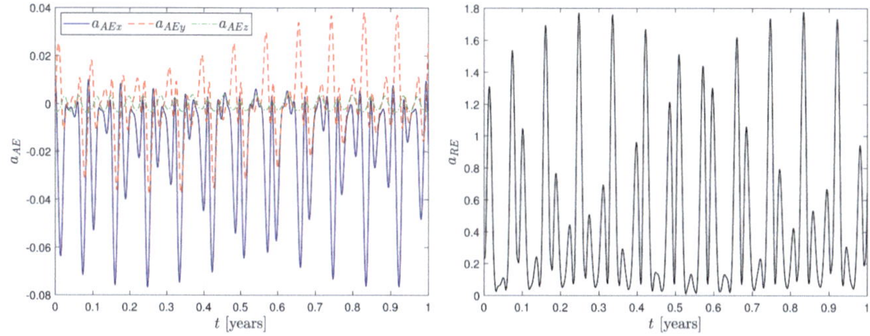

(a) Absolute and relative errors of displaced orbit based on the ERTBP

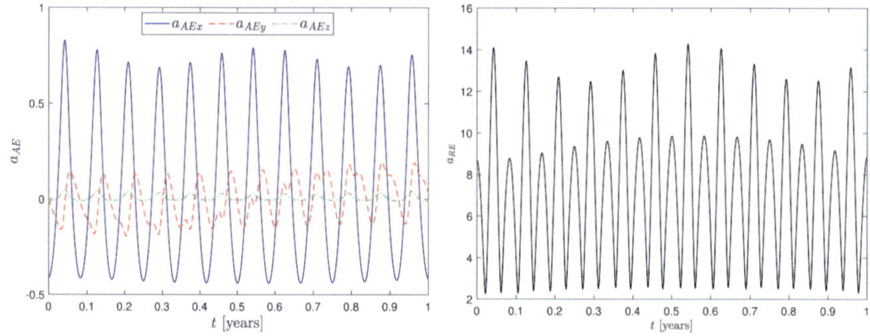

(b) Absolute and relative errors of displaced orbit based on the CRTBP

Fig. 2.4 Absolute and relative errors for two displaced orbits

2.5 Backstepping Station-Keeping Controller

2.5.1 Station-Keeping Controller Design

The error analysis indicates that an active orbital control is required to keep the sail in the reference orbit, whose position, velocity and acceleration are characterized by

$$[\boldsymbol{r}_{ref}]_{\mathcal{T}_R} = \begin{bmatrix} x_{L_2} + \delta x(f_{EM}) \\ \delta y(f_{EM}) \\ \delta z(f_{EM}) \end{bmatrix}, \quad [\boldsymbol{v}_{ref}]_{\mathcal{T}_R} = \begin{bmatrix} \delta x'(f_{EM}) \\ \delta y'(f_{EM}) \\ \delta z'(f_{EM}) \end{bmatrix}, \quad [\boldsymbol{a}_{ref}]_{\mathcal{T}_R} = \begin{bmatrix} \delta x''(f_{EM}) \\ \delta y''(f_{EM}) \\ \delta z''(f_{EM}) \end{bmatrix} \tag{2.51}$$

To this end, a backstepping controller is adopted here for orbit stabilization. Denote $[\boldsymbol{r}]_{\mathcal{T}_R} = [x, y, z]^{\mathrm{T}}$ and $[\boldsymbol{v}]_{\mathcal{T}_R} = [x', y', z']^{\mathrm{T}}$, Eq. (2.20) can be rewritten as

$$\begin{cases} \boldsymbol{r}' = \boldsymbol{v} \\ \boldsymbol{v}' = \begin{bmatrix} 2y' \\ -2x' \\ -z \end{bmatrix} + \dfrac{1}{1 + e_{EM} \cos f_{EM}} \nabla \Omega + \dfrac{1}{1 + e_{EM} \cos f_{EM}} \boldsymbol{a} \end{cases} \tag{2.52}$$

where $\nabla\Omega = [\Omega_x, \Omega_y, \Omega_z]^{\mathrm{T}}$ is the gradient of Ω.

The backstepping control variables are defined as

$$z_1 = \boldsymbol{r} - \boldsymbol{r}_{ref}, \qquad z_2 = \boldsymbol{v} - \boldsymbol{v}_d \tag{2.53}$$

where $\boldsymbol{v}_d$ is the desired command of the next subsystem, designed as

$$\boldsymbol{v}_d = -k_1 z_1 + \boldsymbol{v}_{ref} \tag{2.54}$$

where $k_1 \in \mathbb{R}^+$.

Consider the first candidate Lyapunov function

$$V_1 = \frac{z_1^{\mathrm{T}} z_1}{2} \tag{2.55}$$

whose derivative can be drawn from Eq. (2.54), as

$$V_1' = z_1^{\mathrm{T}}(z_2 + \boldsymbol{v}_d - \boldsymbol{v}_{ref}) = -k_1 z_1^{\mathrm{T}} z_1 + z_1^{\mathrm{T}} z_2 \tag{2.56}$$

From the definition of z_2 and the desired command of $\boldsymbol{v}_d$ in Eq. (2.54), the dynamic of z_2 can be written as

$$z_2' = \boldsymbol{v}' - \boldsymbol{v}'_d = \begin{bmatrix} 2y' \\ -2x' \\ -z \end{bmatrix} + \frac{\nabla\Omega}{1 + e_{EM} \cos f_{EM}} + \frac{\boldsymbol{a}}{1 + e_{EM} \cos f_{EM}} + k_1 z_1' - \boldsymbol{a}_{ref} \tag{2.57}$$

The second candidate Lyapunov function is selected as

$$V_2 = \frac{z_2^{\mathrm{T}} z_2}{2} + V_1 \tag{2.58}$$

with its derivative given by

$$V_2' = z_2^{\mathrm{T}} z_2' + V_1' = z_2^{\mathrm{T}}(\boldsymbol{v}' + k_1 z_1' - \boldsymbol{a}_{ref}) - k_1 z_1^{\mathrm{T}} z_1 + z_1^{\mathrm{T}} z_2 \tag{2.59}$$

To ensure system stability, the control law of the sail is designed as

$$\boldsymbol{a} = (1 + e_{EM}\cos f_{EM})(-k_2 z_2 - z_1^{\mathrm{T}} + \boldsymbol{a}_{ref} - k_1 z_1') - \nabla\Omega - (1 + e_{EM}\cos f_{EM})\begin{bmatrix} 2y' \\ -2x' \\ -z \end{bmatrix} \tag{2.60}$$

where $k_2 \in \mathbb{R}^+$.

Thus, the derivative of V_2 is

$$V_2' = -k_1 z_1^{\mathrm{T}} z_1 - k_2 z_2^{\mathrm{T}} z_2 < 0 \tag{2.61}$$

which indicates the asymptotic stability of the system.

2.5.2 SEP Acceleration Optimization

According to Eq. (2.13), $\boldsymbol{a}_{SRP}$ is determined by sail reflectivity rate u and attitude angles γ and φ. For mission lifetime extension, the sail reflectivity rate u and attitude angles γ and φ should be optimized, so that the magnitude of the acceleration required from the SEP system is minimum. To avoid possible mechanical instabilities, it is reasonable to limit the attitude angles within some range, e.g. $\gamma \in [30, 60]$ deg and $\varphi \in [-45, 45]$ deg. Consequently, the static constrained optimization problem can be stated as follows:

Find the optimal sail reflectivity rate $u^\star$ and attitude angles $\gamma^\star$ and $\varphi^\star$, such that

$$\left\{u^\star, \gamma^\star, \varphi^\star\right\} = \arg\min_{u,\gamma,\varphi} \|\boldsymbol{a}_{SEP}\| = \arg\min_{u,\gamma,\varphi} \|\boldsymbol{a} - \boldsymbol{a}_{SRP}\| \tag{2.62}$$

s.t.

$$\gamma \in [30, 60]\,\text{deg}, \quad \varphi \in [-45, 45]\,\text{deg}, \quad u \in [0, 1] \tag{2.63}$$

where the total required acceleration $\boldsymbol{a}$ is given by Eq. (2.60). In particular, a particle swarm optimization (PSO) method is adopted here to address the constrained optimization problem represented by Eqs. (2.62) and (2.63).

2.6 Numerical Simulations

To examine the performance of proposed method, two mission scenarios are emphasized here, in which the quasi-periodic displaced orbits around L_1 point and L_2 point are taken as reference orbits respectively, see Table 2.1. The nonlinear dynamic equations represented by Eq. (2.20) are numerically simulated and propagated for one sidereal year, with the orbital controller given by Eq. (2.60). The dimensionless characteristic acceleration $\bar{\kappa}$ is set to 0.002 for displaced orbit around L_1 point, and 0.1 for displaced orbit around L_2 point. The initial position errors in each directions are set equal to 10 km, and the feedback gains in Eq. (2.60) are $k_1 = 1$ and $k_2 = 5$.

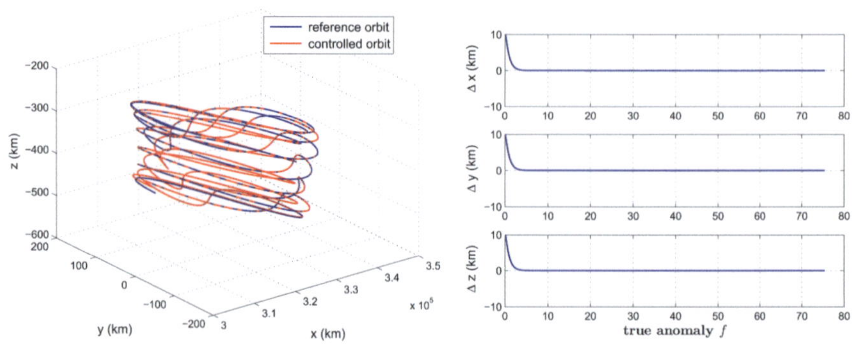

(a) Orbital tracking result (b) Position errors

Fig. 2.5 Orbital maintenance results of the L_1 displaced orbit

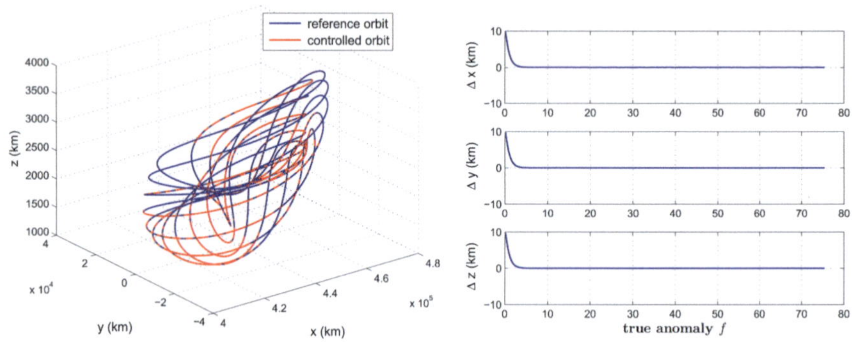

(a) Orbital tracking result (b) Position errors

Fig. 2.6 Orbital maintenance results of the L_2 displaced orbit

Figures 2.5 and 2.6 illustrate the orbital maintenance results of the displaced orbit around L_1 and L_2 point respectively. It can be seen that the position errors diminish asymptotically with the designed controller, indicating that precise orbital tracking can be successfully achieved using the hybrid propulsion system.

Figure 2.7 shows the time histories of the reflectivity rate and attitude angles of the solar sail after optimization for the two cases. In particular, the control strategy resembles a Bang-Bang switching logic for displaced orbit keeping around L_1 point.

Figure 2.8 shows the magnitude and direction angles of the acceleration provided by the SEP system before (solid line) and after (dashed line) optimization. To maintain the displaced orbit around L_1 point, the magnitude of the required thrust lowers slightly while the orientation keeps almost unchanged after optimization. Whereas for the displaced orbit around L_2 point, there exists remarkable difference in both attitude angles and thrust magnitude. It is noteworthy from Fig. 2.8b that, the time intervals during which $a_{SEP} \equiv 0$ indicate the capacity of the solar sail itself to counteract the model errors.

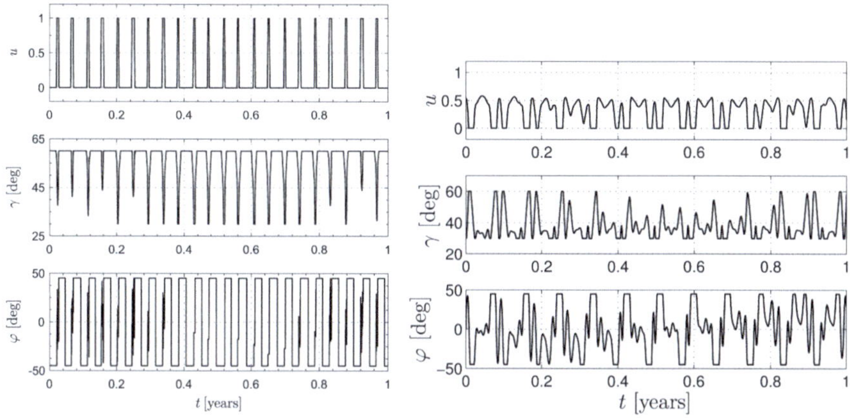

Fig. 2.7 Reflectivity rate and attitude angles of solar sail after optimization. Left: L_1 case; Right: L_2 case

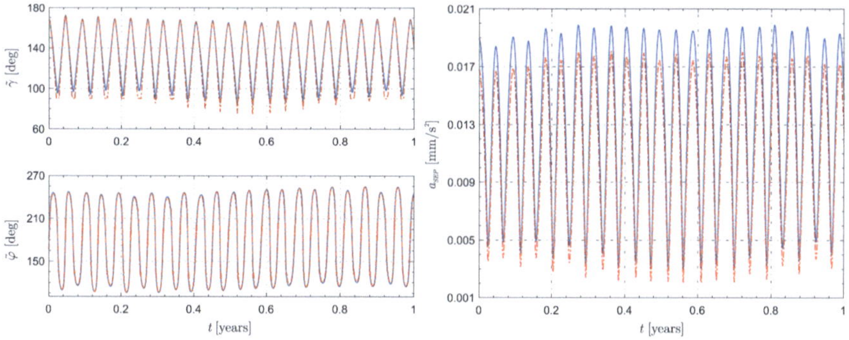

(a) Displaced orbit around L_1 point

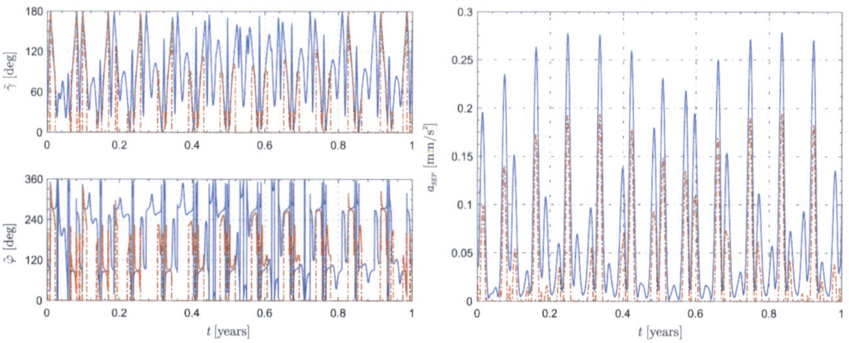

(b) Displaced orbit around L_2 point

Fig. 2.8 Magnitude and direction of SEP acceleration for two cases

Table 2.3 Comparison of fuel consumption

Point	Before optimization (kg)	After optimization (kg)	Reduction rate (%)
L_1	6.2347	5.4382	12.78
L_2	31.8278	12.5625	60.53

To evaluate the propellant consumption of the sail, introduce the mass rate of change

$$\dot{m} = -\frac{a_{SEP}\, m}{I_{sp}\, g_0}, \tag{2.64}$$

where m is the mass of the hybrid sail, I_{sp} is the specific impulse of the SEP system, and $g_0 = 9.8\,\mathrm{m/s^2}$ is the Earth's standard gravity acceleration. It is assumed that the initial mass of the sail $m_0 = 500\,\mathrm{kg}$, and the specific impulse $I_{sp} = 3000\,\mathrm{s}$.

Table 2.3 illustrates the one-year total propellant consumption of the SEP system in the cases where the thrust is optimized and not. It is found that the fuel consumption has reduced by 12.78 % for displaced orbit around L_1 point, whereas for displaced orbit around L_2 point, the fuel reduction is up to 60.53 %. Therefore, it can be concluded that exploiting the most benefit from SRP acceleration by adjusting the sail reflectivity rate and attitude angles, the necessary fuel consumption provided by SEP system can be effectively reduced, especially for the mission applications around L_2 point.

2.7 Conclusions

This chapter investigates the motion of a hybrid sail in the vicinity of cis-lunar libration points within the Earth-Moon ERTBP. Assuming small eccentricities for both the Earth-Moon and Sun-Earth/Moon systems, a new family of quasi-periodic displaced orbits has been constructed. A backstepping orbital controller is proposed to ensure precise reference orbit tracking, while the reflectivity rate and orientation of the solar sail are optimized to minimize the overall propellant consumption of the auxiliary SEP system. In particular, the methodology formulated within the ERTBP framework extends existing results in the literature and has been shown to require less control effort compared to the CRTBP.

Appendix A

The coefficients in s_x component are

$$
\begin{aligned}
c_{x1} &= -\bar{\kappa}\sin i(1-u)\cos^2\gamma\sin\gamma\cos\varphi \\
c_{x2} &= 0 \\
c_{x3} &= -\bar{\kappa}e_m C\frac{u\cos\gamma+2(1-u)\cos^3\gamma}{2} - \bar{\kappa}e_m\frac{3(u\cos\gamma+2(1-u)\cos^3\gamma)}{4} \\
c_{x4} &= \bar{\kappa}e_m C((1-u)\cos^2\gamma\sin\gamma\sin\phi) + \bar{\kappa}e_m\frac{3(1-u)\cos^2\gamma\sin\gamma\sin\varphi}{2} \\
c_{x5} &= \bar{\kappa}\frac{u\cos\gamma+2(1-u)\cos^3\gamma}{2} \\
c_{x6} &= \bar{\kappa}(1-u)\cos^2\gamma\sin\gamma\sin\varphi \\
c_{x7} &= \bar{\kappa}e_{em}(u\cos\gamma+2(1-u)\cos^3\gamma) \\
c_{x8} &= 2\bar{\kappa}e_{em}(1-u)\cos^2\gamma\sin\gamma\sin\varphi \\
c_{x9} &= \bar{\kappa}e_m C\frac{u\cos\gamma+2(1-u)\cos^3\gamma}{2} - \bar{\kappa}e_m\frac{3(u\cos\gamma+2(1-u)\cos^3\gamma)}{4} \\
c_{x10} &= \bar{\kappa}e_m C(1-u)\cos^2\gamma\sin\gamma\sin\varphi - \bar{\kappa}e_m\frac{3(1-u)\cos^2\gamma\sin\gamma\sin\varphi}{2}
\end{aligned}
\tag{A-1}
$$

The coefficients in s_y component are

$$
\begin{aligned}
c_{y1} &= 0 \\
c_{y2} &= \bar{\kappa}\sin i(1-u)\cos^2\gamma\sin\gamma\cos\varphi \\
c_{y3} &= -\bar{\kappa}e_m C(1-u)\cos^2\gamma\sin\gamma\sin\varphi - \bar{\kappa}e_m\frac{3(1-u)\cos^2\gamma\sin\gamma\sin\varphi}{2} \\
c_{y4} &= -\bar{\kappa}e_m C\frac{u\cos\gamma+2(1-u)\cos^3\gamma}{2} - \bar{\kappa}e_m\frac{3(u\cos\gamma+2(1-u)\cos^3\gamma)}{4} \\
c_{y5} &= \bar{\kappa}(1-u)\cos^2\gamma\sin\gamma\sin\varphi \\
c_{y6} &= -\bar{\kappa}\frac{u\cos\gamma+2(1-u)\cos^3\gamma}{2} \\
c_{y7} &= 2\bar{\kappa}e_{em}(1-u)\cos^2\gamma\sin\gamma\sin\varphi \\
c_{y8} &= -\bar{\kappa}e_{em}(u\cos\gamma+2(1-u)\cos^3\gamma) \\
c_{y9} &= \bar{\kappa}e_m C(1-u)\cos^2\gamma\sin\gamma\sin\varphi - \bar{\kappa}e_m\frac{3(1-u)\cos^2\gamma\sin\gamma\sin\varphi}{2} \\
c_{y10} &= -\bar{\kappa}e_m C\frac{u\cos\gamma+2(1-u)\cos^3\gamma}{2} + \bar{\kappa}e_m\frac{3(u\cos\gamma+2(1-u)\cos^3\gamma)}{4}
\end{aligned}
\tag{A-2}
$$

The coefficients in s_z component are

$$
\begin{aligned}
c_{z0} &= \bar{\kappa}(1-u)\cos^2\gamma\sin\gamma\cos\varphi \\
c_{z2} &= \bar{\kappa}\sin i\frac{u\cos\gamma + 2(1-u)\cos^3\gamma}{2} + 2\bar{\kappa}e_{em}(1-u)\cos^2\gamma\sin\gamma\cos\varphi \\
c_{z3} &= -\bar{\kappa}\sin i(1-u)\cos^2\gamma\sin\gamma\sin\varphi \\
c_{z4} &= -3\bar{\kappa}e_m(1-u)\cos^2\gamma\sin\gamma\cos\varphi
\end{aligned}
\tag{A-3}
$$

References

1. McKay R J, Macdonald M, Biggs J, et al. Survey of highly non-Keplerian orbits with low-thrust propulsion[J]. Journal of Guidance, Control, and Dynamics, 2011, 34(3), 645–666.
2. Gong S, Li J. Solar sail heliocentric elliptic displaced orbits[J]. Journal of Guidance, Control, and Dynamics, 2014, 37(6), 2021–2026.
3. Baoyin H, Mcinnes C R. Solar sail halo orbits at the Sun-Earth artificial L_1 point[J]. Celestial Mechanics and Dynamical Astronomy, 2006, 94, 155–171.
4. Macdonald M, McKay R J, Vasile M, et al. Low-thrust-enabled highly-non-Keplerian orbits in support of future Mars exploration[J]. Journal of Guidance, Control, and Dynamics, 2011, 34(5), 1396–1411.
5. Bobskill M R, Lupisella M L. The role of cis-lunar space in future global space exploration[C]. In Global Space Exploration Conference, GLEX-2012.05. 5.4 x12270, Washington D C, USA, 2012.
6. Simo J, McInnes C R. Solar sail orbits at the Earth–Moon libration points[J]. Communications in Nonlinear Science and Numerical Simulation, 2009, 14(12): 4191–4196.
7. Simo J, McInnes C R. Asymptotic analysis of displaced lunar orbits[J]. Journal of Guidance, Control, and Dynamics, 2009, 32(5): 1666–1671.
8. Gong S, Li J, Simo J. Orbital motions of a solar sail around the L_2 Earth–Moon libration point[J]. Journal of Guidance, Control, and Dynamics, 2014, 37(4): 1349–1356.
9. Ozimek M T, Grebow D J, Howell K C. Design of solar sail trajectories with applications to lunar south pole coverage[J]. Journal of Guidance, Control, and Dynamics, 2009, 32(6): 1884–1897.
10. Simo J, McInnes C R. Designing displaced lunar orbits using low-thrust propulsion[J]. Journal of Guidance, Control, and Dynamics, 2010, 33(1): 259–265.
11. Tamakoshi D, Kojima H. Solar sail orbital control using reflectivity variations near the Earth–Moon L_2 point[J]. Journal of Guidance, Control, and Dynamics, 2018, 41(2): 417–430.
12. Tsuda Y, Mori O, Funase R, et al. Achievement of IKAROS-Japanese deep space solar sail demonstration mission[J]. Acta Astronautica, 2013, 82(2): 183–8.
13. Johnson L, Young R, Montgomery E, et al. Status of solar sail technology within NASA[J]. Advances in Space Research, 2011, 48(11): 1687–1694.
14. Betts B, Nye B, Vaughn J, et al. Lightsail 1 mission results and public outreach strategies[C]. 4th International Symposium on Solar Sailing, Kyoto, Japan, 2017.
15. Heiligers J, Diedrich B, Derbes B, et al. Sunjammer: PreliminaryPreliminary end-to-end mission design[C]. AIAA/AAS Astrodynamics Specialist Conference, San Diego, USA, 2014.

Chapter 3
Motion of a Hybrid Sail in the Earth-Moon Circular Restricted Three Body Problem

This chapter investigates the problem of propellant-efficient station-keeping using a hybrid sail in the Earth-Moon system. To achieve high-precision station-keeping and minimize propellant consumption, the problem is approached from both reference orbit design and control strategy perspectives. A high-fidelity model of a hybrid sail, consisting of a solar electric propulsion (SEP) system and a solar sail equipped with reflectivity control devices (RCD), is utilized for reference orbit design in the Earth-Moon system through numerical methods. These hybrid-sail perturbed halo and Lyapunov orbits are parameterized by the sail's reflectivity and are inherently unstable. An orbit-attitude control strategy is proposed for station-keeping, which consists of three components: a nonlinear disturbance observer (NDO)-based optimal periodic orbital controller, SEP acceleration optimization, and an NDO-based robust backstepping attitude controller. Notably, RCD is employed in both orbital and attitude control. Numerical results demonstrate that the proposed control strategy ensures high-precision station-keeping while effectively reducing propellant consumption.

3.1 Introduction

Families of periodic orbits exist around the collinear libration points in the Earth-Moon Circular Restricted Three Body Problem (CRTBP) such as Lyapunov orbits, halo orbits, and vertical Lyapunov orbits. These orbits are of significant importance for exploring and developing cislunar space as they support unique observation and communication missions. Moreover, these natural orbits can be greatly extended by adding low-thrust propulsion into the CRTBP. In particular, the solar-sail libration point orbits (LPOs) have shown potential applications in lunar far-side communications as well as high-latitude observations of the Earth and the Moon [1–4]. Nonetheless, extra propulsion systems are necessary to assist a solar sail for long-term station keeping due to the significant perturbation environments in the Earth-Moon system.

© The Author(s), under exclusive license to Springer Nature Singapore Pte Ltd. 2026
C. Gao and W. Wang, *Solar Sailing for Deep-Space Explorations*,
https://doi.org/10.1007/978-981-95-0109-0_3

Therefore, the purpose of this chapter is to study the problem of station-keeping in the Earth-Moon CRTBP using a hybrid sail propulsion in which a solar sail is coupled with a solar electronic propulsion (SEP) system. To begin with, a high-fidelity model of a hybrid sail is exploited to study the effect of solar radiation pressure (SRP) on the LPOs. In particular, families of hybrid-sail resonant periodic orbits, which are then taken as reference mission orbits for station keeping, are obtained numerically. Finally, a propellant-efficient station-keeping strategy is then proposed for long-term maintenance of the reference orbits.

This chapter aims to achieve high-precision station-keeping while minimizing the propellant consumption of the SEP system, taking into account the propellant efficiency of a hybrid sail. This station-keeping problem can be addressed from two perspectives: reference orbit design and control strategy design. Shirobokov et al. have emphasized that investigations focused on the most precise motion models are crucial for station-keeping, particularly in the Earth-Moon system [5]. Additionally, if the reference orbit is developed using a low-fidelity motion model, the station-keeping method typically incurs higher costs when applied to high-fidelity models. For a hybrid sail, part of the sail area is covered by flexible thin-film solar cells (TFSC), which serve as a power source for the SEP system and other subsystems [6]. Because the optical properties of TFSC differ from those of the standard sail area, the force model for a hybrid sail is more complex than that for a pure solar sail. Therefore, reference orbits for station-keeping should first be designed in the Earth-Moon system using a high-fidelity hybrid sail model.

Regarding the control strategy, orbital controllers with propellant-efficient properties are prioritized. Considering the periodic nature of reference orbits, an optimal periodic orbital controller is designed for station-keeping in this chapter. In fact, optimal controllers, such as the linear quadratic regulator (LQR) and optimal periodic controller, are widely applied in the station-keeping domain due to their simple structure, effectiveness, and robustness [7]. Furthermore, the optimal periodic controller offers faster convergence than the classical LQR [8]. Additionally, the optimal periodic controller should be developed using a disturbance observer for high-precision station-keeping in the presence of external disturbances. Disturbance observers are commonly used to estimate external disturbances and compensate for them via feedforward, thus improving tracking accuracy [9–11]. For example, the extended state observer (ESO)-based proportional control scheme in [12] and the ESO-based LQR scheme in [11] were proposed for low-thrust station-keeping on a halo orbit in the Earth-Moon spacecraft system, where the ESO was employed to estimate the total disturbances. In this chapter, a nonlinear disturbance observer (NDO) is integrated with the optimal periodic controller to enhance station-keeping precision.

In hybrid sail missions, attitude control is as crucial as orbital control, as it directly influences the solar radiation pressure (SRP) force acting on the sail. Due to mass limitations and the large moment of inertia of the sail, conventional actuators like thrusters and control moment gyroscopes are not suitable for attitude control. The reflectivity control device (RCD) is the preferred choice for attitude control in hybrid sail missions, owing to its small mass and lack of propellant consumption. Attitude control using RCD can significantly reduce propellant consumption and extend the mission duration of a hybrid sail. Mu et al. developed the torque model for a spinning disk sail equipped with RCD and explored its attitude control capabilities [13]. They noted that the control torque generated by the RCD is highly constrained and depends on factors such as the area ratio, distribution of the RCD, and the sail's attitude angles. Liu et al. demonstrated the pitch motion control of a sailcraft in Earth orbits using control vanes with RCDs [14]. Tamakoshi et al. investigated the orbital control of displaced orbits above the L_2 point in the Earth-Moon system using a hybrid sail [15].

The chapter is organized as follows. Section 3.2 establishes the force and torque models for a high-fidelity hybrid sail, with the RCD acting as the control actuator. In Sect. 3.3, the orbital and rotational equations of motion for a hybrid sail in the Earth-Moon CRTBP are derived. Section 3.4 discusses the parallel shooting and combination methods used to obtain periodic orbits in the CRTBP. Section 3.5 investigates the evolution of families of hybrid-sail resonant halo and Lyapunov orbits through the application of the parallel shooting method combined with continuation on the SRP acceleration. In Sect. 3.6, a propellant-efficient control strategy for station-keeping is proposed, which includes an NDO-based optimal periodic orbital controller, SEP acceleration optimization, and an NDO-based robust backstepping attitude controller. Section 3.7 presents numerical simulations that demonstrate high-precision station-keeping and effective reduction in propellant consumption using the proposed control strategy. Finally, Sect. 3.8 concludes the chapter.

3.2 Force and Torque Models of a Hybrid Sail

The basic configuration of the hybrid sail considered in this chapter is shown in Fig. 3.1. The hybrid sail is composed of a payload and a SEP thruster at the center as well as a sail covered by TFSC and electrochromic material panels (EMP) respectively. The TFSC acts as a power source for the SEP system and other subsystems

Fig. 3.1 Configuration of a hybrid sail covered by EMP

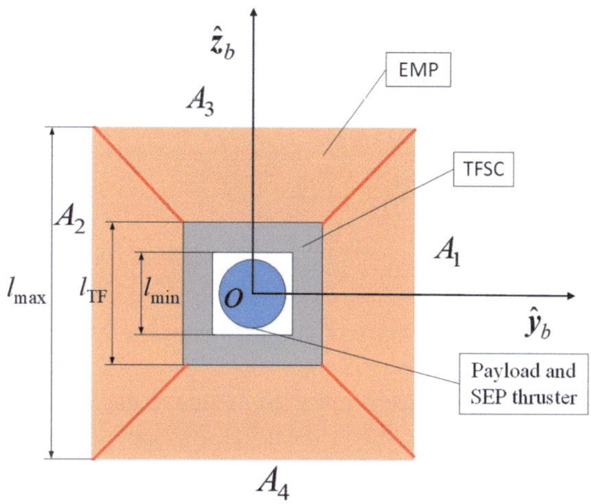

while the EMP is used as RCD for orbit-attitude control. The body-fixed reference frame $\mathcal{T}_B(O;\ \hat{\boldsymbol{x}}_b, \hat{\boldsymbol{y}}_b, \hat{\boldsymbol{z}}_b)$ is defined to describe the attitude of the sail: the origins O is at the center of the sail, the $\hat{\boldsymbol{y}}_b$ axis and the $\hat{\boldsymbol{z}}_b$ axis are in the plane of sail with the $\hat{\boldsymbol{x}}_b$ axis normal to the sail. The area covered by EMPs is divided into four equal parts A_1, A_2, A_3 and A_4 in frame $\mathcal{T}_B$ as shown in Fig. 3.1.

The optical property of the EMP can be switched by power on and off. According to Eqs. (2.10) and (2.11), the SRP forces exerted on the switch-on and switch-off EMP are

$$\begin{cases} \boldsymbol{F}_c^{\text{on}} = 2PS_c^{\text{on}}(\hat{\boldsymbol{n}} \cdot \hat{\boldsymbol{n}}_S)^2 \hat{\boldsymbol{n}} \\ \boldsymbol{F}_c^{\text{off}} = PS_c^{\text{off}}(\hat{\boldsymbol{n}} \cdot \hat{\boldsymbol{n}}_S)\hat{\boldsymbol{n}}_S \end{cases} \tag{3.1}$$

where P is the SRP exerted on the sail surface; $\hat{\boldsymbol{n}}_S$ and $\hat{\boldsymbol{n}}$ are sunlight unit vector and sail unit normal vector, respectively; S_c^{on} and S_c^{off} mean the sail area covered by the switch-on and switch-off EMP.

In addition, the SRP force for the area covered by the TFSC is (see Eq. (2.1))

$$\boldsymbol{F}_{\text{TF}} = PS_{\text{TF}}\left(\hat{\boldsymbol{n}} \cdot \hat{\boldsymbol{n}}_S\right)\left[\left(c_a^{\text{TF}} + c_d^{\text{TF}}\right)\hat{\boldsymbol{n}}_S + \left(\frac{2}{3}c_d^{\text{TF}} + 2c_s^{\text{TF}}\left(\hat{\boldsymbol{n}} \cdot \hat{\boldsymbol{n}}_S\right)\right)\hat{\boldsymbol{n}}\right] \tag{3.2}$$

where c_a^{TF}, c_d^{TF} and c_s^{TF} represent the coefficients of absorption, diffusion, and specular reflection of the TFSC, respectively. According to [6], this chapter sets $c_a^{\text{TF}} = 0.6$, $c_d^{\text{TF}} = 0.0$ and $c_s^{\text{TF}} = 0.4$.

Denote S as the total area of the sail, S_{TF} and S_c as the sail area covered by TFSC and EMP respectively. In addition, let S_c^{off} and S_c^{on} represent the sail area covered by the switch-off and switch-on EMP, then $S_c = S_c^{\text{off}} + S_c^{\text{on}}$ and $S = S_{\text{TF}} + S_c$. Two ratios of area are defined as

$$u_c = \frac{S_c}{S}, \qquad u_c^{\text{off}} = \frac{S_c^{\text{off}}}{S_c} \tag{3.3}$$

where u_c means the ratio of the sail area covered by the EMP to the total area. u_c^{off}, the reflectivity rate, means the sail area covered by switched-off EMP to the area covered by the EMP.

To this end, the acceleration $\boldsymbol{a}_{SRP}$ due to the SRP force for the hybrid sail is

$$\begin{aligned} \boldsymbol{a}_{SRP} &= \frac{\boldsymbol{F}_c^{\text{on}} + \boldsymbol{F}_c^{\text{off}} + \boldsymbol{F}_{\text{TF}}}{m} \\ &= \kappa\left(u_c(1 - u_c^{\text{off}}) + \frac{2}{5}(1 - u_c)\right)(\hat{\boldsymbol{n}} \cdot \hat{\boldsymbol{n}}_S)^2\hat{\boldsymbol{n}} + \kappa\left(\frac{1}{2}u_c u_c^{\text{off}} + \frac{3}{10}(1 - u_c)\right)(\hat{\boldsymbol{n}} \cdot \hat{\boldsymbol{n}}_S)\hat{\boldsymbol{n}}_S \end{aligned} \tag{3.4}$$

where m is the mass of the sail and $\kappa = 2PS/m$ is known as the characteristic acceleration.

Besides, the acceleration $\boldsymbol{a}_{SEP}$ due to the SEP system is

$$\boldsymbol{a}_{SEP} = a_{SEP}\hat{\boldsymbol{m}} \tag{3.5}$$

where a_{SEP} means the magnitude and $\hat{\boldsymbol{m}}$ means the unit thrust vector.

The total acceleration $\boldsymbol{a}$ provided by the hybrid sail due to the SRP force and SEP system can be obtained from the sum of Eqs. (3.4) and (3.5), i.e.,

$$\boldsymbol{a} = \boldsymbol{a}_{SRP} + \boldsymbol{a}_{SEP} \tag{3.6}$$

Assuming that the switched-off EMP are evenly distributed in four parts and area ratios are denoted as u_{y+}, u_{y-}, u_{z+} and u_{z-}, respectively. Ignoring the very small torque generated by the SRP force transverse to the sail surface, the torque about the $\hat{z}_b$ axis due to the SRP force in part A_1 can be expressed as

$$\tau_{z+} = -\int_{\frac{l_{\mathrm{TF}}}{2}}^{\frac{l_{\max}}{2}} 8\frac{u_{y+}}{u_{c\ \max}^{\mathrm{off}}} y^2 \mathrm{d}y \boldsymbol{P}_c^{\mathrm{off}} \cdot \hat{\boldsymbol{n}} - \int_{\frac{l_{\mathrm{TF}}}{2}}^{\frac{l_{\max}}{2}} 2(1 - \frac{4u_{y+}}{u_{c\ \max}^{\mathrm{off}}}) y^2 \mathrm{d}y \boldsymbol{P}_c^{\mathrm{on}} \cdot \hat{\boldsymbol{n}} \tag{3.7}$$

with

$$\begin{aligned} \boldsymbol{P}_c^{\mathrm{on}} &= 2P(\hat{\boldsymbol{n}} \cdot \hat{\boldsymbol{n}}_S)^2 \hat{\boldsymbol{n}} \\ \boldsymbol{P}_c^{\mathrm{off}} &= P(\hat{\boldsymbol{n}} \cdot \hat{\boldsymbol{n}}_S) \hat{\boldsymbol{n}}_S \end{aligned} \tag{3.8}$$

where $u_{c\ \max}^{\mathrm{off}}$ means the maximum value of u_c^{off}, and l_{TF} and $l_{\max}$ represent the length of square areas covered by TFSC and FMP, respectively, as shown in Fig. 3.1.

Similarly, the torque about the $\hat{z}_b$ axis due to the SRP force in part A_2 can be expressed as

$$\tau_{z-} = \int_{\frac{l_{\mathrm{TF}}}{2}}^{\frac{l_{\max}}{2}} 8\frac{u_{y-}}{u_{c\ \max}^{\mathrm{off}}} y^2 \mathrm{d}y \boldsymbol{P}_c^{\mathrm{off}} \cdot \hat{\boldsymbol{n}} + \int_{\frac{l_{\mathrm{TF}}}{2}}^{\frac{l_{\max}}{2}} 2(1 - \frac{4u_{y-}}{u_{c\ \max}^{\mathrm{off}}}) y^2 \mathrm{d}y \boldsymbol{P}_c^{\mathrm{on}} \cdot \hat{\boldsymbol{n}} \tag{3.9}$$

Hence, the total torque about the $\hat{z}_b$ axis is

$$\tau_z = \tau_{z+} + \tau_{z-} = -u_y l_{eq} (\boldsymbol{P}_c^{\mathrm{off}} - \boldsymbol{P}_c^{\mathrm{on}}) \cdot \hat{\boldsymbol{n}} \tag{3.10}$$

with

$$\begin{aligned} u_y &= u_{y+} - u_{y-} \\ l_{eq} &= \int_{\frac{l_{\mathrm{TF}}}{2}}^{\frac{l_{\max}}{2}} \frac{8y^2}{u_{c\ \max}^{\mathrm{off}}} \mathrm{d}y \end{aligned} \tag{3.11}$$

In the same way, the following relation holds for the torque about the $\hat{\boldsymbol{y}}_b$ axis:

$$\tau_y = \tau_{y+} + \tau_{y-} = u_z l_{eq} (\boldsymbol{P}_c^{\mathrm{off}} - \boldsymbol{P}_c^{\mathrm{on}}) \cdot \hat{\boldsymbol{n}} \tag{3.12}$$

with

$$u_z = u_{z+} - u_{z-} \tag{3.13}$$

According to Eqs. (3.10) and (3.12), it can be seen that the torque exerted on the $\hat{\boldsymbol{y}}_b$ axis is independent of u_y while the torque exerted on the $\hat{z}_b$ axis is independent of u_z. This is because A_1 and A_2 are split into half by the $\hat{\boldsymbol{y}}_b$ axis while A_3 and A_4 are split into half by the $\hat{\boldsymbol{y}}_b$ axis as well as the switched-off EMP are distributed evenly in four parts [13].

In addition, the control torque about the $\hat{\boldsymbol{x}}_b$ axis, τ_x, is assumed to be provided by a reaction wheel in this chapter [15]. To this end, in frame $\mathcal{T}_B$ the control torque acting on the hybrid sail is given as

$$[\boldsymbol{\tau}]_{\mathcal{T}_B} = \left[\tau_x, \tau_y, \tau_z\right]^{\mathrm{T}} \tag{3.14}$$

Bearing in mind Eqs. (3.4), (3.10) and (3.12), u_c^{off} is orbit control variable while u_y and u_z are attitude control variables. These three independent parameters are constrained by [13]:

$$\begin{cases} 0 \le u_c^{\mathrm{off}} \le u_{c\ \max}^{\mathrm{off}} \\ -\dfrac{u_{c\ \max}^{\mathrm{off}}}{4} \le u_y \le \dfrac{u_{c\ \max}^{\mathrm{off}}}{4} \\ -\dfrac{u_{c\ \max}^{\mathrm{off}}}{4} \le u_z \le \dfrac{u_{c\ \max}^{\mathrm{off}}}{4} \\ |u_y| + |u_z| \le \min(u_c^{\mathrm{off}}, u_{c\ \max}^{\mathrm{off}} - u_c^{\mathrm{off}}) \end{cases} \tag{3.15}$$

where $u_{c\ \max}^{\mathrm{off}} = 1$ means that all EMP can be switched-off in this chapter.

Equation (3.15) mean that the value of u_y, the difference in ratios of the area of the switched-off EMPs between A_1 and A_2, varies between the maximum value $u_{c\ \max}^{\mathrm{off}}/4$ and the minimum value $-u_{c\ \max}^{\mathrm{off}}/4$. This constraint also holds for u_z. When $u_c^{\mathrm{off}} \le u_{c\ \max}^{\mathrm{off}}/2$, the sum of the absolute values of u_y and u_z is constrained by u_c^{off}; while when $u_c^{\mathrm{off}} > u_{c\ \max}^{\mathrm{off}}/2$, the sum is constrained by $\left(u_{c\ \max}^{\mathrm{off}} - u_c^{\mathrm{off}}\right)$ which is the ratio of the area of the switched-on EMP [13].

3.3 Equations of Motion

In this Section, the orbit and attitude dynamics of a hybrid sail in the Earth-Moon system are established within the framework of CRTBP. In addition, the relative orbital motion dynamics are derived and disturbance acceleration models for station-keeping are established.

3.3.1 Orbit Dynamics of a Hybrid Sail in the Earth-Moon System

The motion of a hybrid sail in the Earth-Moon synodic reference frame $\mathcal{T}_R(O;\ \hat{\boldsymbol{x}}_r, \hat{\boldsymbol{y}}_r, \hat{\boldsymbol{z}}_r)$ is shown in Fig. 3.2. Considering the hybrid sail acceleration, the equations of orbital motion in frame $\mathcal{T}_R$ can be written as

$$\begin{aligned} \ddot{x} - 2\dot{y} &= \Omega_x + a_x \\ \ddot{y} + 2\dot{x} &= \Omega_y + a_y \\ \ddot{z} &= \Omega_z + a_z \end{aligned} \tag{3.16}$$

where $[\boldsymbol{r}]_{\mathcal{T}_R} = [x, y, z]^{\mathrm{T}}$ is the position vector, $[\boldsymbol{v}]_{\mathcal{T}_R} = [\dot{x}, \dot{y}, \dot{z}]^{\mathrm{T}}$ is the velocity vector, $[\boldsymbol{a}]_{\mathcal{T}_R} = [a_x, a_y, a_z]^{\mathrm{T}}$ is the hybrid sail acceleration, and Ω is the potential expressed as

$$\Omega = \frac{1}{2}\left(x^2 + y^2\right) + \frac{1-\mu}{r_1} + \frac{\mu}{r_2} \tag{3.17}$$

$$r_1 = \sqrt{(x+\mu)^2 + y^2 + z^2} \tag{3.18}$$

$$r_2 = \sqrt{(x-1+\mu)^2 + y^2 + z^2} \tag{3.19}$$

where μ is the mass parameter of the Earth-Moon system, and r_1 and r_2 are non-dimensional Earth-sail distance and Moon-sail distance, respectively.

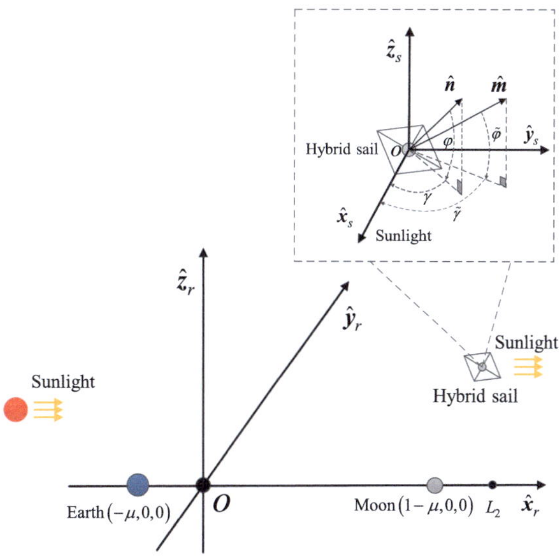

Fig. 3.2 Schematic of a hybrid sail in the Earth-Moon system

Denoting $[\boldsymbol{r}_n]_{\mathcal{T}_R} = [x_n, y_n, z_n]^{\mathrm{T}}$ the nominal periodic orbit, and $[\delta\boldsymbol{r}]_{\mathcal{T}_R} = [\delta x, \delta y, \delta z]^{\mathrm{T}}$ the small deviation of the controlled trajectory $\boldsymbol{r}_c$ from $\boldsymbol{r}_n$. Besides, $\boldsymbol{a}_c$ and $\boldsymbol{a}_n$ are hybrid sail accelerations of the controlled trajectory and the nominal orbit, respectively. Then the nonlinear relative motion dynamics in frame $\mathcal{T}_R$ can be written as

$$\begin{aligned}
\delta\ddot{x} &= 2\delta\dot{y} + \delta x - (1-\mu)\left(\frac{\mu + x_n + \delta x}{r_{1c}^3} - \frac{\mu + x_n}{r_{1n}^3}\right) - \mu\left(\frac{\mu + x_n + \delta x - 1}{r_{2c}^3} - \frac{\mu + x_n - 1}{r_{2n}^3}\right) + \delta a_x \\
\delta\ddot{y} &= -2\delta\dot{x} + \delta y - (1-\mu)\left(\frac{y_n + \delta y}{r_{1c}^3} - \frac{y_n}{r_{1n}^3}\right) - \mu\left(\frac{y_n + \delta y}{r_{2c}^3} - \frac{y_n}{r_{2n}^3}\right) + \delta a_y \\
\delta\ddot{z} &= -(1-\mu)\left(\frac{z_n + \delta z}{r_{1c}^3} - \frac{z_n}{r_{1n}^3}\right) - \mu\left(\frac{z_n + \delta z}{r_{2c}^3} - \frac{z_n}{r_{2n}^3}\right) + \delta a_z
\end{aligned} \tag{3.20}$$

where $[\delta\boldsymbol{a}]_{\mathcal{T}_R} = [\boldsymbol{a}_c]_{\mathcal{T}_R} - [\boldsymbol{a}_n]_{\mathcal{T}_R} = [\delta a_x, \delta a_y, \delta a_z]^{\mathrm{T}}$ means the difference in the hybrid sail acceleration between the controlled trajectory and the nominal orbit, and

$$\begin{aligned}
r_{1n} &= \sqrt{(x_n + \mu)^2 + y_n^2 + z_n^2} \\
r_{2n} &= \sqrt{(x_n + \mu - 1)^2 + y_n^2 + z_n^2} \\
r_{1c} &= \sqrt{(x_n + \delta x + \mu)^2 + (y_n + \delta y)^2 + (z_n + \delta z)^2} \\
r_{2c} &= \sqrt{(x_n + \delta x + \mu - 1)^2 + (y_n + \delta y)^2 + (z_n + \delta z)^2}
\end{aligned} \tag{3.21}$$

Typically for station-keeping problems the dynamics are highly nonlinear as seen in Eq. (3.20), and a traditional approach first linearizes the nonlinear system about the nominal orbit [5, 7]. Hence, the linear relative motion dynamics about $\boldsymbol{r}_n$ can be obtained as

$$\delta\dot{\boldsymbol{x}} = \mathbf{A}(t)\delta\boldsymbol{x} + \mathbf{B}\delta\boldsymbol{a} \tag{3.22}$$

where $[\delta\boldsymbol{x}]_{\mathcal{T}_R} = [\delta x, \delta y, \delta z, \delta\dot{x}, \delta\dot{y}, \delta\dot{z}]^{\mathrm{T}}$ is the error state and

$$\mathbf{A}(t) = \begin{bmatrix} \mathbf{O}_{3\times3} & \mathbf{I}_{3\times3} \\ \Omega_{XX}|_{\boldsymbol{r}=\boldsymbol{r}_n} & \mathbf{C} \end{bmatrix}, \quad \mathbf{B} = \begin{bmatrix} \mathbf{O}_{3\times3} \\ \mathbf{I}_{3\times3} \end{bmatrix} \tag{3.23}$$

with $\mathbf{O}_{3\times3}$ and $\mathbf{I}_{3\times3}$ as zeros matrix and identity matrix, and

$$\Omega_{XX} = \begin{bmatrix} \Omega_{xx} & \Omega_{xy} & \Omega_{xz} \\ \Omega_{yx} & \Omega_{yy} & \Omega_{yz} \\ \Omega_{zx} & \Omega_{zy} & \Omega_{zz} \end{bmatrix}, \quad \mathbf{C} = \begin{bmatrix} 0 & 2 & 0 \\ -2 & 0 & 0 \\ 0 & 0 & 0 \end{bmatrix} \tag{3.24}$$

where

$$
\begin{cases}
\Omega_{xx} = 1 - \dfrac{1-\mu}{r_1^3} - \dfrac{\mu}{r_2^3} + \dfrac{3(1-\mu)(x+\mu)^2}{r_1^5} + \dfrac{3\mu(x+\mu-1)^2}{r_2^5} \\
\Omega_{xy} = \dfrac{3(1-\mu)(x+\mu)y}{r_1^5} + \dfrac{3\mu(x+\mu-1)y}{r_2^5} \\
\Omega_{xz} = \dfrac{3(1-\mu)(x+\mu)z}{r_1^5} + \dfrac{3\mu(x+\mu-1)z}{r_2^5} \\
\Omega_{yx} = \Omega_{xy} \\
\Omega_{yy} = 1 - \dfrac{1-\mu}{r_1^3} - \dfrac{\mu}{r_2^3} + \dfrac{3(1-\mu)y^2}{r_1^5} + \dfrac{3\mu y^2}{r_2^5} \\
\Omega_{yz} = \dfrac{3(1-\mu)yz}{r_1^5} + \dfrac{3\mu yz}{r_2^5} \\
\Omega_{zx} = \Omega_{xz} \\
\Omega_{zy} = \Omega_{yz} \\
\Omega_{zz} = -\dfrac{1-\mu}{r_1^3} - \dfrac{\mu}{r_2^3} + \dfrac{3(1-\mu)z^2}{r_1^5} + \dfrac{3\mu z^2}{r_2^5}
\end{cases}
\tag{3.25}
$$

Since the coefficient matrix $\mathbf{A}(t)$ is the function of the nominal periodic orbit $\boldsymbol{r}_n$, Eq. (3.22) is a linear periodic ordinary differential equation. Hence, an optimal periodic orbital controller can be designed exploiting this inherent periodic property. Once the optimal periodic orbital controller is obtained, it is worth to note that simulations are still implemented using the nonlinear Eq. (3.20).

3.3.2 *Attitude Dynamics of a Hybrid Sail in the Earth-Moon System*

To describe the rotational motion of the hybrid sail, Fig. 3.2 introduces the sunlight reference frame $\mathcal{T}_S(O;\ \hat{\boldsymbol{x}}_s, \hat{\boldsymbol{y}}_s, \hat{\boldsymbol{z}}_s)$. The origin O of frame is at the center of the sail, the $\hat{\boldsymbol{x}}_s$ axis is in the direction of the sunlight, the $\hat{\boldsymbol{z}}_s$ axis is in same direction with the $\hat{\boldsymbol{z}}_r$ axis, and the $\hat{\boldsymbol{y}}_s$ axis satisfies the right-handed rule. Ignoring the small inclination angle between the Earth-Moon plane and the ecliptic plane, the transformation matrix $\mathbf{T}_{RS}$ from $\mathcal{T}_S$ to $\mathcal{T}_R$ is

$$
\mathbf{T}_{RS} = \mathbf{T}_z(-\omega_S t) \tag{3.26}
$$

where $\omega_S = 0.9252$ is the orbital angular velocity of the Sun around the barycenter of the Earth-Moon system, minus symbol means the orbital motion of the Sun is clockwise, and $\mathbf{T}_z$ is the rotation matrix about the $\hat{z}$ axis.

With $3-2-1$ Euler angles, the transformation matrix from $\mathcal{T}_S$ to $\mathcal{T}_B$ is given as

$$\mathbf{T}_{BS} = \mathbf{T}_x(\psi)\mathbf{T}_y(\varphi)\mathbf{T}_z(\gamma) \tag{3.27}$$

where ψ, φ and γ are roll, pitch and yaw angles of the hybrid sail, respectively. $\mathbf{T}_x$ and $\mathbf{T}_y$ are the rotation matrices about the $\hat{\boldsymbol{x}}$ axis and $\hat{\boldsymbol{y}}$ axis.

Bearing in mind relations among $\mathcal{T}_R$, $\mathcal{T}_S$, $\mathcal{T}_B$ and the inertial reference frame, the kinematic equations of the hybrid sail can be obtained as

$$\begin{bmatrix} \dot{\psi} \\ \dot{\varphi} \\ \dot{\gamma} \end{bmatrix} = \begin{bmatrix} 1 & \tan\varphi\sin\psi & \tan\varphi\cos\psi \\ 0 & \cos\psi & -\sin\psi \\ 0 & \sin\psi/\cos\varphi & \cos\psi/\cos\varphi \end{bmatrix} \begin{bmatrix} \omega_x + \sin\varphi(1-\omega_S) \\ \omega_y - \sin\psi\cos\varphi(1-\omega_S) \\ \omega_z - \cos\psi\cos\varphi(1-\omega_S) \end{bmatrix} \tag{3.28}$$

where $[\boldsymbol{\omega}]_{\mathcal{T}_B} = [\omega_x, \omega_y, \omega_z]^{\mathrm{T}}$ is the angular velocity vector of the sail relative to the inertial reference frame expressed in frame $\mathcal{T}_B$.

The attitude dynamics of the hybrid sail can be expressed as

$$\begin{bmatrix} I_x\dot{\omega}_x + (I_z - I_y)\omega_y\omega_z \\ I_y\dot{\omega}_y + (I_x - I_z)\omega_z\omega_x \\ I_z\dot{\omega}_z + (I_y - I_x)\omega_y\omega_x \end{bmatrix} = \begin{bmatrix} \tau_x \\ \tau_y \\ \tau_z \end{bmatrix} + \begin{bmatrix} \tau_{gx} \\ \tau_{gy} \\ \tau_{gz} \end{bmatrix} \tag{3.29}$$

where $[\mathbf{I}]_{\mathcal{T}_B} = \mathrm{diag}([I_x, I_y, I_z])$ is the inertial matrix of the hybrid sail in frame $\mathcal{T}_B$, $[\boldsymbol{\tau}]_{\mathcal{T}_B} = [\tau_x, \tau_y, \tau_z]^{\mathrm{T}}$ is the control torque in frame $\mathcal{T}_B$, and $[\boldsymbol{\tau}_g]_{\mathcal{T}_B} = [\tau_{gx}, \tau_{gy}, \tau_{gz}]^{\mathrm{T}}$ is the gravity gradient torque in frame $\mathcal{T}_B$ due to the primaries which is regarded as disturbance torque in this chapter. According to [16], $\boldsymbol{\tau}_g$ can be written as

$$\begin{bmatrix} \tau_{gx} \\ \tau_{gy} \\ \tau_{gz} \end{bmatrix} = \sum_{i=1}^{2} \frac{\mu_i}{r_i^3} \begin{bmatrix} (I_z - I_y)u_{iyB}u_{izB} \\ (I_x - I_z)u_{ixB}u_{izB} \\ (I_y - I_x)u_{ixB}u_{iyB} \end{bmatrix} \tag{3.30}$$

where $\mu_1 = 1 - \mu$ and $\mu_2 = \mu$, and u_{ixB}, u_{iyB} and u_{izB} are the components of the unit vector $\hat{\boldsymbol{r}}_i$ $(i = 1, 2)$ in frame $\mathcal{T}_B$; $\boldsymbol{r}_1$ and $\boldsymbol{r}_2$ are vectors from the Earth and the Moon to the sail, respectively.

3.3.3 Disturbance Accelerations

In the Earth-Moon system, the eccentric nature of the Moon's orbit and gravity force of the Sun are considered as the main perturbations for orbital motion of spacecraft. Therefore, only these two disturbance forces are considered in this chapter. The disturbance acceleration due to the eccentric nature of the Moon's orbit $\boldsymbol{d}_e$ is mainly in the $\hat{\boldsymbol{x}}_r$ axis direction, and it can be expressed approximately as [17]

$$d_{ex} = \mu\left|\frac{1}{r_c^2} - \frac{1}{r_e^2}\right| \tag{3.31}$$

where r_c is the distance between the libration point and the Moon on Moon's circular orbit, and r_e is the distance between the libration point and the Moon at the pericenter of Moon's elliptical orbit.

The disturbance acceleration due to the gravity force of the Sun is [18]

$$\boldsymbol{d}_s = m_S \left(\frac{(1-\mu)\boldsymbol{r}_{SE}}{\|\boldsymbol{r}_{SE}\|^3} + \frac{\mu \boldsymbol{r}_{SM}}{\|\boldsymbol{r}_{SM}\|^3} - \frac{\boldsymbol{r}_S}{\|\boldsymbol{r}_S\|^3} \right) \tag{3.32}$$

where m_S is mass of the Sun, and $\boldsymbol{r}_{SE}$, $\boldsymbol{r}_{SM}$ and $\boldsymbol{r}_S$ represent the position vectors from the Sun to the Earth, the Moon and the sail, respectively

3.4 Computation Methods

3.4.1 The Parallel Shooting Method

Due to the highly unstable characteristics of the collinear libration points, the computation of periodic orbits of the families emanating from them needs of a parallel shooting method. The idea of shooting method is best described as a process of targeting. Suppose we want to find a periodic orbit. For a trajectory from $x(t_0)$ to $x(t_f)$ under the dynamical system

$$\dot{x} = f(x, t) \tag{3.33}$$

we want to make slight adjustments, $\delta x(t_0)$, of the initial state $x(t_0)$ so that the adjusted trajectory will end up at a desired final state $x_d(t_f)$, near $x(t_f)$. The idea of the "parallel" method is to discretize the trajectory into several parallel shooting processes which can effectively increase the domain of convergence. Taking the CRTBP as example, the procedure is briefly described in the following.

In the parallel shooting procedures, the total time interval (in this case the period T of the system) is divided into a number of shorter time subintervals (even that most of the times the length of the subintervals are chosen to be equal in order to simplify the implementation of the method. In this book sometimes, due to performance reasons, the length of the time-subintervals have been chosen in a non-uniform way). Selecting, for instance, N equally spaced epochs $t_0, \ldots, t_N$, with $t_0 = 0$ the initial one, and $t_N = T$, then $\Delta t = t_{i+1} - t_i$ denotes the length of all the subintervals and

$$\boldsymbol{Q}_i = (t_i, x_i, y_i, z_i, \dot{x}_i, \dot{y}_i, \dot{z}_i)^{\mathrm{T}}, \quad i = 0, 1, \ldots, N \tag{3.34}$$

are the states on the orbit that must be determined in order to satisfy the periodicity condition $\boldsymbol{Q}_0 = \boldsymbol{Q}_N$ of the CRTBP. Let $\Phi_{\Delta t}(\boldsymbol{Q}_i)$ be the image of the state $\boldsymbol{Q}_i$ under the flow Φ associated to Eq. (3.16) after the time interval Δt. Then, the matching conditions of the parallel shooting method require that

$$\boldsymbol{F}(\boldsymbol{Q}_0, \boldsymbol{Q}_1, \dots, \boldsymbol{Q}_{N-1}) = \begin{bmatrix} \Phi_{\Delta t}(\boldsymbol{Q}_0) \\ \Phi_{\Delta t}(\boldsymbol{Q}_1) \\ \vdots \\ \Phi_{\Delta t}(\boldsymbol{Q}_{N-1}) \end{bmatrix} - \begin{bmatrix} \boldsymbol{Q}_1 \\ \boldsymbol{Q}_2 \\ \vdots \\ \boldsymbol{Q}_0 \end{bmatrix} = \boldsymbol{0} \tag{3.35}$$

Note that, since the epochs t_i are fixed ($t_i = i\,\Delta t$) they are not considered as unknowns in the $\boldsymbol{Q}_i$ vectors and, consequently, Eq. (3.35) is a non-linear system of $6 \times N$ equations with $6 \times N$ unknowns. This system is solved using a Newton-Raphson method until a certain accuracy is achieved. In our computations the accuracy requirement has been set to $\|\boldsymbol{F}(\boldsymbol{Q}_0, \boldsymbol{Q}_1, \dots, \boldsymbol{Q}_{N-1})\| \leq 10^{-12}$.

3.4.2 The Numerical Continuation Method

Consider a parametric nonlinear function

$$\boldsymbol{F}(\boldsymbol{X}, p) = 0 \tag{3.36}$$

where $\boldsymbol{X}$ is the state and p represents the parameter. Assume that for a certain value of $p = p_0$ a solution $\boldsymbol{X}(p_0)$ of this system is known. Starting at $\boldsymbol{X}(p_0)$, the goal is to compute the curve $\boldsymbol{X}(p)$, such that $\boldsymbol{F}(\boldsymbol{X}(p), p) = 0$, up to a certain final value of the parameter.

To simplify the notation, the parameter p is introduced as an additional coordinate in $\boldsymbol{X}$, and the augmented state is denoted by $\widetilde{\boldsymbol{X}} = (\boldsymbol{X}, p)$, with $\widetilde{\boldsymbol{X}}_0 = (\boldsymbol{X}(p_0), p_0)$. When the rank of the Jacobian matrix $D\boldsymbol{F}(\widetilde{\boldsymbol{X}})$ is maximal, the implicit function theorem ensures the existence of a solution curve $\boldsymbol{X}(p)$ such that $\boldsymbol{F}(\boldsymbol{X}(p), p) = 0$.

$$\frac{d\boldsymbol{X}}{dp} = -\left(D_X \boldsymbol{F}(\widetilde{\boldsymbol{X}})\right)^{-1} D_p \boldsymbol{F}(\widetilde{\boldsymbol{X}}),$$

In this case the dimension of the kernel of $D\boldsymbol{F}(\widetilde{\boldsymbol{X}}_0)$ is one. Let $\boldsymbol{v} \in \ker(D\boldsymbol{F}(\widetilde{\boldsymbol{X}}_0))$, and consider as initial guess for the computation of the "next" point $\widetilde{\boldsymbol{X}}_1$ in the characteristic curve, $\widetilde{\boldsymbol{X}}_1^0 = \widetilde{\boldsymbol{X}}_0 + \epsilon \boldsymbol{v}$, being ϵ a continuation step size. The refinement towards a solution of the under-determined system $\boldsymbol{F}(\widetilde{\boldsymbol{X}}_1) = 0$, using $\widetilde{\boldsymbol{X}}_1^0$ as initial guess, is obtained solving

$$\boldsymbol{F}(\widetilde{\boldsymbol{X}}_1^k) + D\boldsymbol{F}(\widetilde{\boldsymbol{X}}_1^k)\, \Delta\widetilde{\boldsymbol{X}}_1^k = 0, \quad k = 0, 1, \dots \tag{3.37}$$

where $\widetilde{\boldsymbol{X}}_1^{k+1} = \widetilde{\boldsymbol{X}}_1^k + \Delta\widetilde{\boldsymbol{X}}_1^k$, and requiring $\Delta\widetilde{\boldsymbol{X}}_1^k$ to be minimum with respect to the Euclidean norm. In this way, one obtains the iterative procedure

$$\widetilde{\boldsymbol{X}}_1^{k+1} = \widetilde{\boldsymbol{X}}_1^k - D\boldsymbol{F}(\widetilde{\boldsymbol{X}}_1^k)^{\mathrm{T}} (D\boldsymbol{F}(\widetilde{\boldsymbol{X}}_1^k) D\boldsymbol{F}(\widetilde{\boldsymbol{X}}_1^k)^{\mathrm{T}})^{-1} \boldsymbol{F}(\widetilde{\boldsymbol{X}}_1^k), \quad k = 0, 1, \dots \tag{3.38}$$

The step size ϵ is controlled depending on the number of iterations that iterative refinement procedure needs to achieve a certain accuracy threshold. Adjusting this step size, the continuation approach also works when turning points appear.

At a bifurcation point $\widetilde{\boldsymbol{X}}_0$, the dimension of the kernel of $D\boldsymbol{F}(\widetilde{\boldsymbol{X}}_0)$ is greater than one and the linearization of $\boldsymbol{F}$ is not enough to determine the continuation direction, since more than one possibility appears. In our case the dimension of the kernel of $D\boldsymbol{F}(\widetilde{\boldsymbol{X}}_0)$ at the bifurcation points has been two. Let $\boldsymbol{v}_1$ and $\boldsymbol{v}_2$ be the unitary vectors in ker $(D\boldsymbol{F}(\widetilde{\boldsymbol{X}}_0))$, and let us consider the expansion of $\boldsymbol{F}$ around the point $\widetilde{\boldsymbol{X}}_0$,

$$\boldsymbol{F}(\widetilde{\boldsymbol{X}}_0 + \Delta\widetilde{\boldsymbol{X}}_0) = \boldsymbol{F}(\widetilde{\boldsymbol{X}}_0) + D\boldsymbol{F}(\widetilde{\boldsymbol{X}}_0)\,\Delta\widetilde{\boldsymbol{X}}_0 + \frac{1}{2}\Delta\widetilde{\boldsymbol{X}}_0^{\mathrm{T}}\, H(\widetilde{\boldsymbol{X}}_0)\,\Delta\widetilde{\boldsymbol{X}}_0 + \cdots. \tag{3.39}$$

A continuation direction $\Delta\widetilde{\boldsymbol{X}}_0$ is again written in the form $\epsilon\boldsymbol{v}$ where $\boldsymbol{v}$ is a linear combination of $\boldsymbol{v}_1$ and $\boldsymbol{v}_2$, $\boldsymbol{v} = \lambda_1\boldsymbol{v}_1 + \lambda_2\boldsymbol{v}_2$, such that substituting $\boldsymbol{v}$ into Eq. (3.39) gives, $\boldsymbol{\lambda}^{\mathrm{T}}M\boldsymbol{\lambda} = 0$, where $\boldsymbol{\lambda} = (\lambda_1, \lambda_2)^{\mathrm{T}}$ and $M = \frac{1}{2}(\boldsymbol{v}_1^{\mathrm{T}}, \boldsymbol{v}_2^{\mathrm{T}})^{\mathrm{T}}\, H(\widetilde{\boldsymbol{X}}_0)\,(\boldsymbol{v}_1^{\mathrm{T}}, \boldsymbol{v}_2^{\mathrm{T}})$. Solving this equation one gets two solutions, $\boldsymbol{\lambda}^1 = (\lambda_1^1, \lambda_2^1)^{\mathrm{T}}$ and $\boldsymbol{\lambda}^2 = (\lambda_1^2, \lambda_2^2)^{\mathrm{T}}$, from which the associated directions $\boldsymbol{v}^1 = \lambda_1^1\boldsymbol{v}_1 + \lambda_2^1\boldsymbol{v}_2$ and $\boldsymbol{v}^2 = \lambda_1^2\boldsymbol{v}_1 + \lambda_2^2\boldsymbol{v}_2$ for the continuation procedure are computed. Note that the Jet Transport technique which is going to be introduced in the following can be used to obtain any arbitray order Taylor expansion of a nonlinear function $\boldsymbol{F}$.

3.5 Families of Hybrid-Sail Resonant Periodic Orbits

In the sunlight reference frame $\mathcal{T}_S$, the sunlight unit vector is given as

$$[\hat{\boldsymbol{n}}_S]_{\mathcal{T}_S} = [1, 0, 0]^{\mathrm{T}} \tag{3.40}$$

To create an out-of-plane acceleration component, the sail can be inclined with respect to the Earth-Moon plane. Denote by φ the pitch angle, the sail unit normal vector in $\mathcal{T}_S$ is

$$[\hat{\boldsymbol{n}}]_{\mathcal{T}_S} = [\cos\varphi, 0, -\sin\varphi]^{\mathrm{T}} \tag{3.41}$$

Bearing in mind the relationship between frames $\mathcal{T}_R$ and $\mathcal{T}_S$ in Eq. (3.26) and substituting Eqs. (3.40) and (3.41) into Eq. (3.4), the three components of the SRP acceleration in frame $\mathcal{T}_R$ become

$$\begin{aligned}
a_x &= \kappa\left(u_c(1-u_c^{\mathrm{off}}) + \frac{2}{5}(1-u_c)\right)\cos^3\varphi\cos(\omega_S t) + \kappa\left(\frac{1}{2}u_c u_c^{\mathrm{off}} + \frac{3}{10}(1-u_c)\right)\cos\varphi\cos(\omega_S t)\\
a_y &= -\kappa\left(u_c(1-u_c^{\mathrm{off}}) + \frac{2}{5}(1-u_c)\right)\cos^3\varphi\sin(\omega_S t) - \kappa\left(\frac{1}{2}u_c u_c^{\mathrm{off}} + \frac{3}{10}(1-u_c)\right)\cos\varphi\sin(\omega_S t)\\
a_z &= -\kappa\left(u_c(1-u_c^{\mathrm{off}}) + \frac{2}{5}(1-u_c)\right)\cos^2\varphi\sin\varphi
\end{aligned} \tag{3.42}$$

According to Eq. (3.42), the SRP acceleration is time-varying in frame $\mathcal{T}_R$ which makes the dynamical system (3.16) a periodic system with frequency ω_S. Therefore, perturbing the natural periodic LPOs in the CRTBP with SRP acceleration will make them gain the frequency ω_S and almost all them become two-dimensional invariant tori. Nevertheless, there are some orbits that remain periodic, they are resonant periodic orbits whose period is a (rational) multiple of the period of the Sun. Figure 3.3 presents the families of a natural Lyapunov and halo orbits around the Earth-Moon L_1 and L_2 points and their orbital periods. The periodic orbits marked with red denote the resonant candidates and $T_S = 2\pi/\Omega_S$ means the period of the dynamical system. According to the figure, the resonance is 1:2 for the resonant L_1-point Lyapunov orbit, 1:3 for the resonant L_1-point halo orbit, 1:2 for the resonant L_2-point Lyapunov orbit, and 1:2 for the resonant L_2-point for halo orbit. Starting from these resonant orbits, the associating SRP perturbed families of orbits can be obtained numerically by using the parallel shooting method and continuation method introduced in the last section. Taking the resonant L_1-point Lyapunov orbit and resonant L_2-point halo orbit as example, this section gives the evolution of their families with sail parameters as continuation parameters. The same procedures can be extended to the resonant L_1-point halo orbit and resonant L_2-point Lyapunov orbit. Furthermore, the initial configuration between the Sun and the sail affects the results of numerical methods leading to different families [1]. For simplicity, two initial Sun-sail configurations ($\min x(y = 0)$) and ($\max x(y = 0)$) in Fig. 3.4 are investigated in this chapter. It is clear that the sail is located at the left of the libration point at the initial time for the ($\min x(y = 0)$) configuration while at the right for the ($\max x(y = 0)$) configuration. Parameters of the hybrid sail employed in this chapter are shown in Table 3.1, where m_0 and $\mathbf{I_0}$ are initial mass and inertial matrix of the hybrid sail, respectively. In addition, the value of SRP in the Earth-Moon system is $P = 4.563 \times 10^{-6}$ Pa. Accordingly, the dimensionless characteristic acceleration κ for the sail 0.05. Besides, the coverage rate of the EMPs for the sail is $u_c = S_c/S = 0.95$.

3.5.1 *Continuation of the L_1-Point Resonant Lyapunov Orbits*

Figure 3.5 shows the results obtained for the evolution of the L_1-point Lyapunov orbit when κ, u_c, φ and u_c^{off} are taken as continuation parameters successively. In particular, the pitch angle φ is set to zero when varying κ and u_c. This means that the sail is perpendicular to the incident photons and the out-of-plane component of the SRP acceleration is zero. Hence, the perturbed families remain planar and the characteristic curves show the value of κ and u_c in front of the initial $\hat{\boldsymbol{x}}_r$–axis condition x_0. According to Fig. 3.5a and b, the value of x_0 changes slightly for $\kappa \in [0, 0.05]$ and $u_c \in [0.9, 1]$. Fixing $\kappa = 0.05$ and $u_c = 0.95$ and varying the pitch angle φ, the characteristic curve is given in Fig. 3.5c. Due to the out-of-plane SRP acceleration

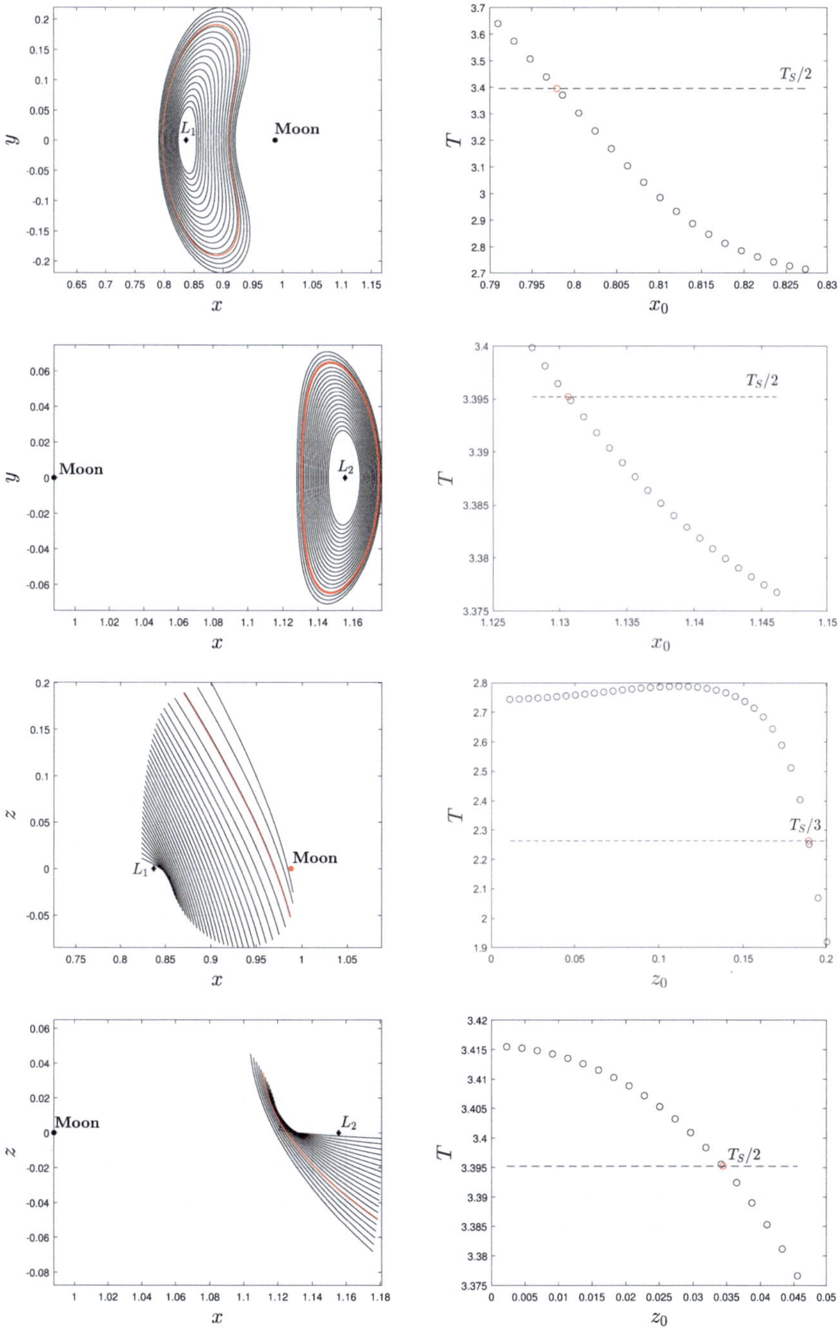

Fig. 3.3 Families of natural LPOs in the Earth–Moon system and their orbital periods. From top to bottom: Lyapunov orbits around the L_1 point, Lyapunov orbits around the L_2 point, halo orbits around the L_1 point, halo orbits around the L_2 point

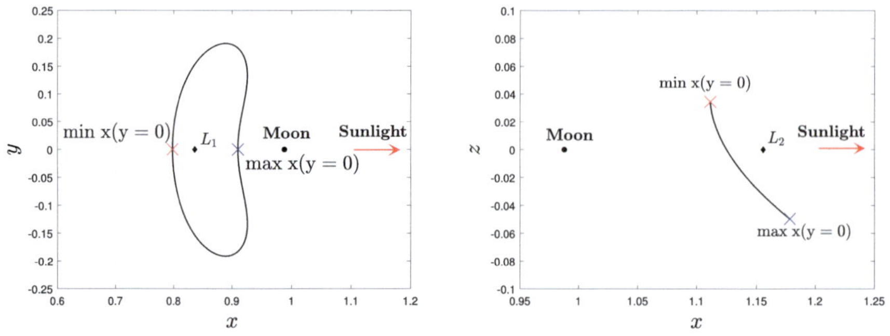

Fig. 3.4 Initial Sun-sail configurations. Left: Lyapunov orbit; Right: halo orbit

Table 3.1 Parameters of the hybrid sail

Symbol	Value
l_{min}	2 m
l_{TF}	7 m
l_{max}	30 m
S	896 m^2
S_c	851 m^2
S_{TF}	45 m^2
m_0	60 kg
$[\mathbf{I_0}]_{\mathcal{T}_B} = \text{diag}\left(\left[I_{x0}, I_{y0}, I_{z0}\right]\right)$	$[20, 3, 3] \times 10^3$ kgm^2

component, the planar orbits become spacial and the initial $\hat{z}_0$–component z_0 is used to characterize the family. For the initial configuration $\min x(y = 0)$, starting from $\varphi = 0$, z_0–value decreases at first and then increase with φ increasing to $\pi/2$. The minimum z_0–value appears at $\varphi \approx 0.603$. The SRP acceleration disappears when $\varphi = \pi/2$ and the perturbed orbit becomes planar again. Due to the symmetry of the system, the curve for $\varphi \in [-\pi/2, 0]$ is symmetric to the one for $\varphi \in [0, \pi/2]$. In addition, for the initial configuration $\max x(y = 0)$ with φ increasing from 0 to $\pi/2$, the value of z_0 increases at first and reaches its maximum at $\varphi \approx 0.7$. Then, z_0–value decreases with the minimum value at $\varphi \approx 0.9$. For $\varphi = \pi/2$, the perturbed orbit becomes planar again. Also, the curve for $\varphi \in [-\pi/2, 0]$ is symmetric to the one for $\varphi \in [0, \pi/2]$. Fixing $\varphi = -0.603$, Fig. 3.5d presents the value of z_0 with respect to u_c^{off}. For both initial configurations, with u_c^{off} increasing from 0 to 1, the z_0 magnitude of the perturbed orbit deceases. When $u_c^{\text{off}} = 1$, the value of z_0 is almost be zero and the orbit is almost planar.

Taking the initial configuration $\min x(y = 0)$ as example, Fig. 3.6 presents the families of perturbed orbits corresponding to the Fig. 3.5. The symbol $\times$ denotes the initial position of the hybrid sail. It can be seen that κ and u_c parameterized families are planar while φ and u_c^{off} parameterized families are spacial. Fixing $\kappa = 0.05$, $u_c = 0.95$ and $\varphi = -0.603$, the initial Lyapunov orbit and three u_c^{off} parameterized

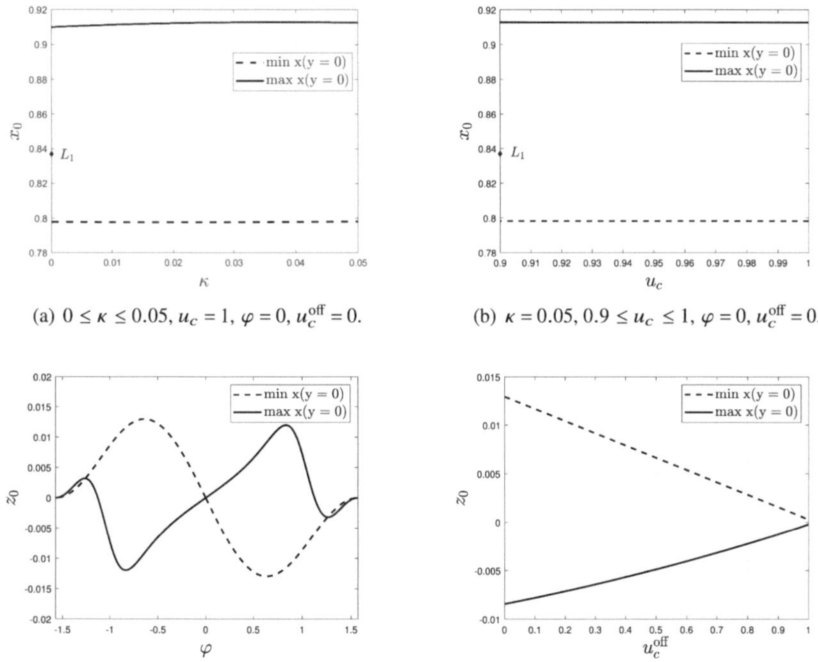

(a) $0 \leq \kappa \leq 0.05$, $u_c = 1$, $\varphi = 0$, $u_c^{\text{off}} = 0$.

(b) $\kappa = 0.05$, $0.9 \leq u_c \leq 1$, $\varphi = 0$, $u_c^{\text{off}} = 0$.

(c) $\kappa = 0.05$, $u_c = 0.95$, $-\pi/2 \leq \varphi \leq \pi/2$, $u_c^{\text{off}} = 0$.

(d) $\kappa = 0.05$, $u_c = 0.95$, $\varphi = -0.603$, $0 \leq u_c^{\text{off}} \leq 1$.

Fig. 3.5 Characteristic curves of hybrid-sail L_1-point Lyapunov orbits

Lyapunov orbits are given in Fig. 3.7. Two orbital revolutions exist within one orbital period and the geometry of 3-dimensional orbit is like a slope which is almost out of the Earth-Moon plane. Such orbits have potential advantages in surveillance and observation if a space station is located in the L_1 point.

3.5.2 Continuation of the L_2-Point Resonant Halo Orbits

The characteristic curves of the perturbed L_2-point halo orbits with respect to parameters κ, u_c and u_c^{off} are shown in Fig. 3.8. Since halo orbits are spacial orbits, the sail normal unit vector $\hat{\boldsymbol{n}}$ is always along the sunlight direction $\hat{\boldsymbol{n}}_S$, i.e. the pitch angle φ is set to zero. The curves with respect to κ in Fig. 3.8a show that z_0–value increases along the continuation process for both $\min x(y = 0)$ and $\max x(y = 0)$ configurations. If $u_c \in [0.9, 1]$ is taken as the continuation parameter with $\kappa = 0$, z_0–value is slightly perturbed and keeps almost unchanged as presented in Fig. 3.8b. Fixing $\kappa = 0.05$ and $u_c = 0.95$, Fig. 3.8c gives the characteristic curve with respect to u_c^{off}. From the figure, it can be seen that the value of z_0 decreases monotonically with u_c^{off} increasing for both $\min x(y = 0)$ and $\max x(y = 0)$ configurations.

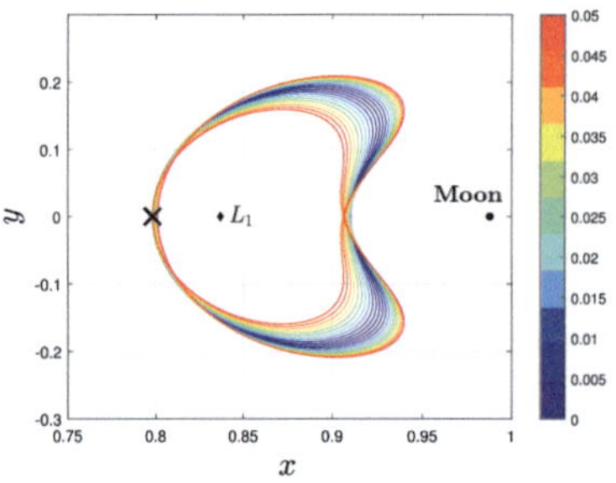

(a) $0 \le \kappa \le 0.05$, $u_c = 1$, $\varphi = 0$, $u_c^{\text{off}} = 0$.

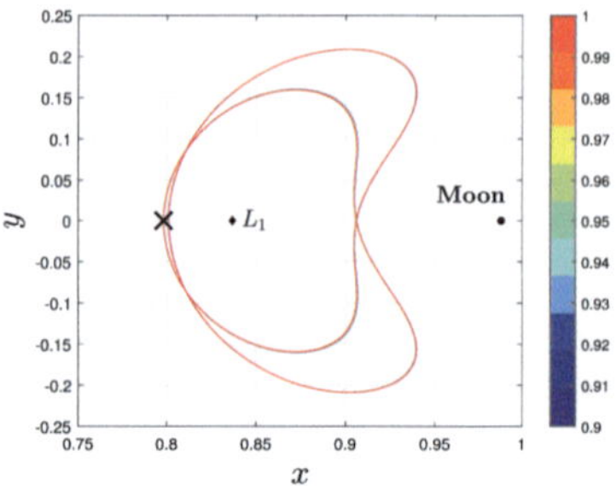

(b) $\kappa = 0.05$, $0.9 \le u_c \le 1$, $\varphi = 0$, $u_c^{\text{off}} = 0$.

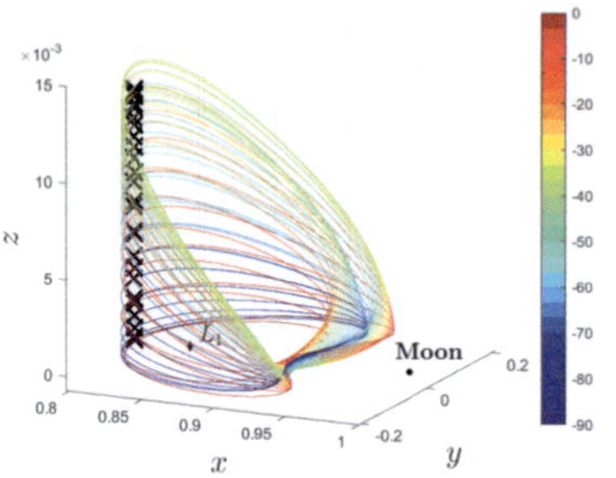

(c) $\kappa = 0.05$, $u_c = 0.95$, $-\pi/2 \le \varphi \le \pi/2$, $u_c^{\text{off}} = 0$.

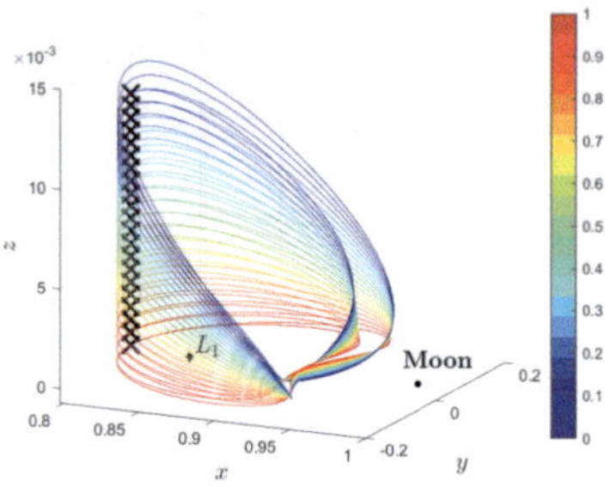

(d) $\kappa = 0.05$, $u_c = 0.95$, $\varphi = -0.603$, $0 \le u_c^{\text{off}} \le 1$.

Fig. 3.6 Families of hybrid-sail L_1-point Lyapunov orbits for the min $x(y = 0)$ configuration

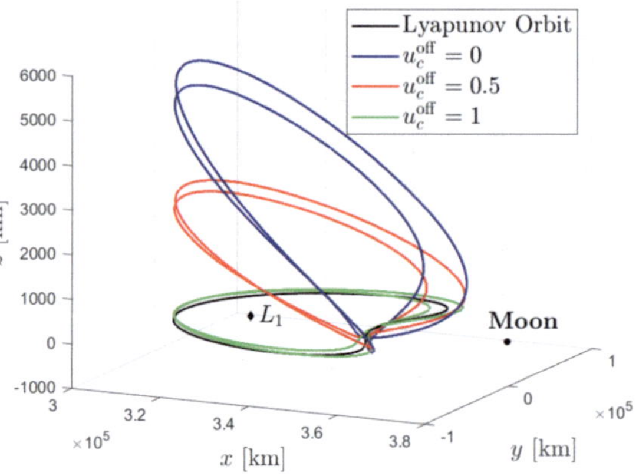

Fig. 3.7 Hybrid-sail L_1-point Lyapunov orbits with $\kappa = 0.05$, $u_c = 0.95$ and $\varphi = -0.603$

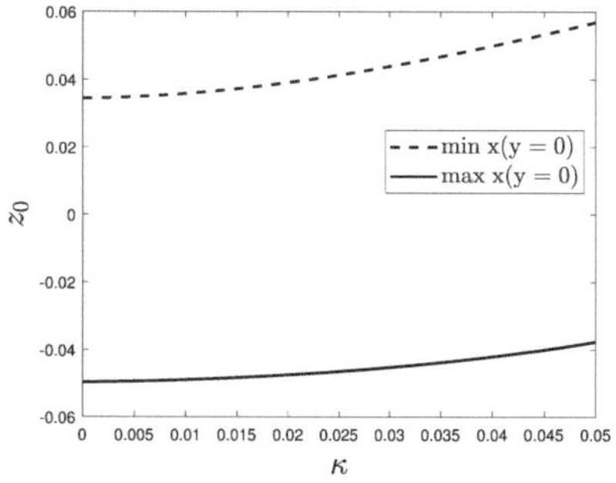

(a) $0 \leq \kappa \leq 0.05$, $u_c = 1$, $\varphi = 0$, $u_c^{\text{off}} = 0$.

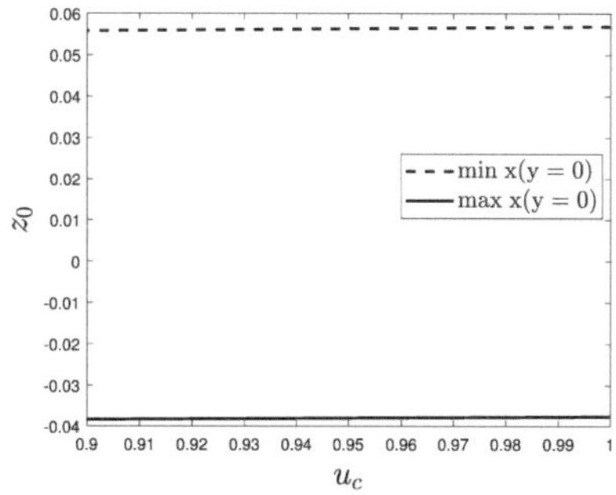

(b) $\kappa = 0.05$, $0.9 \leq u_c \leq 1$, $\varphi = 0$, $u_c^{\text{off}} = 0$.

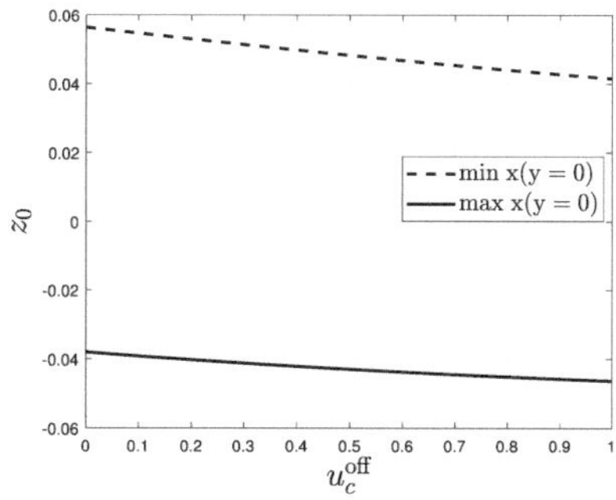

(c) $\kappa = 0.05$, $u_c = 0.95$, $0 \leq u_c^{\text{off}} \leq 1$.

Fig. 3.8 Characteristic curves of hybrid-sail L_2-point halo orbits

Taking the initial configuration $\min x(y = 0)$ as example, the families of hybrid-sail L_2-point halo orbits are shown in Fig. 3.9. Figure 3.10 displays the natural halo orbit and three u_c^{off} parametrized halo orbits with $\kappa = 0.05$ and $u_c = 0.95$. From Fig. 3.10, we can see that these u_c^{off} parametrized orbits are symmetrical about the $\hat{\boldsymbol{x}}_r - \hat{\boldsymbol{z}}_r$ plane and have two orbital revolutions within one orbital period. Adding SRP induced acceleration changes the orbital shape of the natural orbit and enables new types of periodic orbits. Besides, these orbits are always at the outside of the Earth-Moon line which means the far-side of the Moon is visible anytime. Hence, a hybrid sail can employ these orbits to monitor the far-side of the Moon continuously and support its communication with the Earth [1]. In addition, the family of hybrid-sail halo orbits may be employed for the bounded formation flying as each of the orbits are of the same period [19].

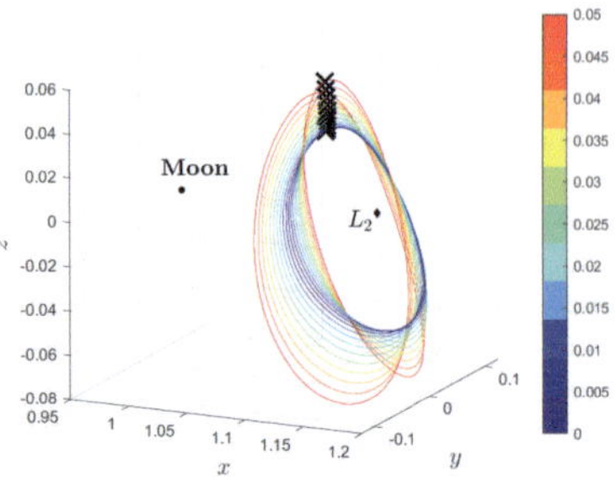

(a) $0 \le \kappa \le 0.05$, $u_c = 1$, $\varphi = 0$, $u_c^{\mathrm{off}} = 0$.

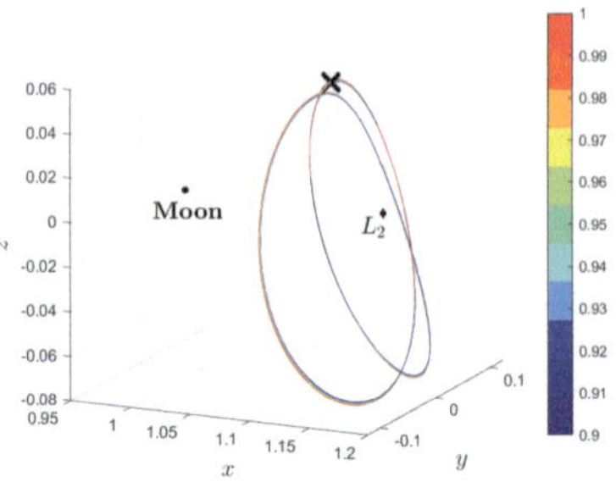

(b) $\kappa = 0.05$, $0.9 \le u_c \le 1$, $\varphi = 0$, $u_c^{\mathrm{off}} = 0$.

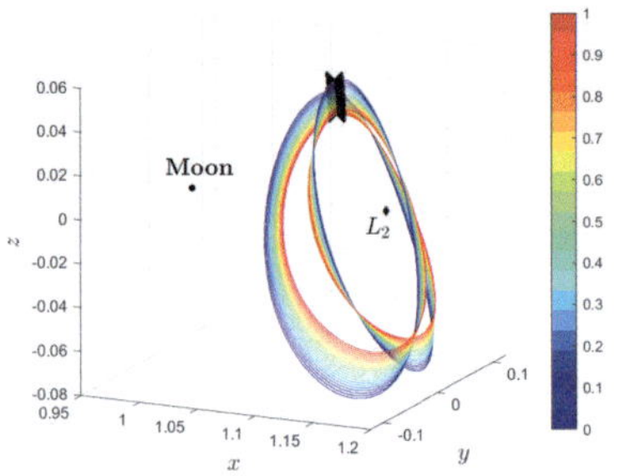

(c) $\kappa = 0.05$, $u_c = 0.95$, $-\pi/2 \le \varphi \le \pi/2$, $u_c^{\mathrm{off}} = 0$.

Fig. 3.9 Families of hybrid-sail L_2-point halo orbits for the min $x(y = 0)$ configuration

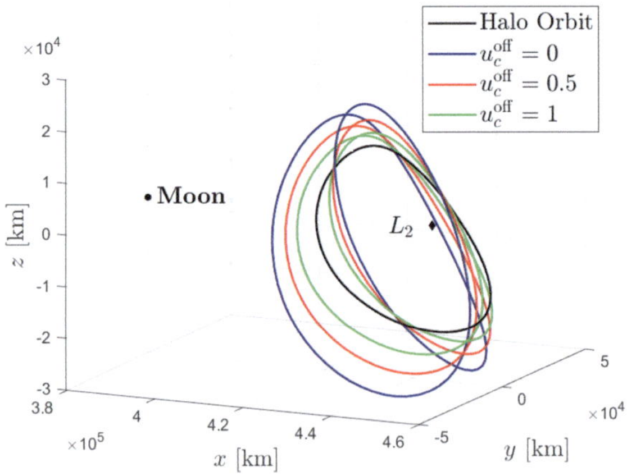

Fig. 3.10 Hybrid-sail L_2-point halo orbits with $\kappa = 0.05$ and $u_c = 0.95$

3.6 Control Strategy Design

3.6.1 NDO-Based Optimal Periodic Orbital Controller

To optimize propellant consumption, an optimal periodic orbital controller is developed exploiting the periodic property of Eq. (3.22). For linear periodic system Eq. (3.22), the optimal control problem is to determine the periodic state feedback control variable

$$\delta \boldsymbol{a} = -\mathbf{R}^{-1}(t)\mathbf{B}^{\mathrm{T}}(t)\mathbf{P}(t)\delta \boldsymbol{x} \tag{3.43}$$

by solving the positive-semidefinite T_S-periodic solution $\mathbf{P}(t)$ of the periodic Riccati differential equation

$$-\dot{\mathbf{P}}(t) = \mathbf{P}(t)\mathbf{A}(t) + \mathbf{A}^{\mathrm{T}}(t)\mathbf{P}(t) - \mathbf{P}(t)\mathbf{B}\mathbf{R}^{-1}(t)\mathbf{B}^{\mathrm{T}}\mathbf{P}(t) + \mathbf{Q}(t) \tag{3.44}$$

which is known to minimize the following performance objective

$$J = \int_{t_0}^{t} \left(\delta \boldsymbol{x}^{\mathrm{T}}\mathbf{Q}(t)\delta \boldsymbol{x} + \delta \boldsymbol{a}^{\mathrm{T}}\mathbf{R}(t)\delta \boldsymbol{a}\right) \mathrm{d}t \tag{3.45}$$

where $\mathbf{Q}(t)$ and $\mathbf{R}(t)$ are both T_S-periodic matrices, i.e. $\mathbf{Q}(t) = \mathbf{Q}(t + T_S)$ and $\mathbf{R}(t) = \mathbf{R}(t + T_S)$ such that $\mathbf{Q}(t) \geq 0$ and $\mathbf{R}(t) > 0$.

In this chapter, the numerical method which is came up with in [7] is adopted to solve Eq. (3.44) and simulation results show that numerical precision can meet mission requirements.

Taking the disturbance accelerations into consideration, Eq. (3.16) can be rewritten as

$$\dot{\boldsymbol{v}} = \boldsymbol{f}(\boldsymbol{v}, \boldsymbol{r}) + \boldsymbol{a} + \boldsymbol{d} \tag{3.46}$$

where $[\boldsymbol{r}]_{\mathcal{T}_R} = [x, y, z]^{\mathrm{T}}$, $[\boldsymbol{v}]_{\mathcal{T}_R} = [\dot{x}, \dot{y}, \dot{z}]^{\mathrm{T}}$, $\boldsymbol{d} = \boldsymbol{d}_e + \boldsymbol{d}_s$, and

$$\boldsymbol{f}(\boldsymbol{v}, \boldsymbol{r}) = \begin{bmatrix} 2\dot{y} + \Omega_x \\ -2\dot{x} + \Omega_y \\ \Omega_z \end{bmatrix} \tag{3.47}$$

Similarly, Eq. (3.22) can be written as

$$\delta \dot{\boldsymbol{x}} = \mathbf{A}(t)\delta \boldsymbol{x} + \mathbf{B}\delta \boldsymbol{a} + \boldsymbol{d} \tag{3.48}$$

A NDO is designed for Eq. (3.48) as

$$\begin{cases} \tilde{\boldsymbol{d}} = \boldsymbol{\varepsilon} + \boldsymbol{p}(\boldsymbol{v}), \\ \dot{\boldsymbol{\varepsilon}} = -\mathbf{L}(\boldsymbol{v})\boldsymbol{\varepsilon} - \mathbf{L}(\boldsymbol{v})\left(\boldsymbol{p}(\boldsymbol{v}) + \boldsymbol{f}(\boldsymbol{v}, \boldsymbol{r}) + \boldsymbol{a}\right) \end{cases} \tag{3.49}$$

where $\tilde{\boldsymbol{d}}$ and $\boldsymbol{\varepsilon}$ are the output and variable of the NDO, and $\mathbf{L}(\boldsymbol{v}) = \partial \boldsymbol{p}(\boldsymbol{v})/\partial \boldsymbol{v}$ where $\boldsymbol{p}(\boldsymbol{v}) = [p_1(\boldsymbol{v}), p_2(\boldsymbol{v}), \ldots, p_n(\boldsymbol{v})]^{\mathrm{T}}$ is the vector-valued function to be designed.

For simplicity, $\boldsymbol{p}(\boldsymbol{v}) = \mathbf{J}_o\boldsymbol{v}$ is employed in this chapter where $\mathbf{J}_o$ is a diagonal matrix with positive elements. Denoting $\boldsymbol{e}_d = \boldsymbol{d} - \tilde{\boldsymbol{d}}$ the estimation error and for disturbance satisfying $\|\dot{\boldsymbol{d}}\| \leq \eta$, $\|\boldsymbol{e}_d\|$ will converge to a closed ball whose radius is

$$\beta_d = \max\left\{\frac{\eta}{\mathbf{J}_{o_i}}, i = 1, 2, \ldots, n\right\} \tag{3.50}$$

The rigorous proof on the convergence of the designed NDO Eq. (3.49) for a nonlinear system is given in [20].

Theorem 1 *Consider systems Eqs.* (3.48) *and* (3.49), *if* $\|\dot{\boldsymbol{d}}\|$ *is bounded and the controller is designed as*

$$\delta \boldsymbol{a} = -\mathbf{K}(t)\delta \boldsymbol{x} - \mathbf{B}^{*}\tilde{\boldsymbol{d}} \tag{3.51}$$

where $\mathbf{K}(t) = \mathbf{R}^{-1}(t)\mathbf{B}^{T}\mathbf{P}(t)$ *and* $\mathbf{B}^{*}$ *is the pseudo-inverse of* $\mathbf{B}$, *then* $\delta \boldsymbol{x} = 0$ *can be asymptotically stable in Eq.* (3.48).

Proof Let a candidate Lyapunov function be

$$V = \frac{1}{2}\delta \boldsymbol{x}^{\mathrm{T}}\delta \boldsymbol{x} \tag{3.52}$$

Then,

$$\begin{aligned} \dot{V} &= \delta \boldsymbol{x}^{\mathrm{T}}\delta \dot{\boldsymbol{x}} \\ &= \delta \boldsymbol{x}^{\mathrm{T}}\left(\mathbf{A}(t)\delta \boldsymbol{x} + \mathbf{B}\delta \boldsymbol{a} + \boldsymbol{d}\right) \\ &= \delta \boldsymbol{x}^{\mathrm{T}}\left(\mathbf{A}(t)\delta \boldsymbol{x} - \mathbf{BK}(t)\delta \boldsymbol{x} - \tilde{\boldsymbol{d}} + \boldsymbol{d}\right) \\ &= \delta \boldsymbol{x}^{\mathrm{T}}\left(\mathbf{A}(t) - \mathbf{BK}(t)\right)\delta \boldsymbol{x} + \delta \boldsymbol{x}^{\mathrm{T}}\boldsymbol{e}_d \\ &\leq \lambda_{\min}\left(\mathbf{A}(t) - \mathbf{BK}(t)\right)\|\delta \boldsymbol{x}\|^2 + \|\delta \boldsymbol{x}\|\|\boldsymbol{e}_d\| \end{aligned} \tag{3.53}$$

According to Eq. (3.50), $\|\boldsymbol{e}_d\|$ is bounded and $\|\boldsymbol{e}_d\| \leq \beta_d$ with $t \to \infty$ and $\lambda_{\min}\left(\mathbf{A}(t) - \mathbf{BK}(t)\right) < 0$, then $\mathbf{Q}(t)$, $\mathbf{R}(t)$ and $\mathbf{J}_o$ can be selected to guarantee $\dot{V} < 0$ which means $\delta \boldsymbol{x} = 0$ can be asymptotically stable. □

3.6.2 SEP Acceleration Optimization

According to Eq. (3.4), $\boldsymbol{a}_{SRP}$ is determined by the sail's reflectivity rate u_c^{off} and the sail unit normal vector $\hat{\boldsymbol{n}}$ which is determined by angles γ and φ in frame $\mathcal{T}_S$. For mission lifetime extension, sail's reflectivity rate u_c^{off} and attitude angles γ and φ should be optimized, such that the magnitude of the acceleration required from the SEP system is minimum. To avoid possible mechanical instabilities, it is reasonable to limit the attitude angles within some ranges, i.e., $\gamma \in [-45, 45]$ deg and $\varphi \in [-45, 45]$ deg in this chapter. Consequently, the static constrained optimization problem can be stated as follows:

Find optimal sail's reflectivity rate $u_c^{\mathrm{off}\star}$ and attitude angles $\alpha^\star$ and $\beta^\star$, satisfying

$$\left\{u_c^{\mathrm{off}\star}, \gamma^\star, \varphi^\star\right\} = \arg \min_{u_c^{\mathrm{off}}, \gamma, \varphi} \|\boldsymbol{a}_{SEP}\| = \arg \min_{u_c^{\mathrm{off}}, \gamma, \varphi} \|\boldsymbol{a}_c - \boldsymbol{a}_{SRP}\| \tag{3.54}$$

s.t.

$$\gamma \in [-45, 45]\,\mathrm{deg}, \quad \varphi \in [-45, 45]\,\mathrm{deg}, \quad u_c^{\mathrm{off}} \in [0, 1], \quad \boldsymbol{a}_c = \boldsymbol{a}_n + \delta\boldsymbol{a} \tag{3.55}$$

where $\boldsymbol{a}_c$ and $\boldsymbol{a}_n$ are the required accelerations for the controlled trajectory and the reference orbit respectively, and $\delta\boldsymbol{a}$ is the control variable obtained through Eq. (3.51). In particular, the particle swarm optimization (PSO) method is adopted here to address the constrained optimization problem represented by Eqs. (3.54) and (3.55). In addition, $\psi^\star$ is set to 0 since $\hat{\boldsymbol{n}}$ is independent of the roll angle.

3.6.3 NDO-Based Robust Backstepping Attitude Controller

According to Eqs. (3.28) and (3.29), the compact form of the rotational motion in frame $\mathcal{T}_B$ can be written as

$$\begin{cases} \dot{\boldsymbol{\sigma}} = \mathbf{G}\boldsymbol{\omega} + (1 - \omega_S)\mathbf{G}\boldsymbol{f}(\boldsymbol{\sigma}) \\ \mathbf{I}\dot{\boldsymbol{\omega}} + \omega^\times \mathbf{I}\boldsymbol{\omega} = \boldsymbol{\tau} + \boldsymbol{\tau}_g \end{cases} \tag{3.56}$$

where

$$\begin{cases} \boldsymbol{\sigma} = [\psi, \varphi, \gamma]^{\mathrm{T}} \\ [\boldsymbol{\omega}]_{\mathcal{T}_B} = [\omega_x, \omega_y, \omega_z]^{\mathrm{T}} \\ \boldsymbol{f}(\boldsymbol{\sigma}) = [\sin\varphi, -\sin\psi\cos\varphi, -\cos\psi\cos\varphi]^{\mathrm{T}} \\ [\boldsymbol{\tau}]_{\mathcal{T}_B} = [\tau_x, \tau_y, \tau_z]^{\mathrm{T}} \\ [\boldsymbol{\tau}_g]_{\mathcal{T}_B} = [\tau_{gx}, \tau_{gy}, \tau_{gz}]^{\mathrm{T}} \\ \mathbf{G} = \begin{bmatrix} 1 & \tan\varphi\sin\psi & \tan\varphi\cos\psi \\ 0 & \cos\psi & -\sin\psi \\ 0 & \sin\psi/\cos\varphi & \cos\psi/\cos\varphi \end{bmatrix} \\ [\mathbf{I}]_{\mathcal{T}_B} = \mathrm{diag}([I_x, I_y, I_z]) \\ \boldsymbol{\omega}^{\times} = \begin{bmatrix} 0 & -\omega_z & \omega_y \\ \omega_z & 0 & -\omega_x \\ -\omega_y & \omega_x & 0 \end{bmatrix} \end{cases} \tag{3.57}$$

Due to the propellant consumption, the deployment of sail and appendages, and the structural flexibility, the inertial matrix $\mathbf{I}$ is uncertain for controller design. Assuming that the inertial matrix is denoted by $\mathbf{I} = \mathbf{I}_0 + \Delta\mathbf{I}$, where $\mathbf{I}_0$ is the nominal one and $\Delta\mathbf{I}$ is the parametric uncertainty which is bounded by $\|\Delta\mathbf{I}\| \le \beta_I$.

Defining backstepping control variables

$$z_1 = \boldsymbol{\sigma} - \boldsymbol{\sigma}_d, \qquad z_2 = \boldsymbol{\omega} - \boldsymbol{\omega}_d \tag{3.58}$$

where $\boldsymbol{\sigma}_d = [\psi^{\star}, \varphi^{\star}, \gamma^{\star}]^{\mathrm{T}}$ means the desired angle obtained by PSO, $\boldsymbol{\omega}_d$ means the reference command of the second subsystem in Eq. (3.56). Then, the dynamics of z_1 is

$$\dot{z}_1 = \mathbf{G}(z_2 + \boldsymbol{\omega}_d) + (1 - \omega_S)\mathbf{G}\boldsymbol{f}(\boldsymbol{\sigma}) \tag{3.59}$$

The reference command $\boldsymbol{\omega}_d$ is designed as

$$\boldsymbol{\omega}_d = -\lambda_1 \mathbf{G}^{-1} z_1 - (1 - \omega_S)\boldsymbol{f}(\boldsymbol{\sigma}) \tag{3.60}$$

where $\lambda_1 \in \mathbb{R}^+$ is a positive constant.

Let a candidate Lyapunov function be

$$V_1 = \frac{1}{2} z_1^{\mathrm{T}} z_1 \tag{3.61}$$

Based on Eqs. (3.59) and (3.60), the derivative of V_1 is

$$\dot{V}_1 = z_1^{\mathrm{T}} \dot{z}_1 = -\lambda_1 z_1^{\mathrm{T}} z_1 + z_1^{\mathrm{T}} \mathbf{G} z_2 \tag{3.62}$$

Bearing in mind the definition of z_2, the reference command Eq. (3.60) and the attitude dynamics Eq. (3.56), the dynamics of z_2 can be written as

$$\begin{aligned}\dot{z}_2 &= \dot{\boldsymbol{\omega}} - \dot{\boldsymbol{\omega}}_d \\ &= \mathbf{I}^{-1}(-\omega^\times \mathbf{I}\boldsymbol{\omega} + \boldsymbol{\tau} + \boldsymbol{\tau}_g) + \lambda_1 \mathbf{G}^{-1}\dot{z}_1 + \lambda_1 \dot{\mathbf{G}}^{-1} z_1 + (1 - \omega_S)\dot{\boldsymbol{f}}(\boldsymbol{\sigma})\dot{\boldsymbol{\sigma}}\end{aligned} \tag{3.63}$$

Considering the inertial matrix uncertainty, Eq. (3.63) can be written as

$$\dot{z}_2 = -\mathbf{I}_0^{-1}\omega^\times \mathbf{I}_0\boldsymbol{\omega} + \lambda_1 \mathbf{G}^{-1}\dot{z}_1 + \mathbf{I}_0^{-1}\boldsymbol{\tau} + \boldsymbol{\tau}_d \tag{3.64}$$

where $\boldsymbol{\tau}_d = -(\mathbf{I}^{-1} - \mathbf{I}_0^{-1})\omega^\times \mathbf{I}_0\boldsymbol{\omega} - \mathbf{I}^{-1}\omega^\times \Delta\mathbf{I}\boldsymbol{\omega} + (\mathbf{I}^{-1} - \mathbf{I}_0^{-1})\boldsymbol{\tau} + \mathbf{I}^{-1}\boldsymbol{\tau}_g + \lambda_1 \dot{\mathbf{G}}^{-1} z_1 + (1 - \omega_S)\dot{\boldsymbol{f}}(\boldsymbol{\sigma})\dot{\boldsymbol{\sigma}}$, and $\boldsymbol{\tau}_d$ is bounded.

Same as Eq. (3.49), a NDO is designed for Eq. (3.64) as

$$\begin{cases} \tilde{\boldsymbol{\tau}}_d = \boldsymbol{\zeta} + \mathbf{J}_a z_2 \\ \dot{\boldsymbol{\zeta}} = -\mathbf{J}_a\boldsymbol{\zeta} - \mathbf{J}_a(\mathbf{J}_a z_2 + \boldsymbol{f}(z_2) + \mathbf{I}_0^{-1}\boldsymbol{\tau}) \end{cases} \tag{3.65}$$

where $\tilde{\boldsymbol{\tau}}_d$ and $\boldsymbol{\zeta}$ are the output and variable of the NDO, $\boldsymbol{f}(z_2) = -\mathbf{I}_0^{-1}\omega^\times \mathbf{I}_0\boldsymbol{\omega} + \lambda_1 \mathbf{G}^{-1}\dot{z}_1$, and $\mathbf{J}_a$ is a diagonal matrix with positive elements.

To this end, the control input can be designed as

$$\boldsymbol{\tau} = \mathbf{I}_0(\mathbf{I}_0^{-1}\omega^\times \mathbf{I}_0\boldsymbol{\omega} - \mathbf{G}^{\mathrm{T}} z_1 - \lambda_2 z_2 - \lambda_1 \mathbf{G}^{-1}\dot{z}_1 - \tilde{\boldsymbol{\tau}}_d) \tag{3.66}$$

Theorem 2 *Consider system Eq. (3.56), if $\|\dot{\boldsymbol{\tau}}_d\|$ is bounded and the controller is designed as Eq. (3.66) with $\lambda_2 > 1/2$ and the minimum eigenvalue of $\mathbf{J_a}$, i.e. $\lambda_{\min}(\mathbf{J_a}) > \mathbf{1/2}$, then the attitude angle $\boldsymbol{\sigma}$ and the angular velocity $\boldsymbol{\omega}$ will converge to small neighborhoods of desired values $\boldsymbol{\sigma}_d$ and $\boldsymbol{\omega}_d$, respectively, with suitable controller parameters.*

Proof Let a candidate Lyapunov function be

$$V_2 = V_1 + \frac{1}{2} z_2^{\mathrm{T}} z_2 + \frac{1}{2}\boldsymbol{e}_{\tau_d}^{\mathrm{T}}\boldsymbol{e}_{\tau_d} \tag{3.67}$$

where $\boldsymbol{e}_{\tau_d} = \boldsymbol{\tau}_d - \tilde{\boldsymbol{\tau}}_d$ means the disturbance torque estimation error, and V_1 is in Eq. (3.61).

The time derivative of V_2 is

$$
\begin{aligned}
\dot{V}_2 &= \dot{V}_1 + z_2^{\mathrm{T}}\dot{z}_2 + e_{\tau_d}^{\mathrm{T}}\dot{e}_{\tau_d} \\
&= -\lambda_1 z_1^{\mathrm{T}} z_1 + z_1^{\mathrm{T}}\mathbf{G}z_2 + z_2^{\mathrm{T}}(-\mathbf{G}^{\mathrm{T}}z_1 - \lambda_2 z_2 + e_{\tau_d}) + e_{\tau_d}^{\mathrm{T}}(\dot{\tau}_d - \mathbf{J}_a e_{\tau_d}) \\
&= -\lambda_1 z_1^{\mathrm{T}} z_1 - \lambda_2 z_2^{\mathrm{T}} z_2 - e_{\tau_d}^{\mathrm{T}}\mathbf{J}_a e_{\tau_d} + z_2^{\mathrm{T}} e_{\tau_d} + e_{\tau_d}^{\mathrm{T}}\dot{\tau}_d \\
&\leq -\lambda_1\|z_1\|^2 - \lambda_2\|z_2\|^2 - \lambda_{\min}(\mathbf{J}_a)\|e_{\tau_d}\|^2 + |\|z_2\|\|e_{\tau_d}\| + \|e_{\tau_d}\|\|\dot{\tau}_d\| \\
&\leq -\lambda_1\|z_1\|^2 - (\lambda_2 - \frac{1}{2})\|z_2\|^2 - (\lambda_{\min}(\mathbf{J}_a) - \frac{1}{2})\|e_{\tau_d}\|^2 + \|e_{\tau_d}\|\|\dot{\tau}_d\| \\
&\leq -2\lambda V_2 + \|e_{\tau_d}\|\|\dot{\tau}_d\|
\end{aligned}
\tag{3.68}
$$

where $\lambda = \min\{\lambda_1, \lambda_2 - \frac{1}{2}, \lambda_{\min}(\mathbf{J_a}) - \mathbf{1/2}\}$.

Therefore, $V_2(t) \leq -V_2(t_0)e^{-2\lambda t} + \|e_{\tau_d}\|\|\dot{\tau}_d\|/(2\lambda)$. Due to the assumption that $\|\dot{\tau}_d\|$ is bounded, $\|e_{\tau_d}\|\|\dot{\tau}_d\|/(2\lambda)$ will converge to the neighborhood of zero quickly with large NDO gains, so that $\lim_{t\to\infty} V_2(t) \leq \xi$ where ξ is a very small number. According to the definition of $V_2(t)$, $\|z_1\|$ and $\|z_2\|$ ultimately converge to the small neighborhood of zero, i.e., $\lim_{t\to\infty} \sigma = \sigma_d$ and $\lim_{t\to\infty} \omega = \omega_d$. □

Bearing in mind Eqs. (3.10), (3.12) and (3.15), lower and upper limits of control torque saturation exist due to the physical constraints on u_y and u_z. When $u_c^{\text{off}} = 0.5$, both u_y and u_z can vary between the maximum value 0.25 and the minimum value -0.25. Hence, u_c^{off} is set to 0.5 along the attitude tracking process to ensure sufficient control torques τ_y and τ_z, i.e.,

$$
\tau_i = \begin{cases} \tau_i^{\max}, & \tau_i > \tau_i^{\max} \\ \tau_i & \\ \tau_i^{\min}, & \tau_i < \tau_i^{\min} \end{cases} \qquad i = y, z
\tag{3.69}
$$

where $\tau_i^{\max} = 0.25 l_{eq}(\boldsymbol{P}_{\text{off}} - \boldsymbol{P}_{\text{on}}) \cdot \hat{\boldsymbol{n}}$ and $\tau_i^{\min} = -0.25 l_{eq}(\boldsymbol{P}_{\text{off}} - \boldsymbol{P}_{\text{on}}) \cdot \hat{\boldsymbol{n}}$.

Then, control variables u_y and u_z can be obtained through Eqs. (3.10), (3.12) and (3.69).

To minimize the SEP acceleration and reduce propellant consumption of the SEP system as much as possible, the value of u_c^{off} is switched to $u_c^{\text{off}\star}$ obtained by Eqs. (3.54) and (3.55) once desired attitude angles have been tracked, i.e., $|\gamma - \gamma^\star| \leq 0.1$ deg and $|\varphi - \varphi^\star| \leq 0.1$ deg.

To this end, the proposed attitude-orbit control strategy for station-keeping is shown in Fig. 3.11.

3.7 Numerical Simulations

To illustrate the proposed control strategy for station-keeping, the L_2-point perturbed halo orbit with $u_c^{\text{off}} = 0.5$ in Fig. 3.10 and the L_1-point perturbed Lyapunov orbit with $u_c^{\text{off}} = 0.5$ in Fig. 3.7 are taken as reference orbits, respectively. Table 3.2 gives the corresponding initial orbital coordinates. The inertial matrix uncertainty is assumed

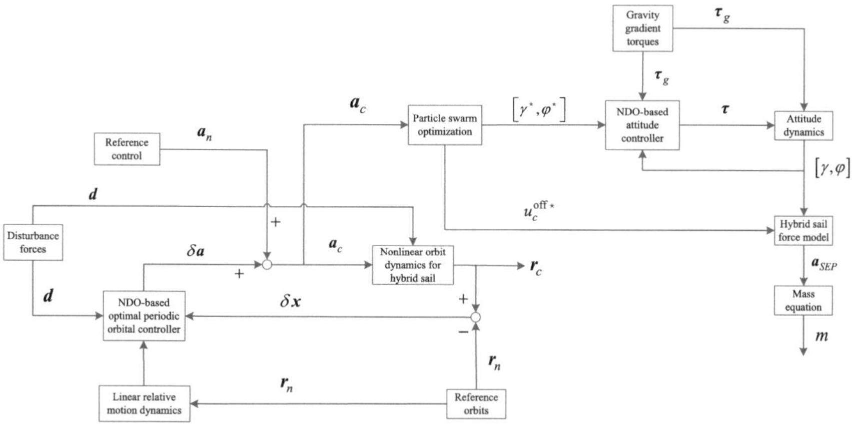

Fig. 3.11 Block diagram of the proposed control strategy for station-keeping

Table 3.2 Initial orbital coordinates for reference orbits

Reference orbit	Parameters	x_0	z_0	$\dot{y}_0$
L_2-point perturbed halo orbit	$\kappa = 0.05$ $u_c = 0.95$ $u_c^{\text{off}} = 0.5$	1.11243777583015	0.0482526769813511	0.174532173018492
L_1-point perturbed Lyapunov orbit	$\kappa = 0.05$ $u_c = 0.95$ $\varphi = -34.55$ deg $u_c^{\text{off}} = 0.5$	0.797845849722269	0.0066369639593008	0.352803791222808

Table 3.3 Parameters in orbital controller and attitude controller

Parameters	Value
$\mathbf{Q}$	$\text{diag}\left([10^5, 10^5, 10^5, 10^3, 10^3, 10^3]\right)$
$\mathbf{R}$	diag([1, 1, 1])
$\mathbf{J_o}$	diag([15000, 17000, 15000])
λ_1	600
λ_2	1000
$\mathbf{J_a}$	$\text{diag}\left([10^5, 10^5, 10^5]\right)$

to be $\Delta I_x = 0.05 I_{x0}$, $\Delta I_y = -0.05 I_{y0}$ and $\Delta I_z = 0.05 I_{z0}$. Initial position errors and initial velocity errors in each directions are set equal to 50 km and 1 m/s, respectively. Parameters in orbital controller and attitude controller are given in Table 3.3 and the simulation is run for one orbital period T_S. To evaluate propellant consumption of the sail, introducing the mass rate of change

$$\dot{m} = -\frac{a_{SEP} m}{I_{sp} g_0} \tag{3.70}$$

where $I_{sp} = 3000\,\mathrm{s}$ is the specific impulse of the SEP system and $g_0 = 9.8\,\mathrm{m/s^2}$ is the Earth's standard gravity acceleration.

In addition, angles $\tilde{\gamma}$ and $\tilde{\varphi}$ are used to define the direction of $\hat{\boldsymbol{m}}$ in frame $\mathcal{T}_S$ according to Fig. 3.2 as

$$\left[\hat{\boldsymbol{m}}\right]_{\mathcal{T}_S} = \left[\cos\tilde{\gamma}\cos\tilde{\varphi},\ \sin\tilde{\gamma}\cos\tilde{\varphi},\ -\sin\tilde{\varphi}\right]^{\mathrm{T}} \tag{3.71}$$

3.7.1 *Example 1: Station-Keeping of the L_2-Point Resonant Halo Orbit*

Figure 3.12 illustrates the station-keeping results where precise orbital tracking can be successfully achieved with the designed NDO-based optimal periodic orbital controller. In particular, Fig. 3.12b shows that position errors along each direction can be kept within 10 m which is a very small quantity compared with the orbital $\hat{z}_r$–axis altitude ($\approx 2 \times 10^4$ km). The disturbance acceleration in orbital control and its estimation are given in Fig. 3.13. The curves of disturbance acceleration and its estimation are almost overlapped in Fig. 3.13a due to the negligible estimation errors in Fig. 3.13b. In particular, the effects of the disturbance acceleration on station-keeping are reduced to the lowest by the NDO through feedforward which guarantees the high-precision station-keeping in Fig. 3.12.

Figure 3.14 gives the time history of sail's attitude angles and control variables when the attitude control is taken into consideration with the designed NDO-based robust backstepping attitude controller. It should be noted that the values of γ and φ are outside the feasible regions in Eq. (3.55) at times which is admissible because constraints on γ and φ in Eq. (3.55) are soft constraints for SEP acceleration optimization. The accurate estimation results lead to the overlap of curves of disturbance torque and its estimation in Fig. 3.15. Figure 3.16 gives the direction angles and magnitude of the acceleration provided by the SEP system in cases of using only SEP and using hybrid SEP/RCDs. Using only SEP means that only SEP system provides the control acceleration $\delta\boldsymbol{a}$ for station-keeping without the assistance of the solar sail. While using hybrid SEP/RCDs means that the sail's reflectivity rate and attitude angles are controlled by RCDs providing the control acceleration together with SEP system. It can be seen that remarkable differences exist in both acceleration magnitude and direction angles using hybrid SEP/RCDs.

3.7.2 *Exmple 2: Station-Keeping of the L_1-Point Resonant Lyapunov Orbit*

Figure 3.17 shows the equivalent station-keeping precision for L_1-point perturbed Lyapunov orbit as halo orbit in Fig. 3.12. Besides, the disturbance acceleration and

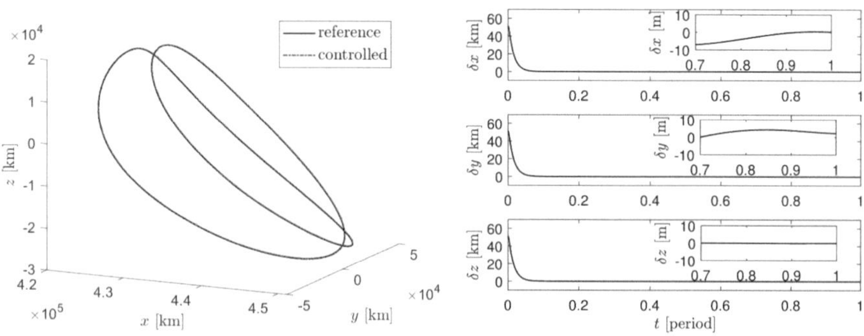

(a) Orbital tracking result (b) Position errors

Fig. 3.12 Example 1: station-keeping results

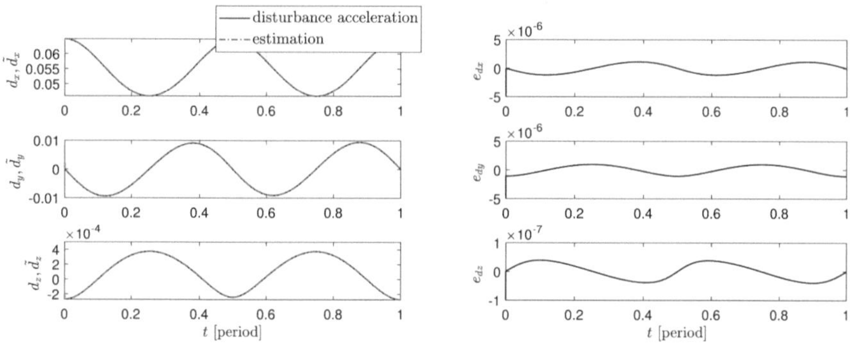

(a) Estimation results (b) Estimation errors

Fig. 3.13 Example 1: disturbance acceleration and estimation

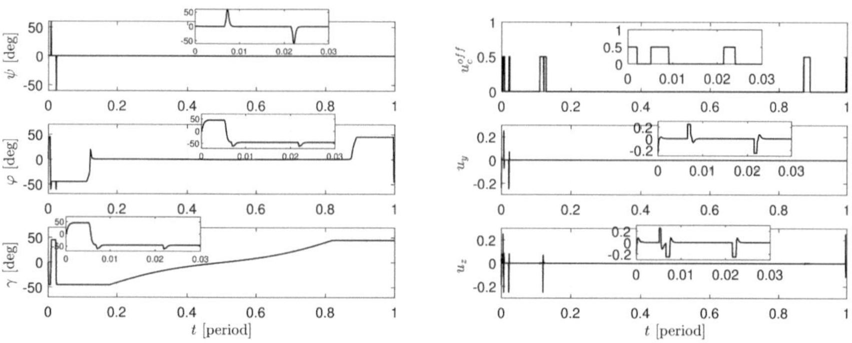

(a) Attitude angles (b) Control variables

Fig. 3.14 Example 1: attitude angles and control variables of the hybrid sail

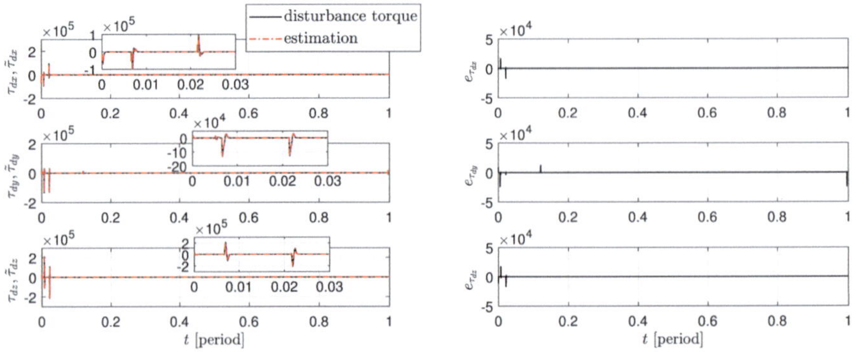

(a) Estimation results (b) Estimation errors

Fig. 3.15 Example 1: disturbance torque and estimation

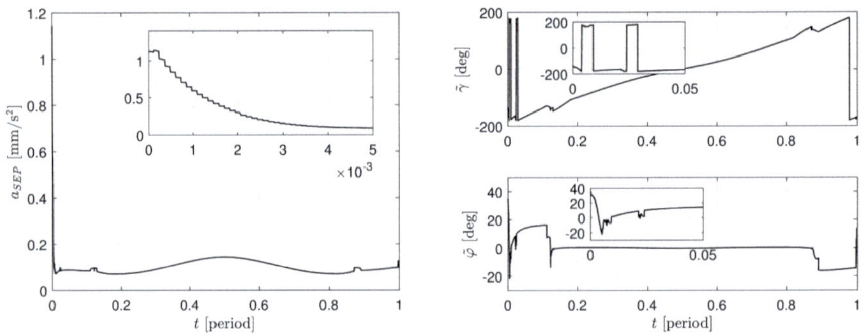

(a) Hybrid SEP/RCDs

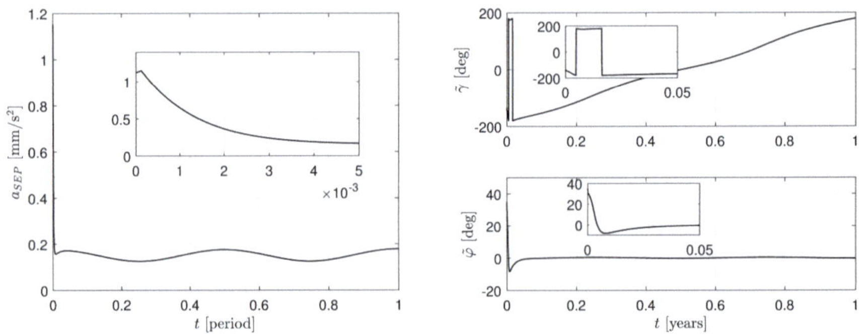

(b) Only SEP

Fig. 3.16 Example 1: direction angles and magnitude of SEP acceleratio

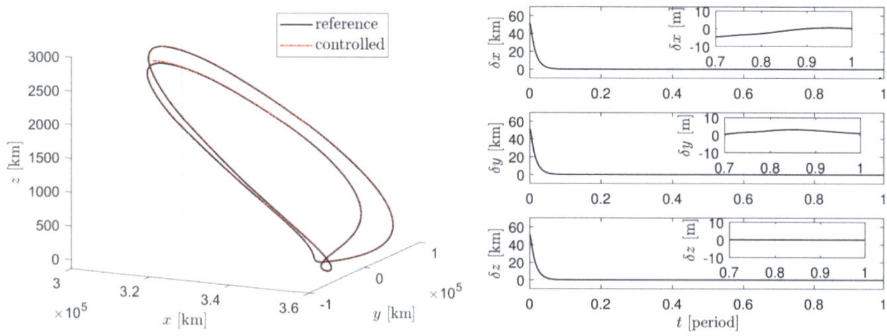

(a) Orbital tracking result (b) Position errors

Fig. 3.17 Example 2: station-keeping results

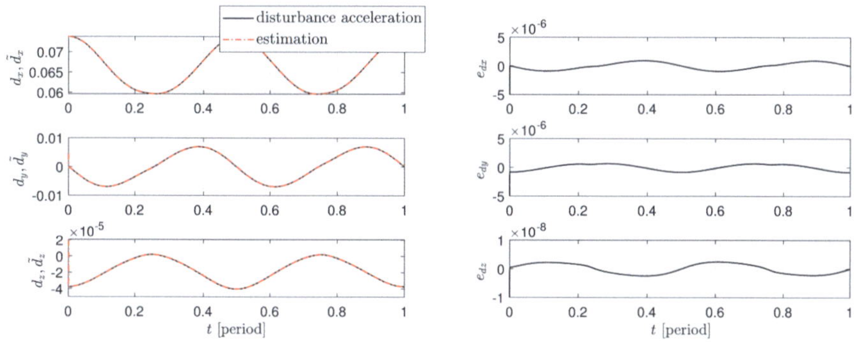

(a) Estimation results (b) Estimation errors

Fig. 3.18 Example 2: disturbance acceleration and estimation

the disturbance torque can be effectively estimated by the NDO with negligible errors as shown in Figs. 3.18 and 3.20, respectively. From Fig. 3.19, we can see that overshoots of attitude angles happen in the process of attitude control as Example 1. In addition, the magnitude of SEP acceleration is reduced due to the control of sail's reflectivity rate and attitude angles as shown in Fig. 3.21.

3.7.3 Comparison of Propellant Consumption

The above simulation results show that the RCD can help optimize the magnitude of required acceleration from SEP system. To further validate the conclusion, the one-period total propellant consumption of the hybrid sail is given in Table 3.4 where station-keeping is carried out using only SEP and hybrid SEP/RCDs. It can be found that the propellant reduction rate is 35.02 % for halo orbit and 28.68 % for Lyapunov orbit using hybrid SEP/RCDs, which means exploiting the most benefit from the

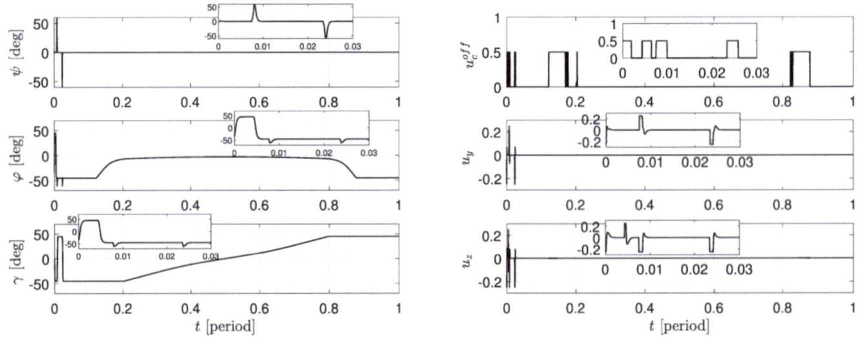

(a) Attitude angles (b) Control variables

Fig. 3.19 Example 2: attitude angles and control variables of the hybrid sail

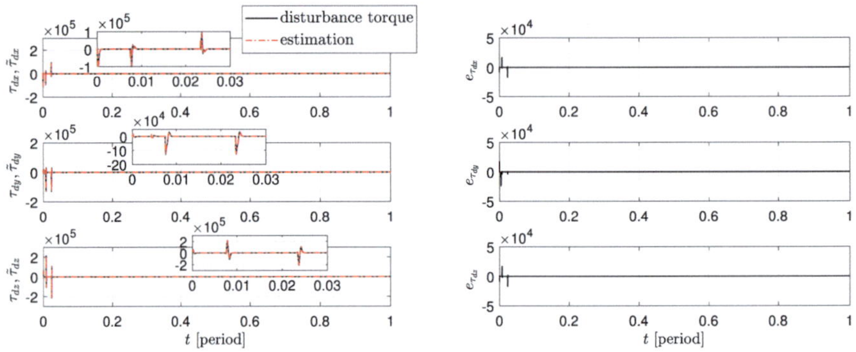

(a) Estimation results (b) Estimation errors

Fig. 3.20 Example 2: disturbance torque and estimation

SRP acceleration by adjusting sail's reflectivity rate and attitude angles can reduce propellant consumption of SEP system effectively.

3.8 Conclusions

In this chapter, the problems of evolution and station-keeping of hybrid-sail resonant periodic orbits in the Earth-Moon system are investigated within the framework of CRTBP. A high-fidelity model of a hybrid sail, which consists of a SEP system and a solar sail covered by RCD, is used. Incorporating the SRP acceleration into the dynamical system, families of resonant periodic orbits are obtained making use of parallel shooting and continuation methods based on the sail's parameters. In particular, the perturbed L_1-point Lyapunov orbits and L_2-point halo orbits are analyzed due to their potential applications in lunar far-side communication and space

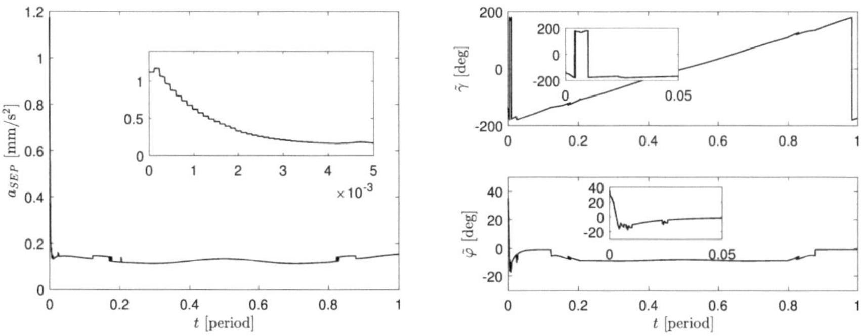

(a) Hybrid SEP/RCDs

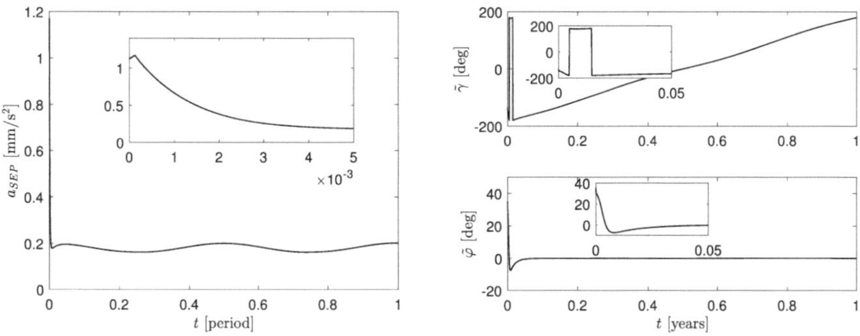

(b) Only SEP

Fig. 3.21 Example 2: direction angles and magnitude of SEP acceleration

station surveillance. To achieve high-precision station-keeping and further reduce propellant consumption, an orbit-attitude control strategy has been proposed for station-keeping by designing a novel NDO-based optimal periodic orbit controller, optimizing SEP acceleration, and designing a novel NDO-based robust backstepping attitude controller. In particular, RCDs are used to control sail's reflectively rate and attitude angles in orbit and attitude control, respectively. Through illustrative examples, we can conclude that: (1) the designed NDO can estimate disturbances with high precision which ensures the effectiveness of the feedforward compensation; (2) high-precision station-keeping is achieved with the designed NDO-based optimal periodic orbital controller in the presence of disturbance accelerations and effective attitude tracking is validated with the designed NDO-based robust backstepping attitude controller in the presence of control torque saturation, inertia uncertainty of the hybrid sail and disturbance torques; (3) compared to using only SEP, using hybrid SEP/RCDs achieves 35.02 % and 28.68 % reduction of propellant consumption for perturbed halo orbit and Lyapunov orbit in one orbital period, respectively.

Table 3.4 Comparison of propellant consumption

Reference orbit	Only SEP	Hybrid SEP/RCDs	Reduction ratios (%)
L_2-point halo orbit	0.7861 kg	0.5108 kg	35.02
L_1-point Lyapunov orbit	0.9385 kg	0.6693 kg	28.68

References

1. Heiligers J, Macdonald M, Parker J S. Extension of Earth-Moon libration point orbits with solar sail propulsion[J]. Astrophysics and Space Science, 2016, 361: 1–20.
2. Heiligers J, Hiddink S, Noomen R, et al. Solar sail Lyapunov and Halo orbits in the Earth-Moon three-body problem[J]. Acta Astronautica, 2015, 116: 25–35.
3. Heiligers J, Parker J S, Macdonald M. Novel solar-sail mission concepts for high-latitude Earth and lunar observation[J]. Journal of Guidance, Control, and Dynamics,2016, 41(1): 212–230.
4. Yuan J, Gao C, Zhang J. Periodic orbits of solar sail equipped with reflectance control device in Earth–Moon system[J]. Astrophysics and Space Science, 2018, 363(2): 1–23.
5. Shirobokov M, Trofimov S, Ovchinnikov M. Survey of station-keeping techniques for libration point orbits[J]. Journal of Guidance, Control, and Dynamics, 2017, 40(5): 1085–1105.
6. Leipold M, G'otz M. Hybrid photonic/electric propulsion[R]. Kayser-Threde GmBH Report: SOL4-TR-KTH-001, Munich, Germany, 2002.
7. Peng H, Zhao J, Wu Z, et al. Optimal periodic controller for formation flying on libration point orbits[J]. Acta Astronautica, 2011, 69(7–8): 537–550.
8. Bryson A E. Time-varying linear-quadratic control[J]. Journal of Optimization Theory & Applications, 1999, 100(3): 515–525.
9. Wang Z, Wu Z. Nonlinear attitude control scheme with disturbance observer for flexible spacecrafts[J]. Nonlinear Dynamics, 2015, 81(1–2): 257–264.
10. Lee D. Nonlinear disturbance observer-based robust control of attitude tracking of rigid spacecraft[J]. Nonlinear Dynamics, 2017, 88(2): 1–12.
11. Narula A, Biggs J. Fault-tolerant station-keeping on libration point orbits[J]. Journal of Guidance, Control, and Dynamics, 2018, 41(4): 879–887.
12. Biggs J D, Henninger H C, Narula A. Enhancing station-keeping control with the use of extended state observers[J]. Frontiers in Applied Mathematics and Statistics, 2018, 4: 1–9.
13. Mu J, Gong S, Li J. Coupled control of reflectivity modulated solar sail for GeoSail formation flying[J]. Journal of Guidance, Control, and Dynamics, 2015, 38(4): 740–751.
14. Liu J, Chen L, Cui N. Solar sail chaotic pitch dynamics and its control in Earth orbits[J]. Nonlinear Dynamics, 2017, 90: 1755–1770.
15. Tamakoshi D, Kojima H. Solar sail orbital control using reflectivity variations near the Earth–Moon L_2 point[J]. Journal of Guidance, Control, and Dynamics, 2018, 41(2): 417–430.
16. Wong B, Patil R, Misra A. Attitude dynamics of rigid bodies in the vicinity of the Lagrangian points[J]. Journal of Guidance, Control, and Dynamics, 2008, 31(1): 252–256.
17. Kulkarni J E, Campbell M E, Dullerud G E. Stabilization of spacecraft flight in halo orbits: An H_∞ approach[J]. IEEE Transactions on Control Systems Technology, 2006, 14(3): 572–578.
18. Gao C, Yuan J, Zhao Y. ADRC for spacecraft attitude and position synchronization in libration point orbits[J]. Acta Astronautica, 2018, 145: 238–249.
19. Biggs J D, McInnes C R. Solar sail formation flying for deep-space remote sensing[J]. Journal of Spacecraft and Rockets, 2009, 46(3): 670–678.
20. Chen W H, Ballance D J, Gawthrop P J, et al. A nonlinear disturbance observer for robotic manipulators. IEEE Transactions on Industrial Electronics, 2000, 47(4): 932–938.

Chapter 4
Solar Sailing in the Earth-Moon Quasi-Bicircular Problem

This chapter focuses on the study of the influence of solar radiation pressure on the dynamics around the dynamical substitutes of the L_1 and L_2 points in the Earth-Moon Quasi-Bicircular Problem. This dynamical model is a periodic perturbation of the Restricted Three Body Problem that incorporates both the gravitational effect of the Sun and the solar radiation pressure acceleration on the sail. Beginning with the simplest invariant objects in the Quasi-Bicircular Problem, namely, the dynamical substitutes of the two equilibrium points and extending to the low-order Sun-resonant periodic orbits with the synodic period of the Sun, we explore the evolution of families of resonant periodic orbits as two sail parameters, defining orientation and efficiency, are varied. The study reveals a complex network of connections between these families. In the analysis, we include their characteristic curves, the maximal Floquet exponent, and the linear normal behavior of the periodic orbits. An interesting observation is that, for certain parameter values, periodic orbits exist that become stable under the influence of solar radiation pressure acceleration.

4.1 Introduction

In the previous two chapters, we discussed solar sailing in the Earth-Moon system based on the Restricted Three-Body Problem (RTBP), without accounting for the gravitational influence of the Sun. However, the Sun's gravity can significantly affect the dynamics around the libration points. Therefore, this chapter takes the analysis a step further by incorporating the Sun's gravitational effect into the Earth-Moon system under the framework of the Quasi-Bicircular Problem (QBCP). Bicircular Problem (BCP) and QBCP are two models that describe the motion of spacecraft in the Sun-Earth-Moon system. Both models are developed accounting for the gravitational effect of the Sun in the Earth-Moon RTBP and they could be seen as periodic perturbations of the RTBP. The QBCP is a coherent version of the BCP. According

© The Author(s), under exclusive license to Springer Nature Singapore Pte Ltd. 2026
C. Gao and W. Wang, *Solar Sailing for Deep-Space Explorations*,
https://doi.org/10.1007/978-981-95-0109-0_4

to [1, 2], the BCP is more adequate for the study of the dynamics near the triangular libration points, while the QBCP is more suitable for the study of the dynamics near the collinear libration points. In both models, the two libration points of the RTBP under consideration (the L_1 and L_2 points) are substituted by small periodic orbits, usually called their dynamical substitutes. In this chapter, we discuss the effect of the solar radiation pressure (SRP) on the dynamics of the QBCP. In particular, we present a complete picture on the evolution of the resonant periodic orbits in the realm of L_1 and L_2 in terms of shape and linear normal behavior. Playing with the parameters of the sail, its lightness number and pitch angle, we find that some of these periodic orbits, together with many newly discovered periodic orbits, belong to families connected via the SRP acceleration. Note that including the Sun's gravity into the Earth-Moon-sail system does not increase the dynamical complexity of the model, since the SRP depends periodically in time with the same period as the Sun around the Earth-Moon system.

The chapter is organized as follows. Section 4.1 is devoted to set the equations of the QBCP including the SRP acceleration. Section 4.2 summarizes some known results about the low order resonant periodic orbits of the QBCP around the two collinear libration points L_1 and L_2. The main contributions of the chapter can be found in Sects. 4.3 and 4.4, in which we detail the intricated web of connections between the families of resonant periodic orbits arising from resonant periodic orbits of the RTBP. Finally, some conclusions are given in Sect. 4.5.

4.2 Equations of Motion

4.2.1 The Quasi-Bicircular Problem

The QBCP is first introduced by Andreu in [3] where the orbits of the Sun, the Earth and the Moon follow a solution of the general three body problem. The quasi-bicircular solution of the three bodies is planar, and is determined by means of a collocation method for the Fourier coefficients of their Jacobi coordinates. As compared to the BCP, the QBCP does not introduce any relevant additional complexity, since it is also a time-periodic system, but it is a better approximation of the true motion of these three Solar System bodies. In particular, it captures the dynamical consequences associated with the 1:2 resonance between the inner frequency of the halo family around the L_2 point and the synodic frequency of the Moon. In fact, near this resonance and using continuation procedures, one can compute long time quasi-periodic orbits close to periodic orbits of the QBCP in the N-body Solar System ephemeris model, while the procedures diverge when the initial seed has been obtained using simpler models, such as the RTBP or the BCP (see [3] for some examples).

Similar to the procedures used in the RTBP, we consider a pulsating rotating reference frame to study the motion of spacecraft in the Earth-Moon QBCP. The key

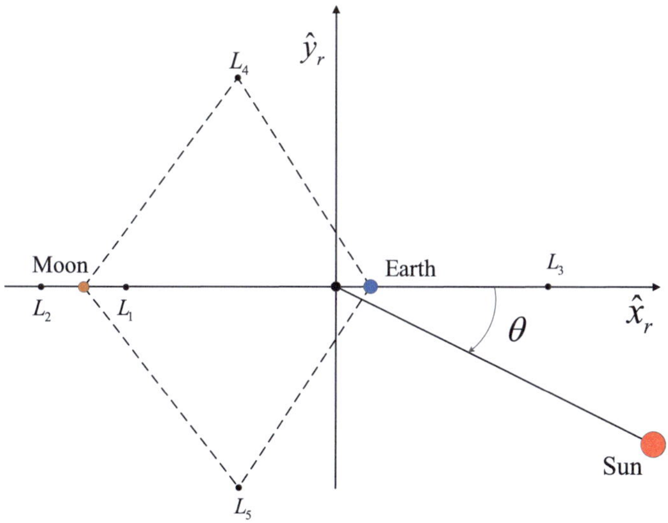

Fig. 4.1 The Quasi-Bicircular Problem

difference is that, following the traditional convention in the QBCP rotating frame, the Earth is located at $(\mu, 0, 0)$ and the Moon is at $(\mu - 1, 0, 0)$ as seen in Fig. 4.1.

The QBCP can be written as a Hamiltonian system with three and a half degrees-of-freedom, and the Hamiltonian function is [3]:

$$\begin{aligned} H_{QBCP}(t, x, y, z, p_x, p_y, p_z) =& \frac{1}{2}\alpha_1(p_x^2 + p_y^2 + p_z^2) + \alpha_2(xp_x + yp_y + zp_z) \\ &+ \alpha_3(yp_x - xp_y) + \alpha_4 x + \alpha_5 y - \alpha_6\left(\frac{1-\mu}{r_{PE}} + \frac{\mu}{r_{PM}} + \frac{m_S}{r_{PS}}\right) \end{aligned} \tag{4.1}$$

The associated equations of motion are

$$\begin{cases} \dot{x} = \alpha_1 p_x + \alpha_2 x + \alpha_3 y \\ \dot{y} = \alpha_1 p_y + \alpha_2 y - \alpha_3 x \\ \dot{z} = \alpha_1 p_z + \alpha_2 z \\ \dot{p}_x = -\alpha_2 p_x + \alpha_3 p_y - \alpha_4 - \alpha_6\left(\dfrac{1-\mu}{r_{PE}^3}(x-\mu) + \dfrac{\mu}{r_{PM}^3}(x-\mu+1) + \dfrac{m_S}{r_{PS}^3}(x-\alpha_7)\right) \\ \dot{p}_y = -\alpha_2 p_y - \alpha_3 p_x - \alpha_5 - \alpha_6\left(\dfrac{1-\mu}{r_{PE}^3}y + \dfrac{\mu}{r_{PM}^3}y + \dfrac{m_S}{r_{PS}^3}(y-\alpha_8)\right) \\ \dot{p}_z = -\alpha_2 p_z - \alpha_6\left(\dfrac{1-\mu}{r_{PE}^3}z + \dfrac{\mu}{r_{PM}^3}z + \dfrac{m_S}{r_{PS}^3}z\right) \end{cases} \tag{4.2}$$

In the above equations $[x, y, z]^{\mathrm{T}}$ denotes the position vector, $[p_x, p_y, p_z]^{\mathrm{T}}$ the associated momenta, m_S the mass of Sun, $r_{PE} = \sqrt{(x-\mu)^2 + y^2 + z^2}$, $r_{PM} =$

Table 4.1 Main parameters of the QBCP. They have been obtained from the JPL-DE405 ephemeris data file. $T_S = 2\pi/\omega_S$ is the period of the Sun and of the non-autonomous system (4.2)

μ	m_S	ω_S	T_S
0.0121505816	328900.5423094043	0.925195985520347	6.791193871907968

$\sqrt{(x-\mu+1)^2+y^2+z^2}$, $r_{PS} = \sqrt{(x-\alpha_7)^2+(y-\alpha_8)^2+z^2}$, and the coefficients $\alpha_1, \ldots, \alpha_8$ are real-valued periodic functions whose Fourier expansions are given by:

$$\alpha_i(\theta) = a_0^i + \sum_{k\geq 0} a_k^i \cos(k\theta) + \sum_{k\geq 0} b_k^i \sin(k\theta), \quad i = 1, \ldots, 8 \tag{4.3}$$

where $\theta = \omega_S t$ and ω_S denotes the frequency of the synodic motion of the Sun with respect to the Earth-Moon line. The numerical values of the Fourier expansions for the α_i functions can be found in Appendix A, while the main remaining parameters in Eq. (4.2) are given in Table 4.1.

Moreover, the α_i are odd functions for $i = 1, 3, 4, 6, 7$, and even ones for $i = 2, 5, 8$. Their physical meaning is the following:

1. $[\alpha_7, \alpha_8, 0]^T$ is the position of the Sun in the plane of motion of the primaries. The modulus of this vector is close to 389, which approximately is the ratio between the Sun-(Earth-Moon barycenter) and the Earth-Moon distances.
2. $\alpha_1, \alpha_2, \alpha_3$, and α_6, capture the fact that the distance between the Earth and Moon pulsates. They slightly vary periodically with time about their respective constant values of the RTBP model. This is, $\alpha_1 \simeq 1$, $\alpha_2 \simeq 0$, $\alpha_3 \simeq 1$ and $\alpha_6 \simeq 1$.
3. α_4 and α_5, related with the coherent non-circular motion of the Sun about the Earth-Moon system, oscillate with zero mean and amplitude close to 2.

Because of the periodic perturbation of the Sun, the equilibrium points of the RTBP do not longer subsist in the QBCP. They are replaced by (small) periodic orbits with the same period of the Sun. These periodic orbits are usually called the *dynamical substitutes* (or equivalents) of the equilibrium points. In this chapter we only consider the L_1 and L_2 points, and we are going to refer to their periodic dynamical substitutes also by L_1 and L_2 which are shown in Fig. 4.2.

It is well known that several families of periodic Lyapunov and halo orbits exist around the above mentioned libration points of the RTBP. When the orbits of these families are continued to the QBCP they 'gain' the frequency of the Sun and almost all them become two-dimensional invariant tori. Nevertheless, there are some orbits that remain periodic in this model, they are *resonant periodic orbits* whose period is a (rational) multiple of the period of the Sun. During the (numerical) continuation of resonant periodic orbits from the RTBP to the QBCP, several bifurcations to two or more periodic orbits of the QBCP appear. According to [3], Table 4.2 shows the low order resonant periodic orbits, the number of bifurcated orbits, and the labels given to them in both the RTBP and the QBCP. The first column shows the labels assigned to the orbits in the RTBP, all of them are with first character '0', the second

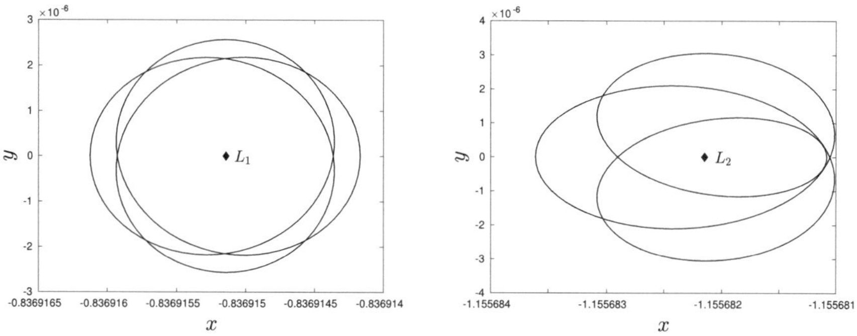

Fig. 4.2 Dynamical substitutes of the L_1 point (left) and the L_2 point (right)

Table 4.2 Low order resonant periodic orbits around the L_1 and L_2 points

RTBP	Resonance	Number of bifurcated orbits	QBCP label
012	1 : 2	2	**12, 13**
014	1 : 1	4	**14, 15, 16, 17**
018	1 : 1	4	**18, 19, 110, 111**
$\mathbf{01}C^{\pm}$	1 : 3	2	$\mathbf{1}C^{\pm}, \mathbf{1}D^{\pm}$
$\mathbf{01}E^{\pm}$	1 : 3	2	$\mathbf{1}E^{\pm}, \mathbf{1}F^{\pm}$
022	1 : 2	4	**22, 23, 24, 25**
026	1 : 1	4	**26, 27, 28, 29**
$\mathbf{02}A^{\pm}$	1 : 2	4	$\mathbf{2}A^{\pm}, \mathbf{2}B^{\pm}, \mathbf{2}C^{\pm}, \mathbf{2}D^{\pm}$
$\mathbf{02}E^{\pm}$	1 : 3	2	$\mathbf{2}E^{\pm}, \mathbf{2}F^{\pm}$
$\mathbf{02}G^{\pm}$	1 : 4	2	$\mathbf{2}G^{\pm}, \mathbf{2}H^{\pm}$

character of the labels indicates the equilibrium point (1 or 2) and the third character is a number when the orbit is a Lyapunov orbit, or a letter when it is a halo one. The second column gives the order of the resonance. The third column displays the number of bifurcated orbits and in the fourth column the corresponding labels are assigned to them in the QBCP (see [3]). In addition, the orbits, whose label is a number, are the continuation of planar Lyapunov orbits, while those with a letter in the label correspond to the continuation of halo orbits. In addition, the superindex $\pm$ denotes if the halo-type orbit is of the north or the south type. The north and south families are each other mirrored counterparts with respect to the $\hat{z}_r = 0$ plane.

4.2.2 The QBCP with the SRP

Denoting by $[\boldsymbol{a}]_{\mathcal{T}_R} = [a_x, a_y, a_z]^{\mathrm{T}}$ the acceleration due to the SRP, it is easy to write the equations of motion according to Eq. (4.2) that include this additional acceleration

$$\left\{\begin{aligned} \ddot{x} - \frac{\dot{\alpha}_1}{\alpha_1}\dot{x} - 2\alpha_3\dot{y} &= \left(\dot{\alpha}_2 + \alpha_2^2 + \alpha_3^2 - \frac{\dot{\alpha}_1\alpha_2}{\alpha_1}\right)x + \left(\dot{\alpha}_3 - \frac{\dot{\alpha}_1\alpha_3}{\alpha_1}\right)y - \alpha_1\alpha_4 \\ &\quad - \alpha_1\alpha_6\left(\frac{1-\mu}{r_{PE}^3}(x-\mu) + \frac{\mu}{r_{PM}^3}(x-\mu+1) + \frac{m_S}{r_{PS}^3}(x-\alpha_7)\right) + a_x \\ \ddot{y} - \frac{\dot{\alpha}_1}{\alpha_1}\dot{y} + 2\alpha_3\dot{y} &= \left(\dot{\alpha}_2 + \alpha_2^2 + \alpha_3^2 - \frac{\dot{\alpha}_1\alpha_2}{\alpha_1}\right)y - \left(\dot{\alpha}_3 - \frac{\dot{\alpha}_1\alpha_3}{\alpha_1}\right)x - \alpha_1\alpha_5 \\ &\quad - \alpha_1\alpha_6\left(\frac{1-\mu}{r_{PE}^3}y + \frac{\mu}{r_{PM}^3}y + \frac{m_S}{r_{PS}^3}(y-\alpha_8)\right) + a_y \\ \ddot{z} - \frac{\dot{\alpha}_1}{\alpha_1}\dot{z} &= \left(\dot{\alpha}_2 + \alpha_2^2 - \frac{\dot{\alpha}_1\alpha_2}{\alpha_1}\right)z - \alpha_1\alpha_6\left(\frac{1-\mu}{r_{PE}^3}z + \frac{\mu}{r_{PM}^3}z + \frac{m_S}{r_{PS}^3}z\right) + a_z \end{aligned}\right. \tag{4.4}$$

where $[\boldsymbol{x}]_{\mathcal{T}_R} = [x, y, z, \dot{x}, \dot{y}, \dot{z}]^{\mathrm{T}}$ denotes the state of spacecraft in synodic reference frame $\mathcal{T}_R$.

According to Eq. (2.6), the SRP acceleration exerted on an ideal solar sail is

$$\beta\frac{m_S}{r_{PS}^2}(\hat{\boldsymbol{n}}\cdot\hat{\boldsymbol{n}}_S)^2\hat{\boldsymbol{n}} \tag{4.5}$$

where β is the lightness number, and $\hat{\boldsymbol{n}}$ and $\hat{\boldsymbol{n}}_S$ are the unitary vectors normal to the sail and along the Sun-sail direction, respectively.

However, since in the QBCP the Earth-Moon distance pulsates, similar to the gravitational terms in Eq. (4.4) the SRP acceleration of Eq. (4.5) has to be modified by means of the periodic term $\alpha_1\alpha_6$. This is,

$$\alpha_1\alpha_6\,\beta\frac{Gm_S}{r_{PS}^2}(\hat{\boldsymbol{n}}\cdot\hat{\boldsymbol{n}}_S)^2\hat{\boldsymbol{n}} \tag{4.6}$$

and taking into account that the functions α_7 and α_8 provide the position of the Sun, the unitary vector $\hat{\boldsymbol{n}}_S$ can be written as

$$[\hat{\boldsymbol{n}}_S]_{\mathcal{T}_R} = \frac{1}{\sqrt{\alpha_7^2 + \alpha_8^2}}[-\alpha_7, -\alpha_8, 0]^{\mathrm{T}} \tag{4.7}$$

Moreover, we are going to consider that the sail can be inclined with respect to the Earth-Moon plane to create an out-of-plane acceleration component. Denoting by φ the pitch angle, and according to Fig. 4.3, the unitary vector normal to the sail $\hat{\boldsymbol{n}}$ can be written as

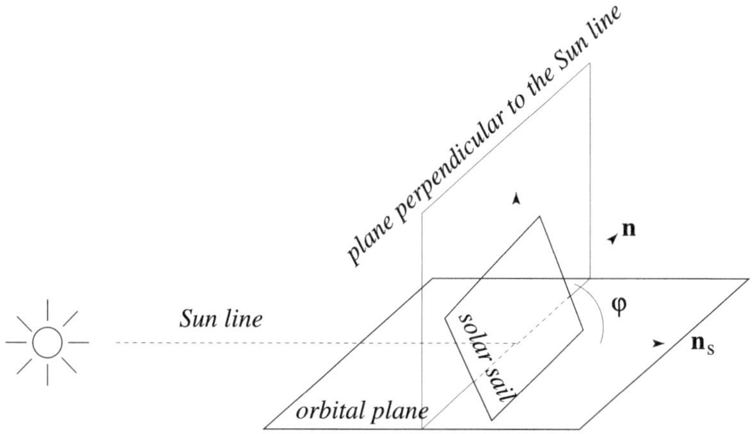

Fig. 4.3 Schematic representation of the pitch angle φ. When $\varphi = 0$ the sail is perpendicular to the Sun line

$$[\hat{\boldsymbol{n}}]_{\mathcal{T}_R} = \left[-\frac{\alpha_7 \cos\varphi}{\sqrt{\alpha_7^2 + \alpha_8^2}}, -\frac{\alpha_8 \cos\varphi}{\sqrt{\alpha_7^2 + \alpha_8^2}}, -\sin\varphi \right]^{\mathrm{T}} \tag{4.8}$$

Note that the in-plane angle of the sail is not considered in this chapter, i.e., the projection of the unitary vector normal to the sail in the Earth-Moon plane is always aligned with the Sun-sail direction.

Assuming that the sail moves in the neighborhood of the Earth-Moon barycenter, the following simplification has been done:

$$r_{PS}^2 \approx \sqrt{\alpha_7^2 + \alpha_8^2} \tag{4.9}$$

Then, according to all the above considerations, the three components of SRP acceleration become

$$\begin{aligned} a_x &= -\alpha_1\alpha_6\beta \frac{m_S}{\alpha_7^2 + \alpha_8^2} \frac{\alpha_7}{\sqrt{\alpha_7^2 + \alpha_8^2}} \cos^3\varphi \\ a_y &= -\alpha_1\alpha_6\beta \frac{m_S}{\alpha_7^2 + \alpha_8^2} \frac{\alpha_8}{\sqrt{\alpha_7^2 + \alpha_8^2}} \cos^3\varphi \\ a_z &= -\alpha_1\alpha_6\beta \frac{m_S}{\alpha_7^2 + \alpha_8^2} \cos^2\varphi \sin\varphi \end{aligned} \tag{4.10}$$

4.3 Resonant Periodic Orbits

Table 4.2 has listed some of the low-order resonant orbits in the QBCP, while Table 4.3 provides their stability properties. The stability properties (linear normal behavior) of these periodic orbits are characterized by the eigenvalues of the associated monodromy matrix, which can be gathered by pairs. Each pair of complex conjugate eigenvalues (λ, $\overline{\lambda} \in \mathbb{C}$ with $|\lambda| = 1$) corresponds to a central behavior, and each pair of real eigenvalues (λ, $\lambda^{-1} \in \mathbb{R}$) to a saddle one. There exist also a complex saddle case involving two pairs of inverse and complex conjugate eigenvalues with non zero real part. In this way, the label $(C \times C \times C)$ stands for Center $\times$ Center $\times$ Center, $(S \times C \times C)$ for Saddle $\times$ Center $\times$ Center, $(S \times CS)$ for Saddle $\times$ Complex Saddle.

The periodic orbits with labels **11** and **21** are the dynamical substitutes of the two libration points L_1 and L_2. There are few cases in Table 4.3, those with labels $\mathbf{2B^{\pm}}$ and $\mathbf{2D^{\pm}}$, for which their linear normal behavior is not given. This is because they can not be continued from the RTBP up to the QBCP when varying the mass of the Sun from zero to its actual value m_S. In fact, starting the continuation at $\mathbf{02A^{\pm}}$, two orbits, $\mathbf{2B^{\pm}}$ and $\mathbf{2D^{\pm}}$, appear after a first pitchfork bifurcation and disappear after a second one. Moreover, the resonant orbits $\mathbf{1N^{\pm}}$, $\mathbf{1H^{\pm}}$, **210**, $\mathbf{2I^{\pm}}$, $\mathbf{2K^{\pm}}$, and $\mathbf{2L^{\pm}}$ do not appear in the Table 4.2. They correspond to orbits of particular bifurcated

Table 4.3 Linear stability characteristics of the periodic orbit of the QBCP

Equilibrium point	Resonance	QBCP (label and linear stability)
	1 : 1	**11** $(S \times C \times C)$
	1 : 2	**12** $(S \times S \times S)$, **13** $(S \times S \times C)$
	1 : 1	**14** $(S \times S \times S)$, **15** $(S \times S \times C)$, **16** $(S \times S \times S)$, **17** $(S \times S \times C)$
L_1	1 : 1	**18** $(S \times S \times C)$, **19** $(S \times S \times S)$, **110** $(S \times S \times C)$, **111** $(S \times S \times S)$
	1 : 3	$\mathbf{1C^{\pm}}$ $(S \times C \times C)$, $\mathbf{1D^{\pm}}$ $(S \times S \times C)$, $\mathbf{1N^{\pm}}$ $(S \times S \times C)$, $\mathbf{1H^{\pm}}$ $(S \times S \times C)$
	1 : 3	$\mathbf{1E^{\pm}}$ $(S \times S \times C)$, $\mathbf{1F^{\pm}}$ $(S \times S \times C)$
	1 : 1	**21** $(S \times C \times C)$
	1 : 2	**22** $(S \times C \times C)$, **23** $(S \times S \times C)$, **24** $(S \times C \times C)$, **25** $(S \times S \times C)$
	1 : 1	**26** $(S \times S \times S)$, **27** $(S \times S \times C)$, **28** $(S \times S \times S)$, **29** $(S \times S \times C)$
	1 : 3	**210** $(S \times C \times C)$
L_2	1 : 2	$\mathbf{2A^{\pm}}$ $(S \times S \times C)$, $\mathbf{2B^{\pm}}$, $\mathbf{2C^{\pm}}$ $(S \times C \times C)$, $\mathbf{2D^{\pm}}$
	1 : 2	$\mathbf{2I^{\pm}}$ $(S \times C \times C)$, $\mathbf{2L^{\pm}}$ $(S \times CS)$, $\mathbf{2K^{\pm}}$ $(S \times S \times S)$
	1 : 3	$\mathbf{2E^{\pm}}$ $(C \times C \times C)$, $\mathbf{2F^{\pm}}$ $(C \times C \times C)$
	1 : 4	$\mathbf{2G^{\pm}}$ $(S \times C \times C)$, $\mathbf{2H^{\pm}}$ $(S \times S \times C)$

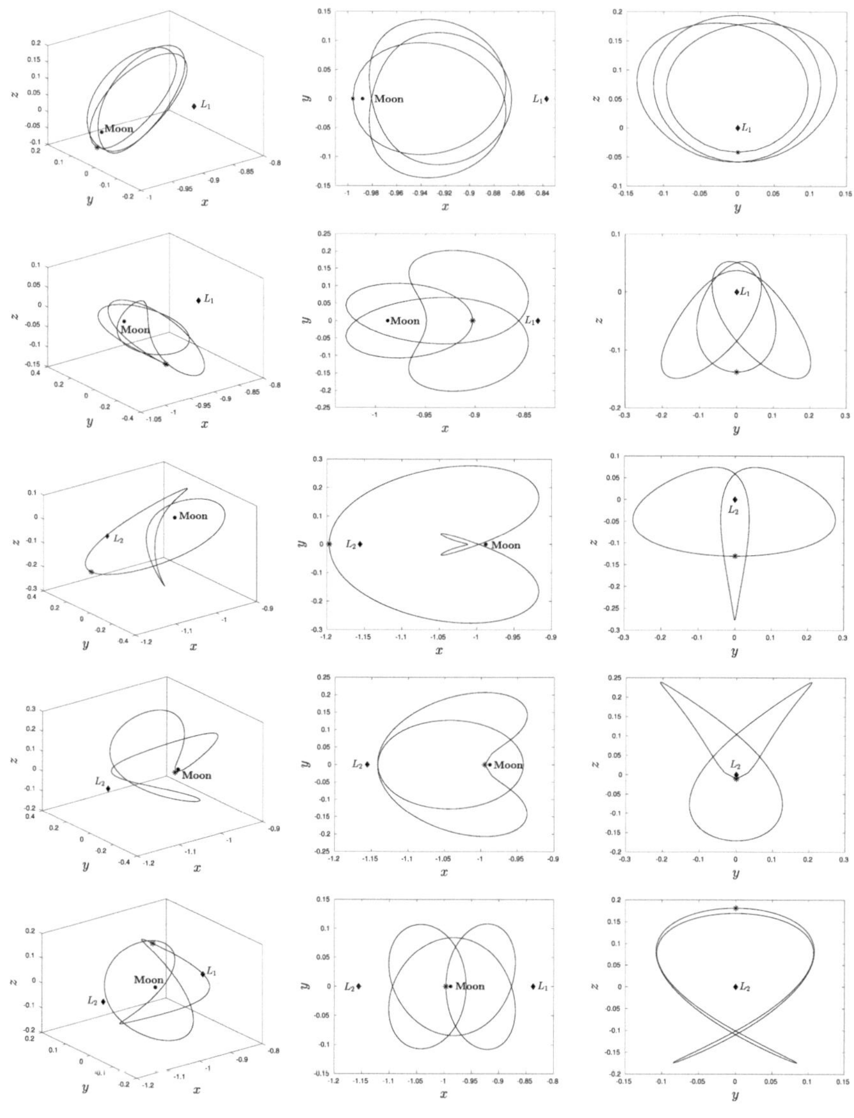

Fig. 4.4 From top to bottom, 3D representation and $\hat{x}_r - \hat{y}_r$ and $\hat{y}_r - \hat{z}_r$ coordinate projections of the halo-type resonant orbits $\mathbf{1N^-}$, $\mathbf{1H^-}$, $\mathbf{2I^-}$, $\mathbf{2K^-}$ and $\mathbf{2L^-}$

families newly determined during the computations as will be discussed in the next two sections. The 3D representation of these new halo-type orbits, together with two of their coordinate projections, are shown in Fig. 4.4. In addition, the planar Lyapunov-type orbit **210** is shown in Fig. 4.5.

The initial conditions of the bifurcated symmetric periodic orbits displayed in Table 4.3 can be found in Tables of the Appendix B. Initial conditions for non-

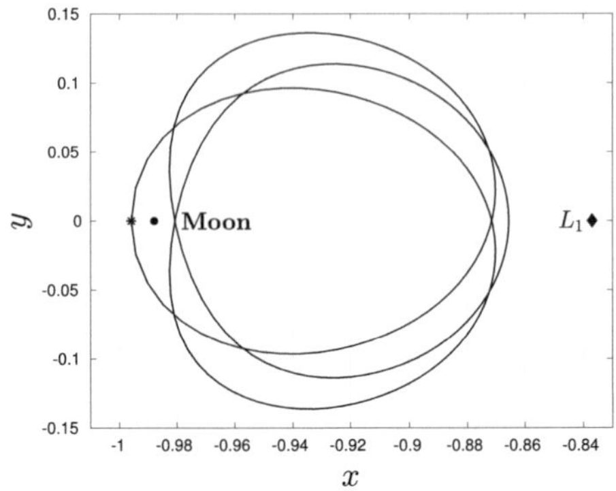

Fig. 4.5 Lyapunov-type orbit **210**

symmetric orbits, which are not discussed in what follows, are not given (a periodic orbit is said to be symmetric if its trajectory $[x(t), y(t), z(t)]^{\mathrm{T}}$is symmetric with respect to the $\hat{x} - \hat{z}$ plane, and their initial conditions are $[x, y = 0, z = 0, \dot{x} = 0, \dot{y}, \dot{z} = 0]^{\mathrm{T}}$). It must be noted that, due to their highly unstable character, the refinement and continuation of the families needs of a parallel shooting method introduced in Chap. 3, so for convenience, the coordinates of three states $\boldsymbol{Q}_0$, $\boldsymbol{Q}_1$, and $\boldsymbol{Q}_2$ are provided in the tables. In the next section we analyze how these resonant symmetric QBCP periodic orbits evolve when including a solar sail.

4.4 The Web of Periodic Orbits Around L_1

In this section we present the results of the continuation of the symmetric resonant orbits **11**, **12**, **13**, **14**, **16**, **18**, **110**, $\mathbf{1}\boldsymbol{C}^-$, $\mathbf{1}\boldsymbol{N}^-$, $\mathbf{1}\boldsymbol{D}^\pm$, and $\mathbf{1}\boldsymbol{H}^\pm$ with respect to the two sail parameters: the lightness number β and the pitch angle φ. As it has already been said, the non-symmetric orbits have been excluded from the explorations, as well as the ones associated to the halo-type orbits $\mathbf{1}\boldsymbol{E}^\pm$ and $\mathbf{1}\boldsymbol{F}^\pm$. The reason for the exclusion of these last two orbits is that a bifurcation happens at the very beginning of both continuations, and the computation of the Hessian matrix becomes very sensitive due to a near collision with a primary. At this point it would be required the use of regularized equations of motion that have not been implemented in this chapter.

For all the families we display their *characteristic curves*. For β-continuation families of Lyapunov-type orbits, the characteristic curves show, as a function of β, the value of x-coordinate of the point $\boldsymbol{Q}_0$, that corresponds to the orthogonal intersection of the orbit with the x-axis (i.e. $y = 0$) at the initial epoch t_0. For the families of Lyapunov-type orbits computed varying the angle φ, and of halo-type orbits computed varying the parameter β or φ, the characteristic curves show the value of z-coordinate of the first orthogonal intersection $\boldsymbol{Q}_0$ of the orbit with the $\hat{x} - \hat{z}$ plane as a function of one of these two parameters.

In both cases and for each family, stability curves are also provided. In them, in addition to the norm of the largest eigenvalue of the monodromy matrix M, the evolution of the linear stability properties of the orbits is given as a function the parameter (this is, we show their linear normal behavior, according to the characteristics of the eigenvalues of M).

For all the families, the lightness number varies within a range $\beta \in [0, \beta_{\max}]$, where $\beta_{\max}$ is always larger than the lightness number of the current solar sail capabilities, which is about 0.04 [4, 5]. For the families of planar Lyapunov-type orbits, the continuation with respect to β has been done setting the pitch angle φ equal to zero (the sail is perpendicular to the Sun direction), while for the halo-type orbits different values of φ are considered within $[-\pi/2, \pi/2]$.

4.4.1 Continuation of Periodic Orbits Around L_1 with Respect to β

Since for the continuations of planar orbits with respect to β the pitch angle φ is set equal to zero, there is no out-of-plane acceleration component due to the SRP, therefore the families of resonant periodic orbits departing from the planar orbits of the QBCP with labels **11**, **12**, **13**, **14**, **16**, **18**, and **110** remain planar for all values of β. Figure 4.6 summarizes the results we obtain for these seven Lyapunov-type orbits, and the main results are the following:

- The top row of Fig. 4.6 shows the characteristic and stability curves for the continuations starting at the orbits labeled as **11** and **12**. It can be seen that the two families of resonant periodic orbits obtained are, in fact, the same one. Starting at any of the two orbits, **11** or **12**, there is a turning point at $\beta \approx 0.096403$ where the two families coincide, and it corresponds to a transition in the linear normal behaviour between $(S \times C \times C)$ and $(S \times S \times C)$. Clearly, all the orbits of the family are highly unstable.
- As we see in Table 4.2, there is a second orbit **13** that bifurcates at a 1:2 resonance, together with orbit **12**, from the same resonant orbit **012**. In fact, both orbits are very close in the configuration space [3], however the behaviour of the two families that emanate from them is quite different under the effect of the SRP. The second row of Fig. 4.6 shows that the **13** – **14** connection curve has a turning point at $\beta \approx 0.1301$, with no changes in the stability properties of the orbits at this point (this is the only case that has been found where the turning point is not associated to a stability bifurcation). Some orbits of the families connecting **11** with **12**, and **13** with **14** are shown in Fig. 4.7.
- The characteristic and stability curves of the family connecting orbits **16** and **110** are given in the third row of Fig. 4.6. Along this connection there is a transition between $(S \times S \times C)$ and $(S \times S \times S)$ at the turning point close to $\beta \approx 0.0307$.
- The last row of Fig. 4.6 shows the results obtained for the resonant orbit **18**. All the orbits of this family are unstable and with a $(S \times S \times C)$ behaviour. Due to

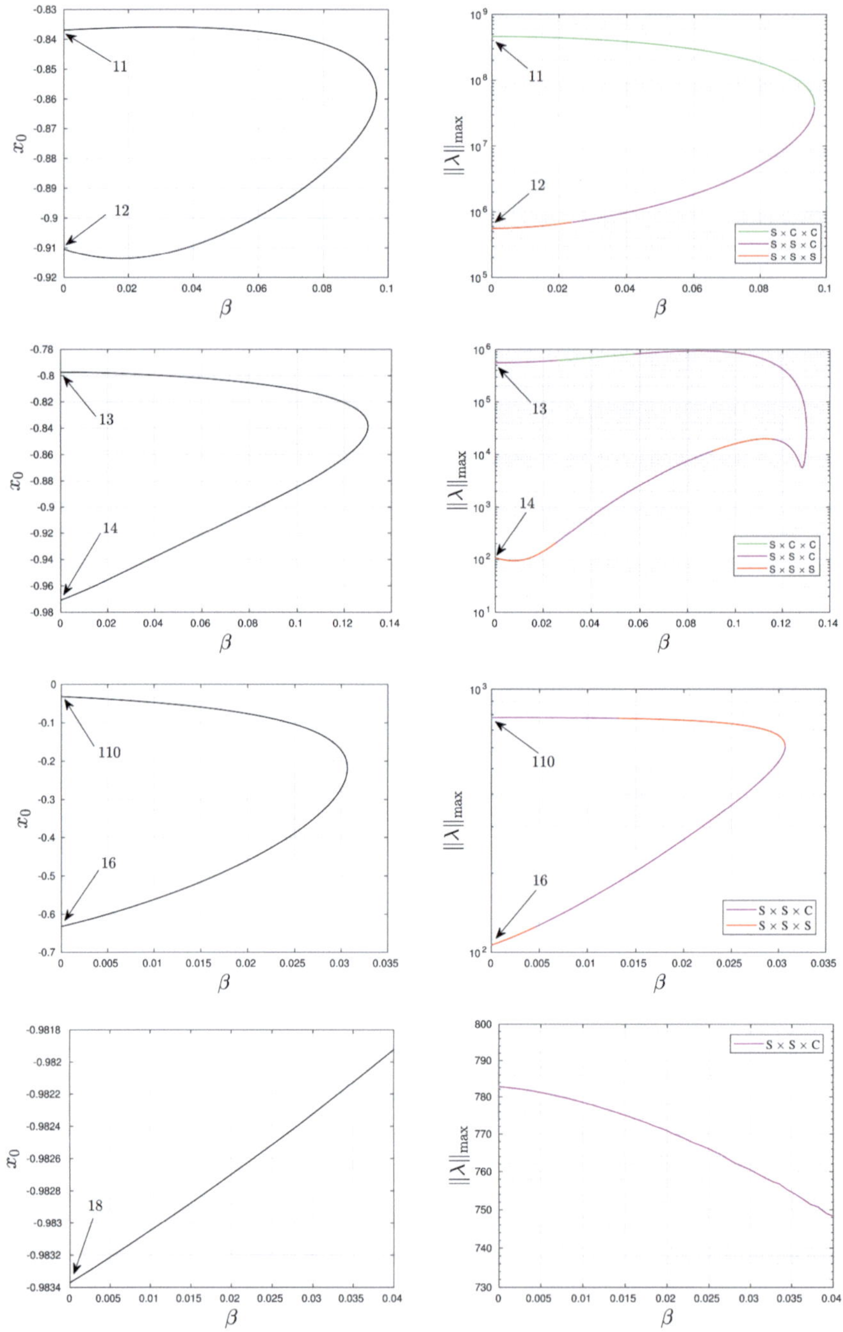

Fig. 4.6 Characteristic (left) and stability (right) curves of the families of connections between several resonant Lyapunov-type orbits. From top to bottom: family connecting orbits **11** and **12**, family connecting orbits **13** and **14**, family connecting orbits **16** and **110**, and family departing from orbit **18**

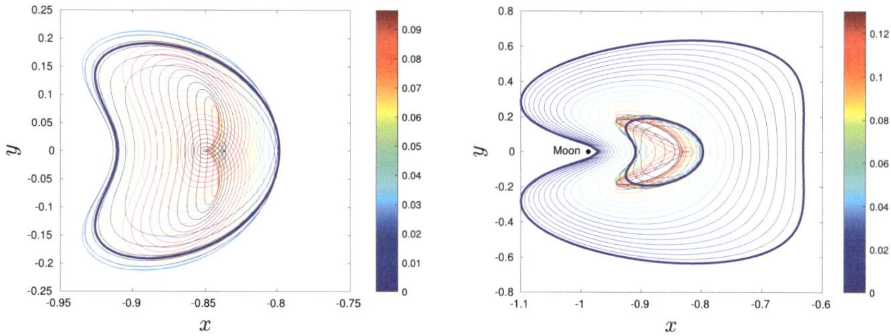

Fig. 4.7 Left: periodic orbits of the family connecting orbit **11** (in the middle of the plot, but not displayed since it is too small) and orbit **12** (thick blue one). Right: periodic orbits of the family connecting orbit **13** (inner thick blue one) and orbit **14** (outer thick blue one). The color indicates the value of the lightness number β

their close approach to the Moon the continuation has been stopped at $\beta = 0.04$, little beyond the maximum current technological feasible value of this parameter.

Figure 4.8 gives the results obtained for the halo-type orbits $\mathbf{1C}^-$, $\mathbf{1N}^-$, $\mathbf{1D}^-$ and $\mathbf{1H}^-$ and are briefly described in what follows.

- The top row of Fig. 4.8 shows how the evolution with β provides a connection between the halo-type orbits $\mathbf{1C}^-$ and $\mathbf{1N}^-$ for different values of the pitch angle φ (see also [1]). Due to the symmetry of the system, there is another connection between $\mathbf{1C}^+$ and $\mathbf{1N}^+$ symmetric to this one, this is: for each value of φ, there are curves connecting $\mathbf{1C}^-$ and $\mathbf{1N}^-$ which are symmetric to the ones that connect $\mathbf{1C}^+$ and $\mathbf{1N}^+$ for $-\varphi$. The value of β for which the turning point of the curve occurs increases as the absolute value of φ does. At the turning points there is a transition in the stability behavior between $(S \times C \times C)$ and $(S \times S \times C)$. According to the stability results it follows that some orbits of the families obtained for different values of φ become stable (the norm of the maximum eigenvalue is equal to one) in certain narrow ranges of β (that depend on φ). This means that unstable periodic orbits in the QBCP can be stabilized by the SRP. For $\varphi = 0$, some orbits of the connection are shown in Fig. 4.9. A stable orbit of this family, with $\beta = 0.00041344$, is shown in the bottom row of this same figure. Note that, due to the 1:3 resonance, the orbits perform three revolutions along their period.
- The bottom row of Fig. 4.8 shows some β-connections between the halo-type orbits $\mathbf{1D}^-$ and $\mathbf{1H}^-$. For values of φ larger than -0.1407 (≈ -8.1 deg) (we have explored up to $\varphi = 1.0$ (≈ 57.3 deg)), the characteristic curves that connect $\mathbf{1D}^-$ and $\mathbf{1H}^-$ are similar in behaviour, and all them have a turning point at certain β values that depend on φ (for instance, when $\varphi = 1.0$ the turning point is at $\beta = 0.0284$). At these points the stability behaviour of the periodic orbits changes from $(S \times C \times C)$ to $(S \times S \times C)$. For $\varphi = -0.1407$ there is a bifurcation at the turning point $\beta = 0.0199$. At this bifurcation point one of the branches connects $\mathbf{1D}^-$ and $\mathbf{1H}^-$, as in the preceding case, and the other branch corresponds to a

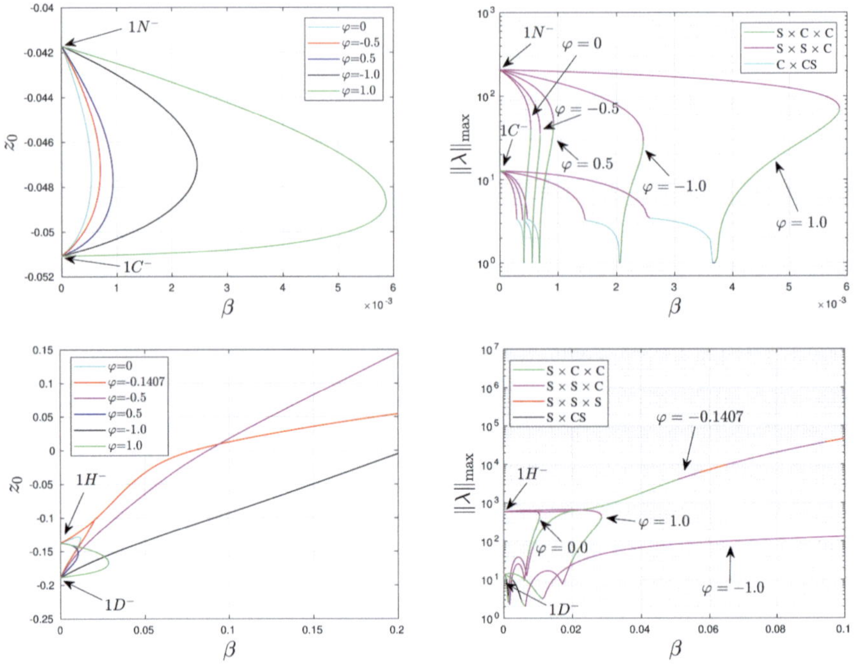

Fig. 4.8 Characteristic (left) and stability (right) curves of the families departing from several resonant halo-type orbits for several fixed values of φ. Top: families connecting orbits $\mathbf{1C^-}$ and $\mathbf{1N^-}$ where stable orbits (with maximum eigenvalue norm equal to one) are found for some narrow ranges of β. Bottom: families departing from orbits $\mathbf{1D^-}$ and $\mathbf{1H^-}$

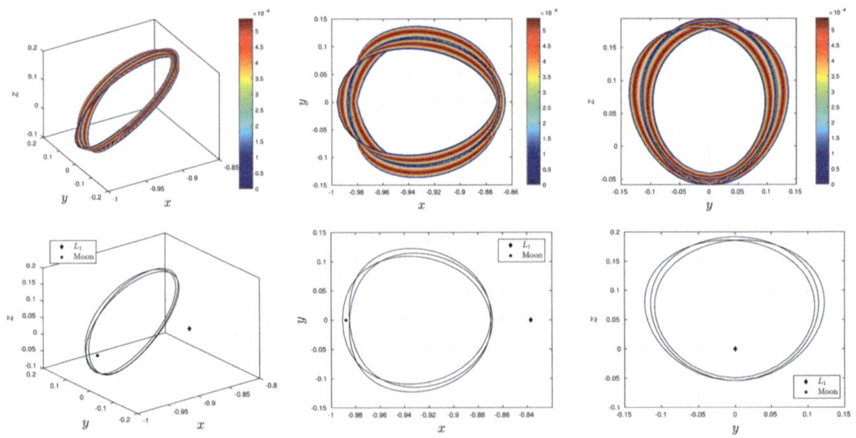

Fig. 4.9 In the top row, 3D representation and $x - y$ and $y - z$ coordinate projections of periodic orbits of the family connecting $\mathbf{1C^-}$ (thick blue one) and $\mathbf{1N^-}$ (thick blue one, see Fig. 4.4) for $\varphi = 0$. The colors indicate the values of lightness number β. The bottom row displays one of the stable periodic orbits of this family for $\beta = 0.00041344$

characteristic curve in which z_0 increases with β. When $\varphi < -0.1407$ only this second branch is detected, and there is no turning point or bifurcation along the characteristic curve. In all cases, the continuation procedure with respect to β has been carried out up to $\beta = 0.2$, which is about 5 times larger than the current maximum technological value.

4.4.2 Continuation of Periodic Orbits Around L_1 with Respect to φ

This section shows the results obtained for the evolution of the same resonant orbits of the preceding section when the parameter β is fixed and φ varies in $[-\pi/2, \pi/2]$. Figures 4.10 and 4.11 show the results obtained for the resonant Lyapunov-type orbits. A brief summary of these results is given in what follows.

- The top row in Fig. 4.10 shows the characteristic and stability curves of the families of periodic orbits departing from the dynamical substitute **11** for some fixed values of β. The behavior of the curves depends on the value of this parameter. For $\beta = 0.096403$ (black curve), at $\varphi = 0$ there is a bifurcation of the characteristic curve into two branches. Along one branch, φ keeps monotonically increasing until the curve reaches again the departure orbit **11** at $\varphi = \pi/2$. Along the other branch φ decreases, and the curve ends at the resonant periodic orbit **12** at $\varphi = -\pi/2$. For $\beta < 0.096403$, the family ends with the departure orbit **11** when φ increases from $-\pi/2$ to $\pi/2$. Note that, in these cases, the maximum z_0-value occurs at $-\varphi_{\max}$, where $\varphi_{\max} = \sin^{-1}(1/\sqrt{3}) \approx 0.6155$ (≈ 35.3 deg) is the value for which the out-of-plane acceleration component due to the SRP is maximum. Due to the symmetry, the minimum z_0-value is at $\varphi_{\max}$. For values of β slightly larger than 0.096403 (we have explored $\beta = 0.1$, which is displayed in green), orbits **11** and **12** are connected by varying the pitch angle φ.
- According to the stability curves associated to these families, when β is less than the bifurcation value 0.096403 (the results are displayed for $\beta = 0.01$), the value of the maximum eigenvalue reaches its minimum at $\varphi = 0$, and all the orbits in the family are of $(S \times C \times C)$ type. For $\beta \geq 0.096403$, i.e. $\beta = 0.1$, the maximum eigenvalue decreases along the family. In addition, there is a first change in the linear stability behaviour at the turning point where the stability changes from $(S \times C \times C)$ to $(S \times S \times C)$. Later it follows a second transition to $(S \times CS)$, and a third one to $(S \times S \times S)$.
- The results obtained for the continuation of the resonant Lyapunov-type orbit **12** are similar to the ones already described for orbit **11**, and are shown in the second row of Fig. 4.10. If $\beta < 0.096403$, the characteristic curves show the connection of orbit **12** with itself; if $\beta = 0.096403$ there is a bifurcation at $\varphi = 0$, as in the preceding case; and for $\beta > 0.096403$, orbits **11** and **12** are connected, as it has already seen.

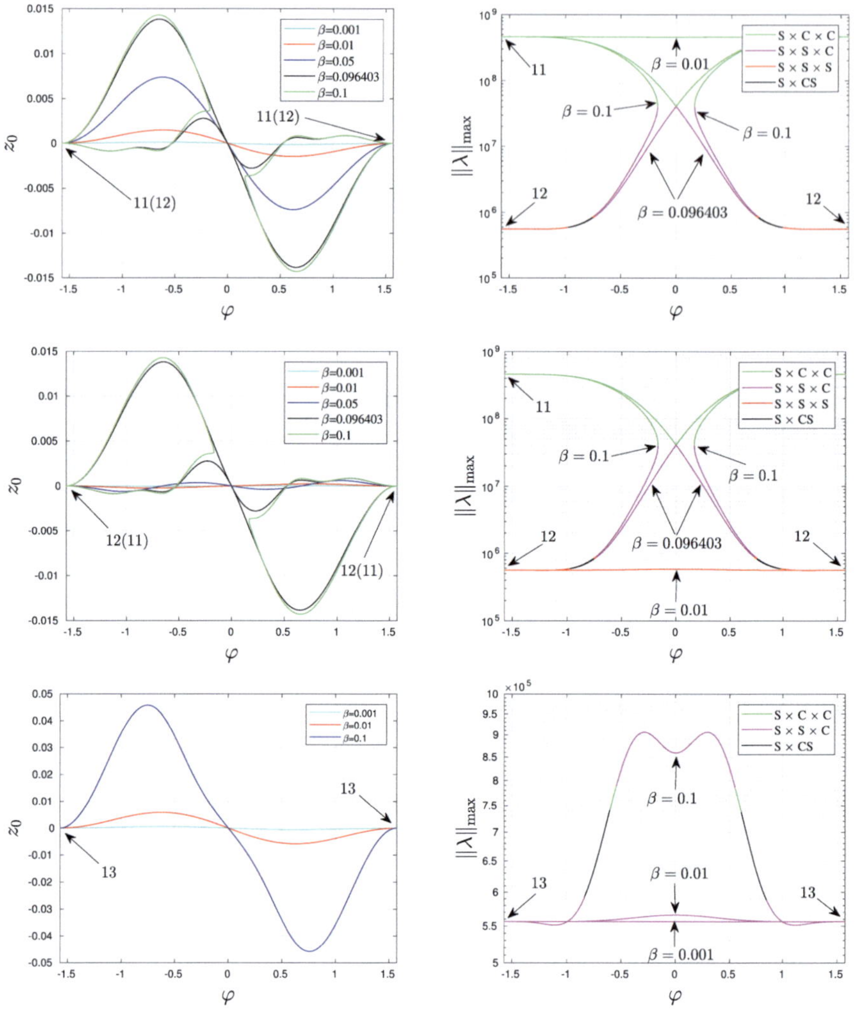

Fig. 4.10 Characteristic (left) and stability (right) curves of the families of several resonant Lyapunov-type orbits obtained by continuation with respect to $\varphi \in [-\pi/2, \pi/2]$ and for several fixed values of β. From top to bottom: families departing from: **11**, **12**, and **13**

- The third row of Fig. 4.10 shows the behaviour of the characteristic and stability curves of the families departing from the resonant orbit **13**. There are no bifurcations along the characteristic curves, and, furthermore, the stability properties of the orbits do not change along the families, they are all $(S \times S \times C)$ except for $\beta \geq 0.1$.
- The behavior of the families departing from the Lyapunov-type orbits **14** and **18** in Fig. 4.11 is completely similar to the one described for the families departing from orbit **13**. The characteristic curves are symmetric, connect each orbit with

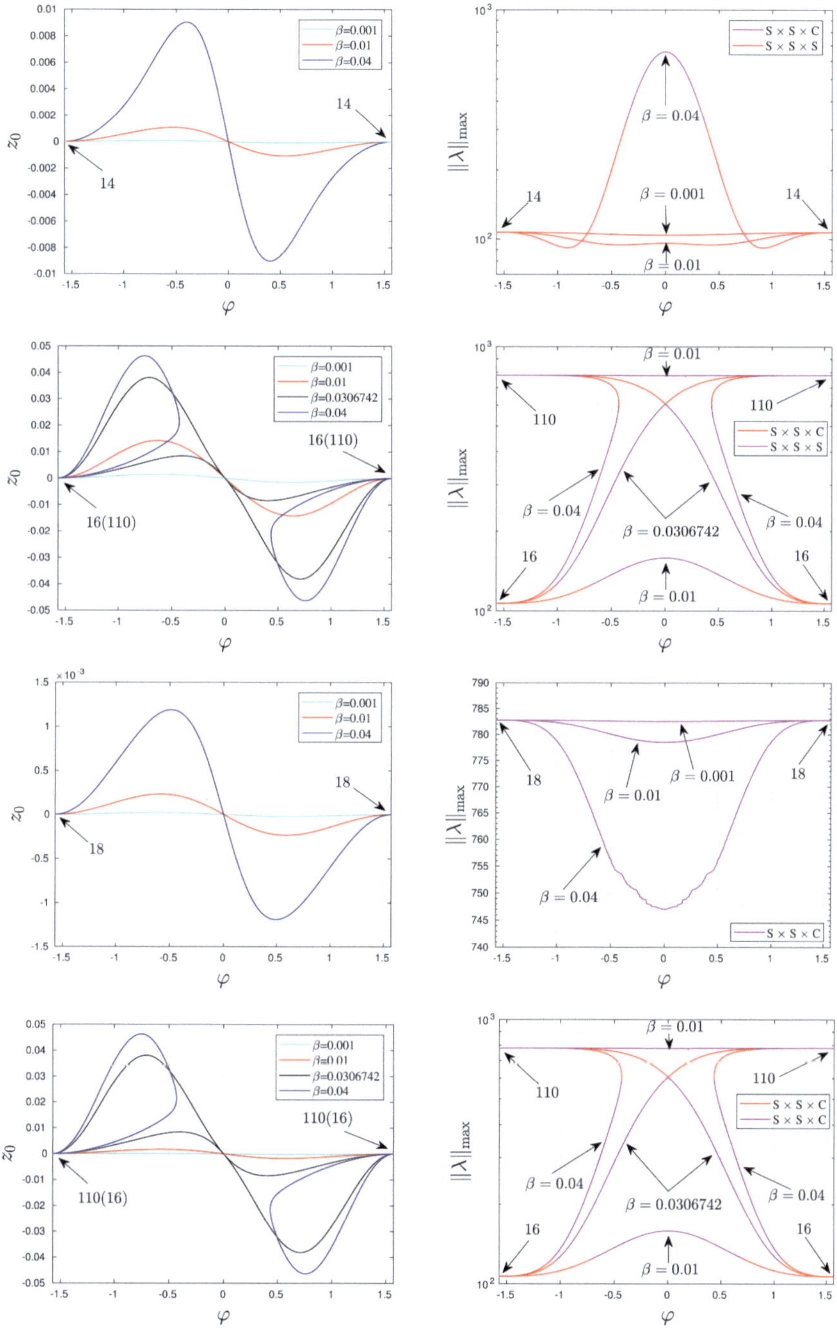

Fig. 4.11 Characteristic (left) and stability (right) curves of the families of several resonant Lyapunov-type orbits obtained by continuation with respect to $\varphi \in [-\pi/2, \pi/2]$ and for several fixed values of β. From top to bottom: families departing from: **14**, **16**, **18**, and **110**

itself, and there are almost no changes in the stability properties of the orbits along the families: they are of $(S \times S \times S)$ type in the **14–14** connection (there are only variations for $\beta \geq 0.04$) and of $(S \times S \times C)$ in the **18–18** one.

- The behavior of the families departing from the Lyapunov-type orbits **16** and **110** is very similar as seen in Fig. 4.11b and d, although the two orbits are different. The second row of Fig. 4.11 shows the characteristic and stability curves for the families departing from **16** for different values of β. The two orbits are connected with themselves for $\beta < 0.0306742$; while for $\beta \geq 0.0306742$, the orbit departing from **16** is connected with the one departing from **110**, and vice versa. In the associated characteristic curves there is a turning point at a value of φ that depends on β, and is associated to a change in the stability properties between $(S \times S \times C)$ and $(S \times S \times S)$.
- Note that the maximum value $\beta_{\max} = 0.04$ has been reached for the orbits **14**, **18**, **16** and **110**, because the associated families approach closer and closer to the Moon as β increases. Since no regularization procedure has been implemented in the equations of motion, the continuations finish at its value.

The results obtained for the families departing from the halo-type resonant orbits $\mathbf{1C^-}$, $\mathbf{1D^\pm}$, $\mathbf{1H^\pm}$, and $\mathbf{1N^-}$ are given in Fig. 4.12, and are briefly described in what follows.

- The top row of Fig. 4.12 shows characteristic and stability curves departing from $\mathbf{1C^-}$ and $\mathbf{1N^-}$ varying $\varphi \in [-\pi/2, \pi/2]$, and for different values of β. Bifurcation occurs at $\varphi \approx -0.08256$ (≈ -4.7 deg) for $\beta = 0.0005321$ where the two characteristic curves (in black), departing from $\mathbf{1C^-}$ and $\mathbf{1N^-}$, are connected. For $\beta > 0.0005321$, the characteristic curves have two components and their turning points correspond to the stability transition between $(S \times C \times C)$ and $(S \times S \times C)$. Furthermore, and according to what was already found in the continuations with respect to β, the linear behaviour of some orbits is $(C \times C \times C)$, which means they are stable. So, as it was already said, the SRP can stabilize orbits.
- The bottom row of Fig. 4.12 shows the results for the families departing from the halo-type orbits $\mathbf{1D^\pm}$ and $\mathbf{1H^\pm}$ for different values of β. For $\beta = 0.0096856$, the characteristic curves (in red) departing from $\mathbf{1D^-}$ and $\mathbf{1H^-}$ are connected at the bifurcation point $\varphi = 0.2935$ (≈ 16.8 deg). For $\beta > 0.0096856$ (i.e. $\beta = 0.01$), the characteristic curve (in dark) associated to the $\mathbf{1D^-}$–$\mathbf{1H^-}$ connection has two components and their turning points correspond to stability changes between $(S \times C \times C)$ and $(S \times S \times C)$. Furthermore, for $\beta = 0.029711$, there is another bifurcation at $\varphi = 0$, where the $\mathbf{1D^-}$–$\mathbf{1H^-}$ characteristic curve (in black) departing at $\varphi = -\pi/2$ is connected with its symmetric counterpart. For $\beta > 0.029711$, the characteristic curves departing from $\mathbf{1D^-}$ and $\mathbf{1H^-}$ at $\varphi = -\pi/2$ reach their symmetric counterparts at $\varphi = \pi/2$, this is: $\mathbf{1D^-}$ and $\mathbf{1D^+}$ are connected, as well as $\mathbf{1H^-}$ and $\mathbf{1H^+}$. See, for instance, the green curve when $\beta = 0.1$.

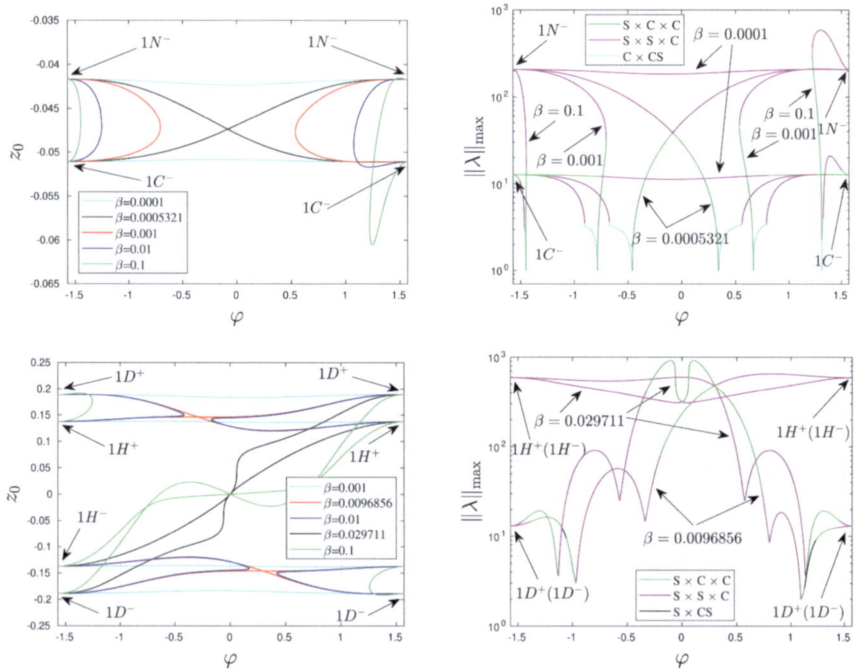

Fig. 4.12 Characteristic (left) and stability (right) curves of the families of several resonant halo-type orbits obtained by continuation with respect to $\varphi \in [-\pi/2, \pi/2]$ and for several fixed values of β. From top to bottom: families departing from: $\mathbf{1C^-}$ and $\mathbf{1N^-}$, and $\mathbf{1D^\pm}$ and $\mathbf{1H^\pm}$

4.5 The Web of Periodic Orbits Around L_2

Similar to the preceding section, now we focus on the continuation of families of symmetric resonant periodic orbits around L_2, again with respect to the sail parameters β and φ. The results correspond to the resonant periodic families emanating from the Lyapunov-type orbits **21**, **22**, **24**, **26**, **28**, and **210**, and from the halo-type orbits $\mathbf{2A^\pm}$, $\mathbf{2C^\pm}$, $\mathbf{2E^\pm}$, $\mathbf{2F^\pm}$, $\mathbf{2G^\pm}$, $\mathbf{2I^\pm}$, $\mathbf{2K^\pm}$, and $\mathbf{2L^\pm}$. It must be noted that some of these orbits, **210**, $\mathbf{2I^\pm}$, $\mathbf{2K^\pm}$, and $\mathbf{2L^\pm}$, are not listed in [1, 3]. They have been obtained from the bifurcations of families emanating from the other resonant orbits.

4.5.1 Continuation of Periodic Orbits Around L_2 with Respect to β

As in the L_1 case, the continuation of Lyapunov-type orbits with respect to β has been done keeping the pitch angle φ equal to zero, so the families of resonant periodic orbits departing from the orbits **21**, **22**, **24**, **26**, and **28** remain planar for all β. Figure 4.13 shows the characteristic and stability curves obtained for these families.

- The first two rows of Fig. 4.13 show that there is a family of resonant orbits connecting orbits **21** and **24**, and another family connecting orbits **22** and **26**. The turning points for the **21** – **24** and **22** – **26** connections are at $\beta = 0.01097$ and $\beta = 0.0996$, respectively, and they correspond to $(S \times C \times C)$ and $(S \times S \times C)$ transition. Figure 4.14 shows some orbits of the **21** – **24** and the **22** – **26** connections, where it can be seen that **26** is a quasi-colliding orbit with the Moon.
- The global behaviour of the family departing from orbit **28** is similar to the one departing from **18** which has been discussed in the previous section. As β increases, the orbits of the family become closer and closer to the Moon, so the computations finish at $\beta = 0.04$ without detecting any turning point.

Figure 4.15 shows the families departing or reaching the halo-type orbits $\boldsymbol{2A^{\pm}}$, $\boldsymbol{2C^{\pm}}$, $\boldsymbol{2I^{\pm}}$, $\boldsymbol{2K^{-}}$, and $\boldsymbol{2L^{+}}$ and Fig. 4.16 the families departing or reaching the halo-type orbits $\boldsymbol{2E^{\pm}}$, $\boldsymbol{2F^{-}}$, $\boldsymbol{2G^{-}}$, as well as the planar Lyapunov-type orbit **210**.

- The top row of Fig. 4.15 shows how the orbits $\boldsymbol{2A^{\pm}}$, $\boldsymbol{2I^{\pm}}$, and $\boldsymbol{2L^{+}}$ evolve with β. For $\varphi < 0$, the turning point of the characteristic curve connecting $\boldsymbol{2A^{-}}$ and the Lyapunov-type orbit **22** corresponds to stability transition between $(S \times C \times C)$ and $(S \times S \times C)$. For $0 < \varphi < 0.0444345$ (≈ 2.5 deg) (red curve), $\boldsymbol{2A^{-}}$ is connected with $\boldsymbol{2I^{+}}$ with a $(S \times S \times C)$ to $(S \times S \times S)$ transition at the turning point of the characteristic curve. For $\varphi = 0.0444345$ (black curve) a bifurcation in the family departing from $\boldsymbol{2A^{-}}$ appears at $\beta \approx 0.09054$, and the new bifurcated family ends at $\boldsymbol{2I^{-}}$. Furthermore, for not too large negative values of φ, i.e. $\varphi = -0.5$ (≈ -28.6 deg), the orbits $\boldsymbol{2I^{-}}$ and $\boldsymbol{2L^{+}}$ are also connected (dotted dark blue curve).
- The second and third rows of Fig. 4.15 show the results for the halo-type orbits $\boldsymbol{2C^{-}}$ and $\boldsymbol{2K^{-}}$. For $\varphi = 0$ (light blue curve), $\boldsymbol{2C^{-}}$ is connected with its symmetric counterpart $\boldsymbol{2C^{+}}$ with a turning point at $\beta \approx 0.1508$. For $\varphi = 0.210475$ (≈ 12.1 deg), there is a bifurcation at $\beta \approx 0.01194$ where along one branch β decreases up to zero and the family tends to the Lyapunov-type orbit **24** and along the other to $\boldsymbol{2K^{-}}$. The $\boldsymbol{2C^{-}}$ – **24** connection has a turning point with a transition between $(S \times C \times C)$ and $(S \times S \times C)$ stability types. For $\varphi = -0.960245$ (≈ -55.0 deg) (cyan curve) there is a bifurcation at $\beta = 0.12607$, and $\boldsymbol{2C^{-}}$ is connected with $\boldsymbol{2K^{-}}$. And the stability properties change between $(S \times C \times C)$ to $(S \times S \times C)$ at the turning point.

Figure 4.16 shows the families departing from $\boldsymbol{2E^{\pm}}$, $\boldsymbol{2F^{-}}$, and $\boldsymbol{2G^{-}}$ for several fixed values of φ. The main results are the following:

- For $\varphi = 0$ (light blue curve), orbit $\boldsymbol{2E^{-}}$ is connected with its symmetric counterpart $\boldsymbol{2E^{+}}$ with a turning point at $\beta \approx 0.10976$. For $\varphi > 0$, $\boldsymbol{2E^{-}}$ and the Lyapunov-type orbit **210** are connected after a turning point whereas for $\varphi < 0$, no bifurcations have been detected in the family departing from $\boldsymbol{2E^{-}}$. Besides, for some ranges

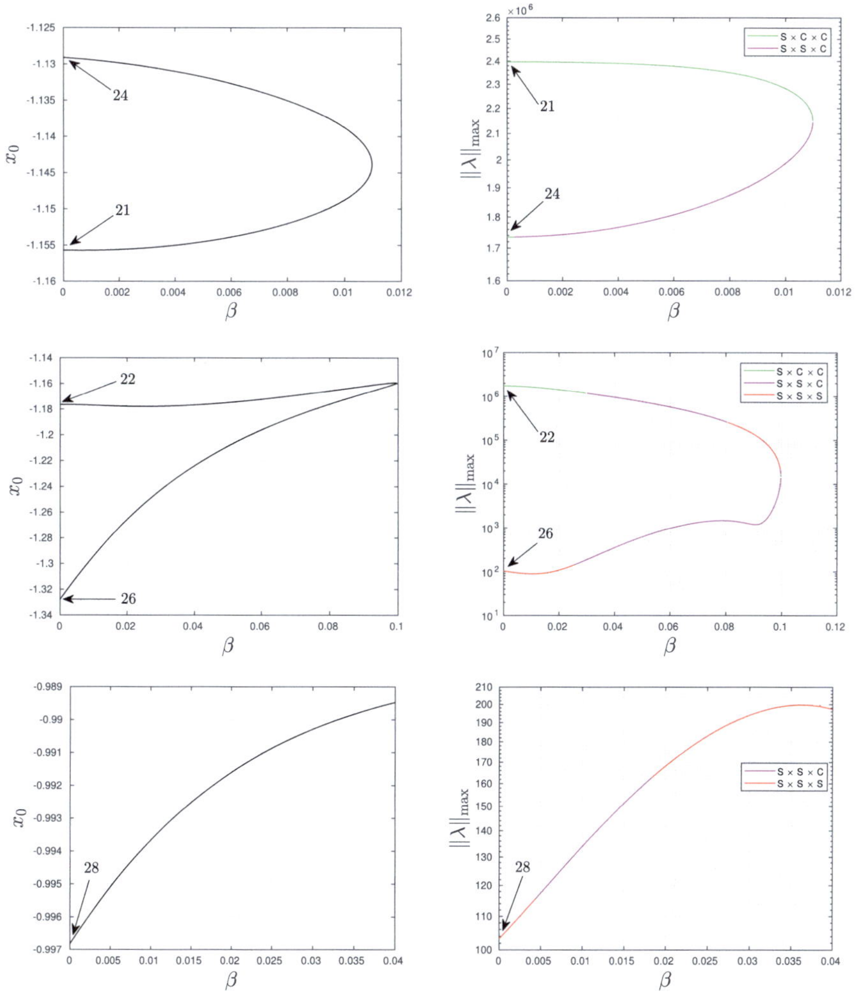

Fig. 4.13 Characteristic (left) and stability (right) curves of the families of connections between several resonant Lyapunov-type orbits. From top to bottom: family connecting orbits **21** and **24**, family connecting orbits **22** and **26**, and family departing from orbit **28**

of β, that depend on φ, the orbits of the families departing from $\mathbf{2E^-}$ have a $(C \times C \times C)$ behaviour and so, they become stable due to the SRP.

- The families departing from $\mathbf{2F^-}$ and $\mathbf{2G^-}$, due to a collision (or close approach) to the Moon, have been continued most only up to $\beta = 0.04$, and the behaviour of the orbits of both families is simple.

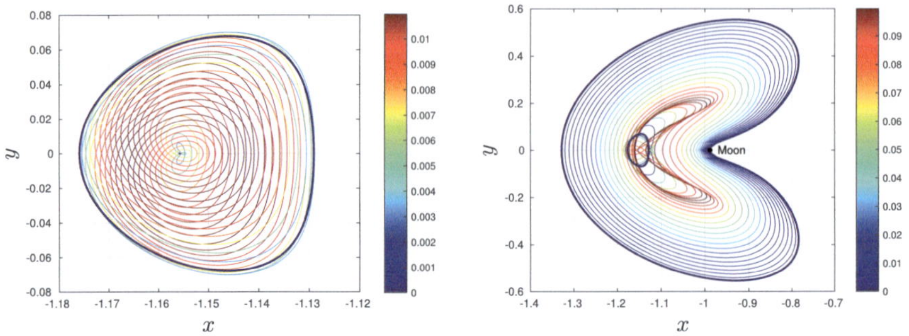

Fig. 4.14 Left: periodic orbits of the family connecting **21** (in the middle of the plot, but not displayed since it is too small) and **24** (thick blue one). Right: periodic orbits of the family connecting **22** (inner thick blue one) and **26** (outer thick blue one). The colors indicate the values of lightness number β

4.5.2 Continuation of Periodic Orbits Around L_2 with Respect to φ

This section shows the evolution of resonant orbits around L_2 when the parameter φ varies in the interval $[-\pi/2, \pi/2]$, and β is fixed. From Figs. 4.17, 4.18, 4.19 and 4.20 we present the characteristic and stability curves obtained for the Lyapunov-type and halo-type families departing from resonant orbits of the QBCP. A summary of the main results is given in what follows.

Figure 4.17 shows the characteristic and stability curves of the families of periodic orbits departing from, or arriving to, the resonant periodic orbits **21**, **22**, $\mathbf{2A^{\pm}}$, $\mathbf{2C^{\pm}}$, $\mathbf{2I^{\pm}}$, and $\mathbf{2L^{\pm}}$. From this figure, the following remarks can be stated:

- In the top row of Fig. 4.17 we see that, departing from orbit **21** a bifurcation appears for $\beta = 0.0109704$ (black curve) at $\varphi = 0$. For $\beta > 0.0109704$, departing from orbit **21**, the family ends at the halo-type orbit $\mathbf{2C^{+}}$ instead of **21** itself, as in the case of $\beta < 0.0109704$.
- The second row of Fig. 4.17 shows the families connecting the Lyapunov-type orbit **22** and the halo-type orbit $\mathbf{2A^{\pm}}$. For $\beta = 0.003678$, the (black) characteristic curves departing from **22** and $\mathbf{2A^{-}}$ meet at $\varphi = -0.6148$ (≈ -35.2 deg). For $\beta > 0.003678$, **22** and $\mathbf{2A^{-}}$ are connected by (red and green) curves crossing twice the value $\varphi = -\pi/2$. Symmetrically, **22** and $\mathbf{2A^{+}}$ are connected by ones crossing twice the value $\varphi = \pi/2$.
- As is shown in the third row of the same figure, another bifurcation appears for $\beta = 0.04832$. The (black) curve connecting $\mathbf{2A^{-}}$ and $\mathbf{2A^{+}}$ meet with the (black) curves starting from $\mathbf{2I^{-}}$ and $\mathbf{2I^{+}}$. Therefore, $\mathbf{2A^{\pm}}$ and $\mathbf{2I^{\pm}}$ are connected with each other. For $\beta = 0.099615$ there is a bifurcation of the curve connecting $\mathbf{2I^{-}}$ with $\mathbf{2I^{+}}$, and $\mathbf{2L^{-}}$ with $\mathbf{2L^{+}}$ at $\varphi = 0$. At this point these four orbits are connected with respect to φ.

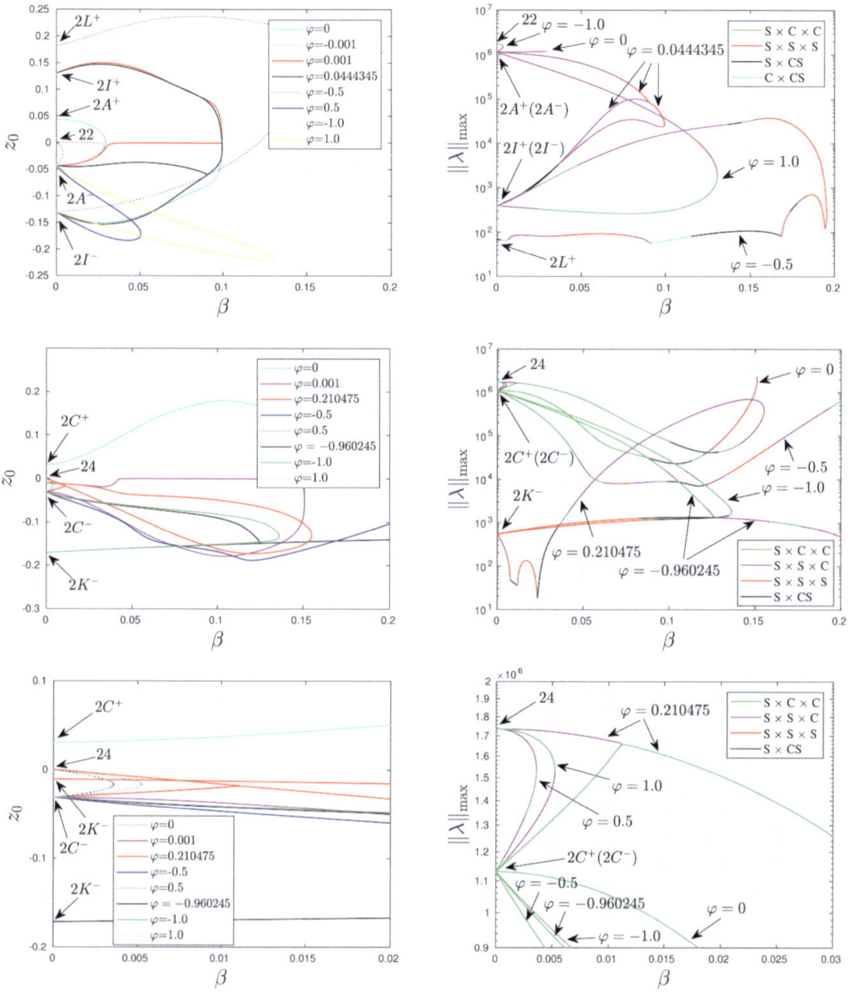

Fig. 4.15 Characteristic (left) and stability (right) curves of the families departing from several resonant halo-type orbits for several fixed values of φ. Top: families departing from orbits $\mathbf{2A}^{\pm}$, $\mathbf{2I}^{\pm}$, and $\mathbf{2L}^{+}$. Middle: families departing from orbits $\mathbf{2C}^{\pm}$ and $\mathbf{2K}^{-}$. Bottom: magnification of the middle one

In Fig. 4.18 we see the connections between the resonant orbits **24** (Lyapunov-type) and $\mathbf{2C}^{\pm}$ (halo-type) for different values of β. The main remarks are the following:

- For $\beta = 0.0034016$, the (black) curve departing from **24** intersects the one coming from $\mathbf{2C}^{+}$ at $\varphi = -0.6296$ (≈ -36.1 deg). Moreover, for $\beta \in [0.0034016, 0.0109704]$, i.e. $\beta = 0.01$ (red curve) , the orbit $\mathbf{2C}^{-}$ is also connected with $\mathbf{2C}^{+}$ after going through two turning points.

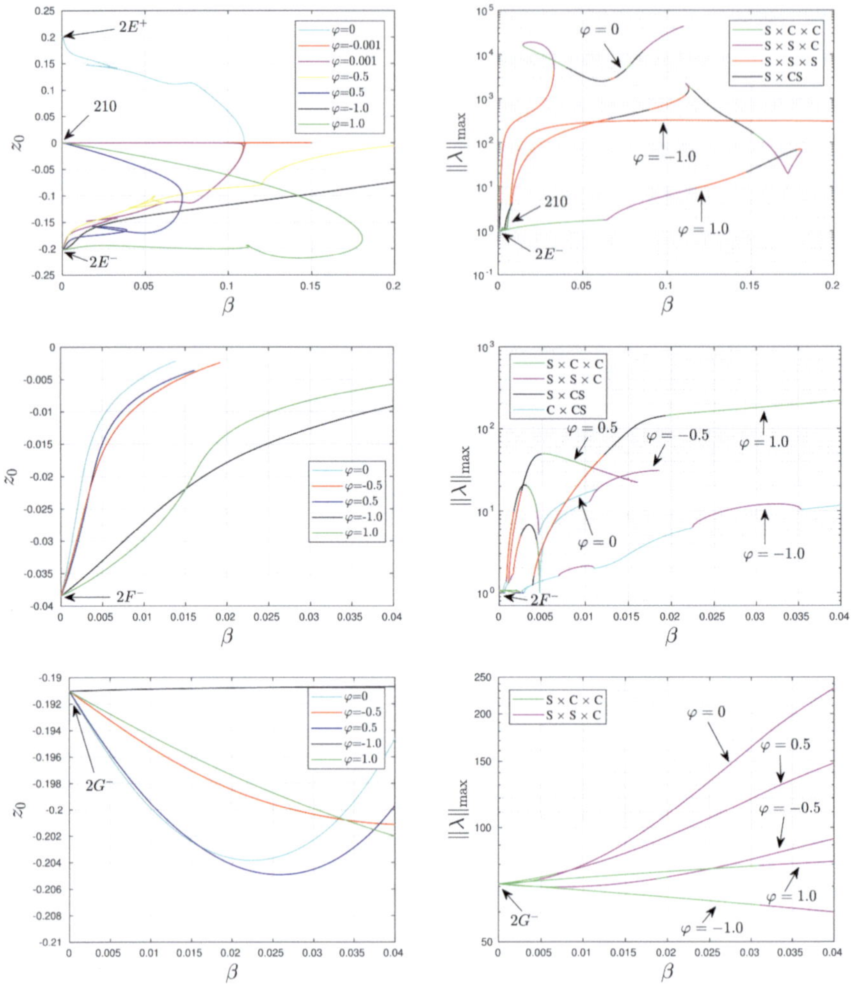

Fig. 4.16 Characteristic (left) and stability (right) curves of the families departing from several resonant halo-type orbits for several fixed values of φ. From top to bottom: family connecting orbits $\mathbf{2E^{\pm}}$ and **210**, family departing from orbit $\mathbf{2F^-}$, family departing from orbit $\mathbf{2G^-}$

- For $\beta = 0.0109704$ there is a bifurcation of the (dark blue) characteristic curves departing from $\mathbf{2C^{\pm}}$ at $\varphi = 0$. In this case there are two families departing from $\mathbf{2C^-}$, one reaches the halo-type orbit $\mathbf{2C^+}$ and the other the small-size dynamical substitute **21**. For $\beta > 0.0109704$ (i.e $\beta = 0.1$), the halo-type orbit $\mathbf{2C^-}$ (or $\mathbf{2C^+}$) and Lyapunov-type orbit **21** are connected by a (green) characteristic curve.

Figure 4.19 shows the results obtained for the φ-continuation families departing from the Lyapunov-type orbits **26** and **28**. Again, due to a near-collision with the Moon in some orbits of both families, the maximum β value that has been explored

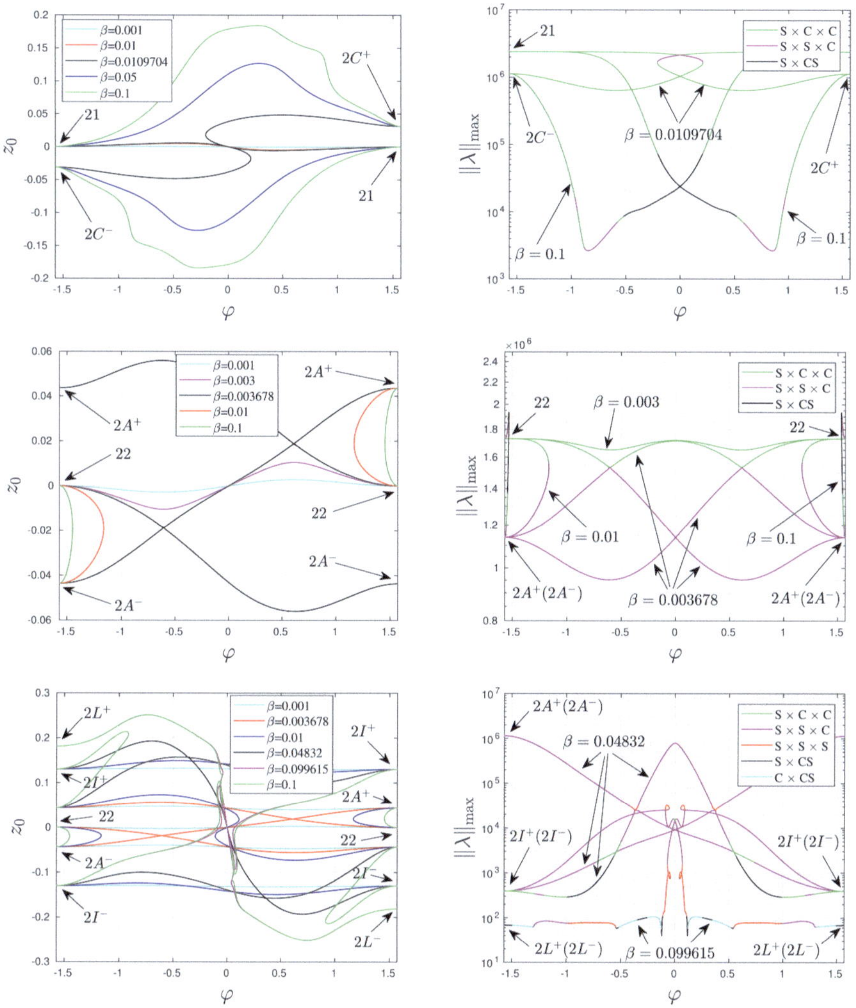

Fig. 4.17 Characteristic (left) and stability (right) curves of the families departing from several resonant periodic orbits obtained by continuation with respect to $\varphi \in [-\pi/2, \pi/2]$ for several fixed values of β. From top to bottom: family departing from orbits **21** and $\boldsymbol{2C^{\pm}}$, family departing from orbits **22** and $\boldsymbol{2A^{\pm}}$, family departing from orbits **22**, $\boldsymbol{2A^{\pm}}$, $\boldsymbol{2I^{\pm}}$, and $\boldsymbol{2L^{\pm}}$

is 0.04. As we see in the figure, the families departing from these two orbits return back to themselves, sometimes experiencing stability transitions in between when β is big enough.

Figure 4.20 shows the results obtained for the families departing from the orbits **210**, $\boldsymbol{2E^{\pm}}$, $\boldsymbol{2F^{-}}$, and $\boldsymbol{2G^{-}}$. The main results obtained are the following:

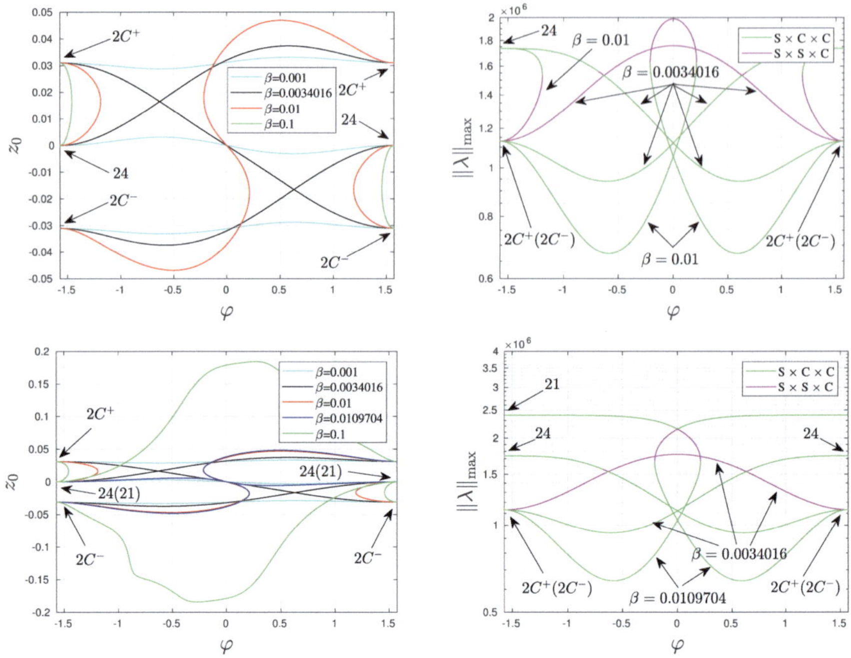

Fig. 4.18 Characteristic (left) and stability (right) curves of the families departing from several resonant periodic orbits with respect to $\varphi \in [-\pi/2, \pi/2]$ for several fixed value of β. Top: families departing from orbits **24** and $\mathbf{2C^{\pm}}$. Bottom: extension of the top one with families departing from orbits $\mathbf{2C^{\pm}}$ for $\beta \geq 0.0109704$

- The top row of Fig. 4.20 shows the connection between the halo-type orbit $\mathbf{2E^{\pm}}$ and Lyapunov-type orbit **210**. For $\beta = 0.0719049$ (black curve), a bifurcation occurs at $\varphi \approx \mp 0.4750$ (≈ 27.2 deg), where the curves connecting $\mathbf{2E^{\pm}}$ and **210** intersect. Furthermore, according to the stability curve, the halo-type families departing from $\mathbf{2E^{\pm}}$ remain stable for a large range of values of φ.
- The second row of Fig. 4.20 shows the evolution of the families departing from orbit $\mathbf{2F^{-}}$. For values of β near to 0.04, the family starting at $\varphi = -\pi/2$ or $\varphi = \pi/2$ approaches to the Moon, and the continuation procedure stops. According to the stability curves, some orbits of the families departing from $\mathbf{2F^{-}}$ are stable for certain ranges of φ that depend on β.
- According to the last row of Fig. 4.20, the characteristic curves departing from $\mathbf{2G^{-}}$ connect with itself for values of β up to 0.04. The families departing from $\mathbf{2G^{-}}$ have stability changes between $(S \times C \times C)$ and $(S \times S \times C)$ for the larger values of β, while for $\beta = 0.001$ all the orbits of the family are of $(S \times C \times C)$ type.

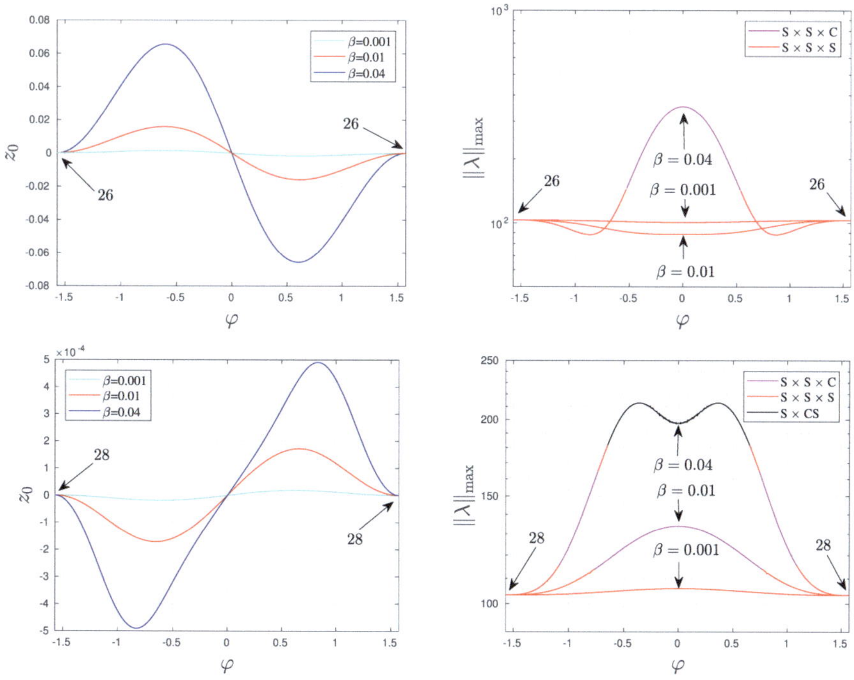

Fig. 4.19 Characteristic (left) and stability (right) curves of the families departing from several resonant Lyapunov-type orbits obtained by continuation with respect to $\varphi \in [-\pi/2, \pi/2]$ and for several fixed values of β. From top to bottom: families departing from orbits **26** and **28**

4.6 Relevant Properties of Some Useful Orbits Found

As we have seen, the neighborhood of L_1 and L_2 in the Earth-Moon system provides a very rich structure from which one could obtain interesting particular applications for missions involving solar sailing. A detailed presentation of the characteristics for all the families and evolution we have found is out of the scope of this chapter, as well as the station keeping control or the optimal transfer procedures that could allow the navigation among them, changing, for instance, the solar sail attitude. However, in this section we show few particular cases that could be of interest.

A main thing to consider is that, due to their geometry, many solar-sail periodic orbits around the L_2 point can be useful as translunar nominal orbits for spacecraft missions. For instance, according to Fig. 4.21, the solar-sail periodic orbits emanating from the halo-type orbit $\mathbf{2C}^-$ are able to support lunar far-side communications, because the Moon does not obstruct the orbit when viewed from the Earth. Some other families with similar properties have also been found, for instance the β–continuation families departing from $\mathbf{2A}^-$ and φ–continuation families departing from **22**. Moreover, when compared to the orbit $\mathbf{2C}^-$, the results in Fig. 4.15 show

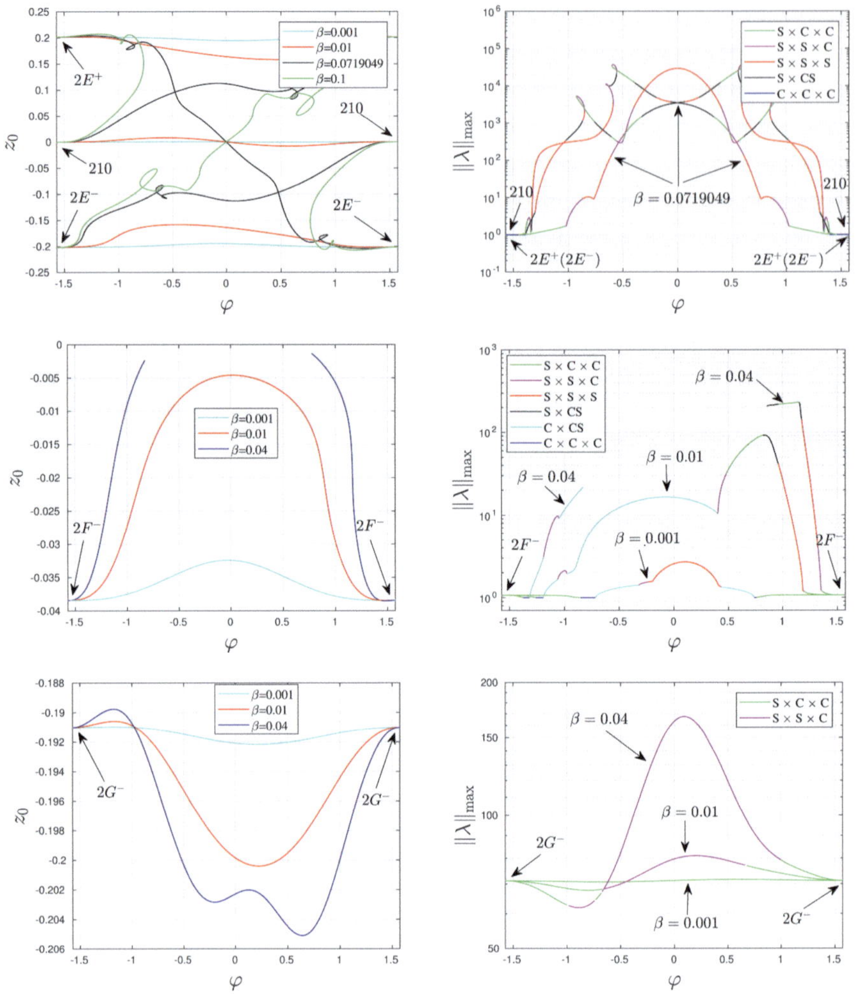

Fig. 4.20 Characteristic (left) and stability curves (right) of the families departing from several resonant halo-type orbits obtained by continuation with respect to $\varphi \in [-\pi/2, \pi/2]$ for several fixed values of β. From top to bottom: families connecting orbits $\mathbf{2E^{\pm}}$ and **210**, families departing from orbit $\mathbf{2F^{-}}$, families departing from orbit $\mathbf{2G^{-}}$

that these solar-sail orbits are much more stable. In addition, they are good candidates to be employed for bounded formation flying, since all the orbits have the same period.

Changing the value of the pitch angle φ, some planar orbits increase in size and transition to 3-dimensional orbits with the z-amplitude tuned by this parameter. Figure 4.22 displays examples of this 3D evolution associated to the families departing from orbits **12** and **22**. A similar behavior happens for the family departing from **11**.

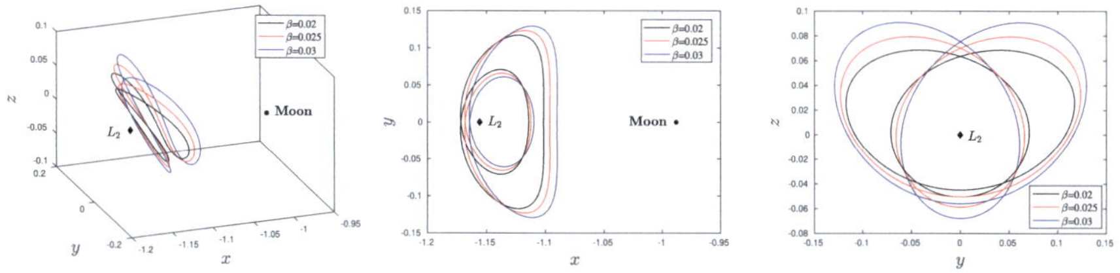

Fig. 4.21 3D representation and $x - y$ and $y - z$ coordinate projections of three solar-sail periodic orbits from the β-continuation family departing from the halo-type orbit $\mathbf{2C}^-$ for $\varphi = 0$

Near rectilinear halo orbits of the RTBP have been considered as strong candidates for the Gateway station. As it is shown in Fig. 4.23, some orbits of the families $\mathbf{2E}^-$ and $\mathbf{2G}^-$ correspond to the dynamical solar sail substitutes of these orbits. They are associated to a 3:1 and a 4:1 resonances, respectively. As it follows from the $x - z$ projection of the $\mathbf{2E}^-$ orbit, some part of it is between L_1 and Moon, while the other part is between Moon and L_2, this property means that spacecraft following these orbits can visit both the near-side and the far-side of the Moon.

Several kinds of orbits approaching the Moon, the Earth, or both can be found in the families computed; orbits with close approaches to the Earth and the Moon are good candidates for transfer purposes between these two bodies. Figure 4.24 shows some examples. Along the continuation with respect to β, the solar-sail periodic orbits departing from orbit **18** approach to the Moon, the ones departing from **28** approach to the Earth, and the ones departing from **16** approach to both and end at orbit **110**. In addition, the near rectilinear halo orbits of the family $\mathbf{2G}^-$ displayed in the bottom rows of Fig. 4.23 can be chosen with close approaches to the Moon. Similar properties are found for another near rectilinear halo orbit $\mathbf{2F}^-$, as shown in Fig. 4.25, by adjusting the pitch angel φ.

Another interesting thing that should be noted is that the SRP could change the resonance order along the family of periodic orbits. For instance, as shown in the Fig. 4.7, when the orbit **13** is continued to **14**, with β as the continuation parameter, the resonance order goes from 1:2 to 1:1 (and vice versa). A similar transition happens between the orbits **22** and **26** displayed in Fig. 4.14.

4.7 Conclusions

By means of a careful numerical computation, in this chapter we investigate the motion of a solar-sail spacecraft near the dynamical substitutes of the L_1 and L_2 points in the Earth-Moon QBCP. To understand the effect of the SRP on the natural dynamics, the QBCP is augmented by means of including the SRP acceleration which depends on the lightness number β and pitch angle φ of the sail. Making use of parallel shooting and continuation methods based on these two parameters, we obtain the characteristic curves and the linear stability character of a web of resonant

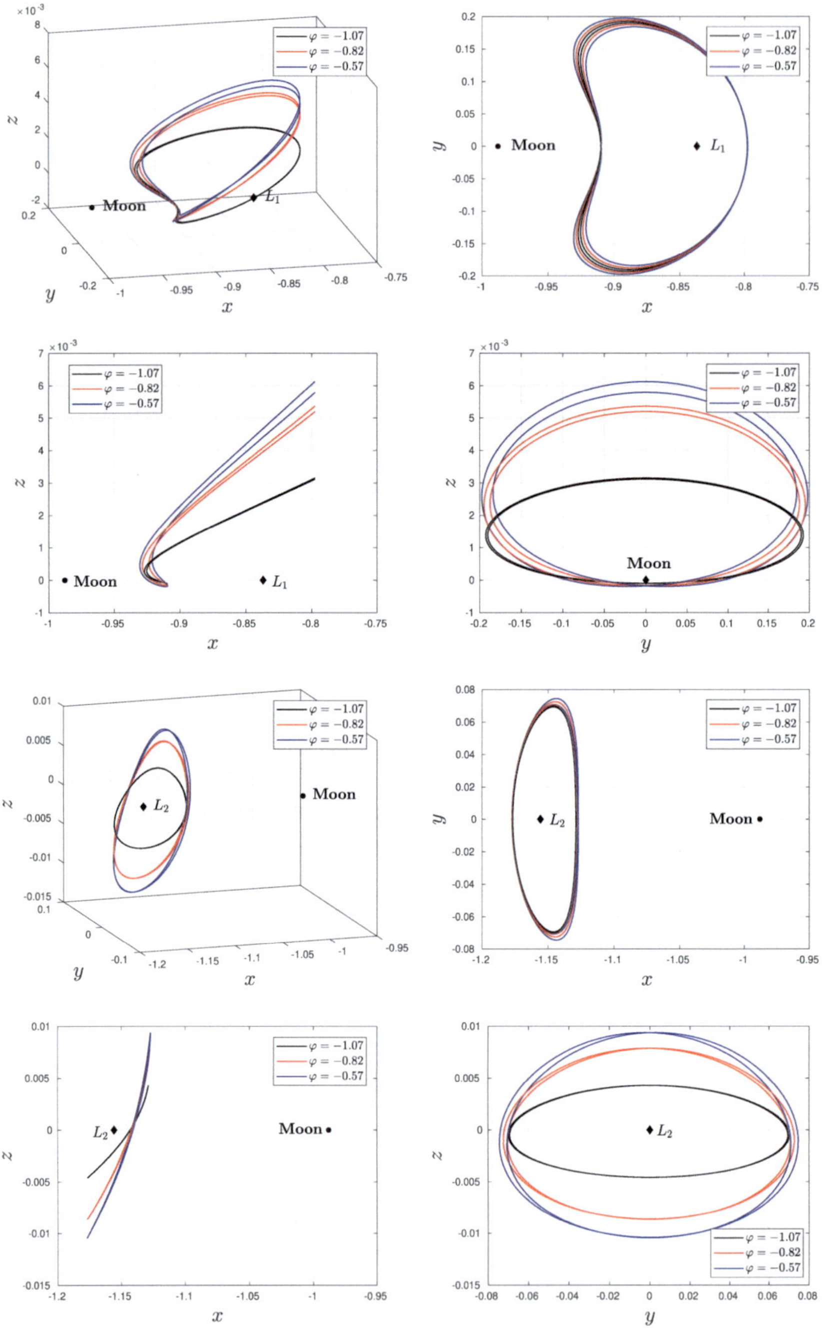

Fig. 4.22 3D representation and coordinate projections of orbits of the families, computed varying φ, departing from the planar orbits **13** (top rows) with $\beta = 0.01$, and **22** (bottom rows) with $\beta = 0.003$

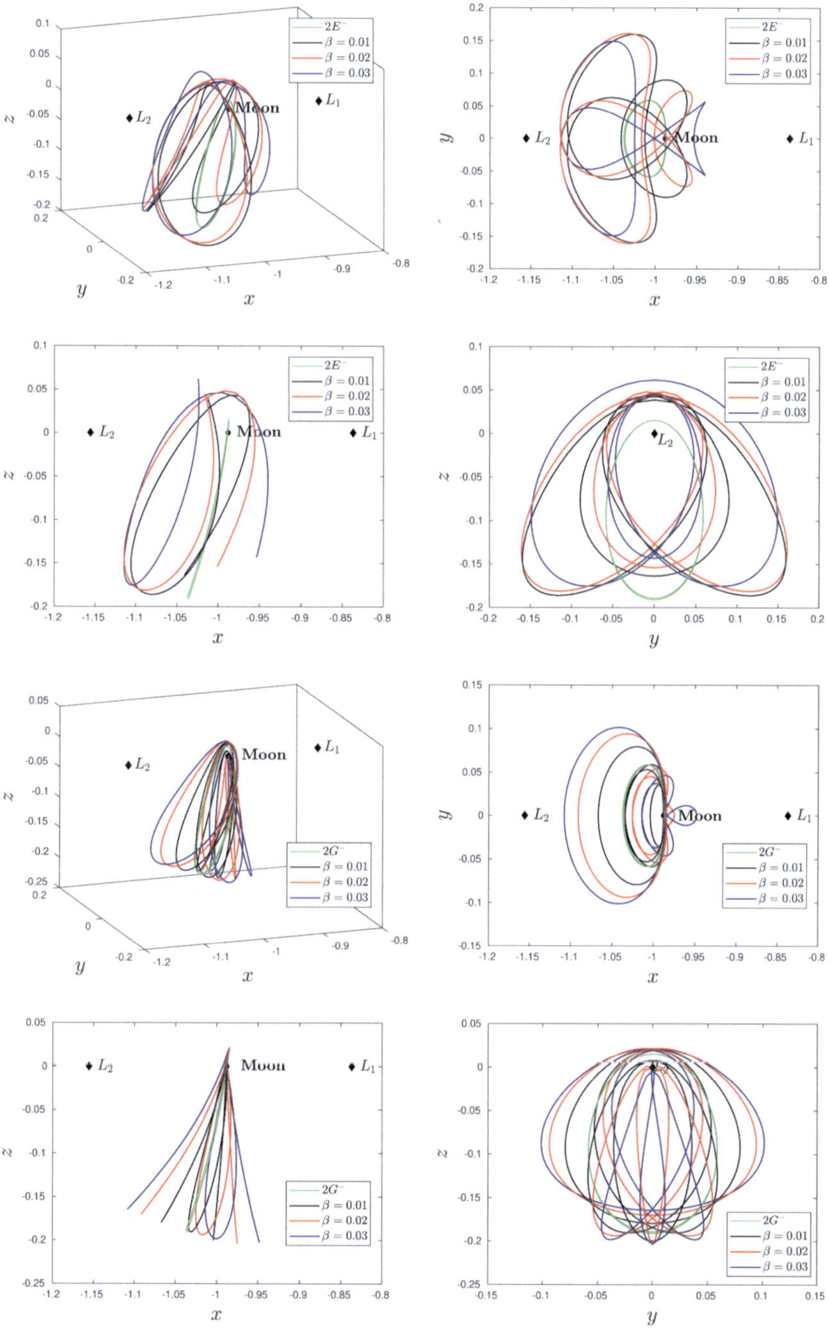

Fig. 4.23 3D representation and coordinate projections of two orbits of the families, computed varying β with $\varphi = 0.0$, departing from $\mathbf{2E}^-$ (top rows) and $\mathbf{2G}^-$ (bottom rows) which are near rectilinear halo orbits associated to a 3:1 and a 4:1 resonance, respectively

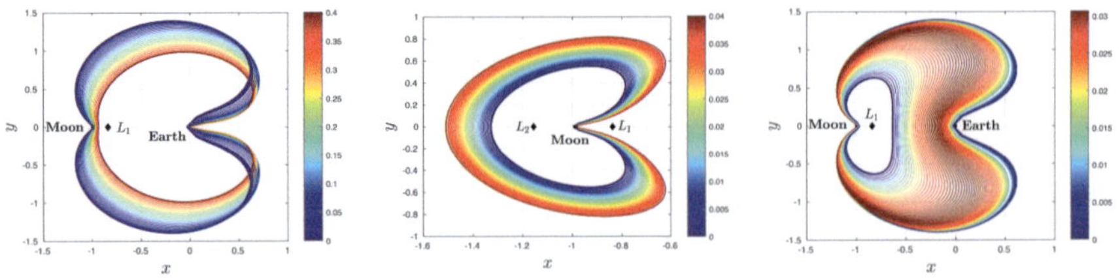

Fig. 4.24 Moon, Earth and both Earth-Moon passages. Left: periodic orbits of the family departing from **18**. Middle: periodic orbits of the family departing from **28**. Right: periodic orbits of the family connecting **16** and **110**. The color indicates the value of the lightness number β

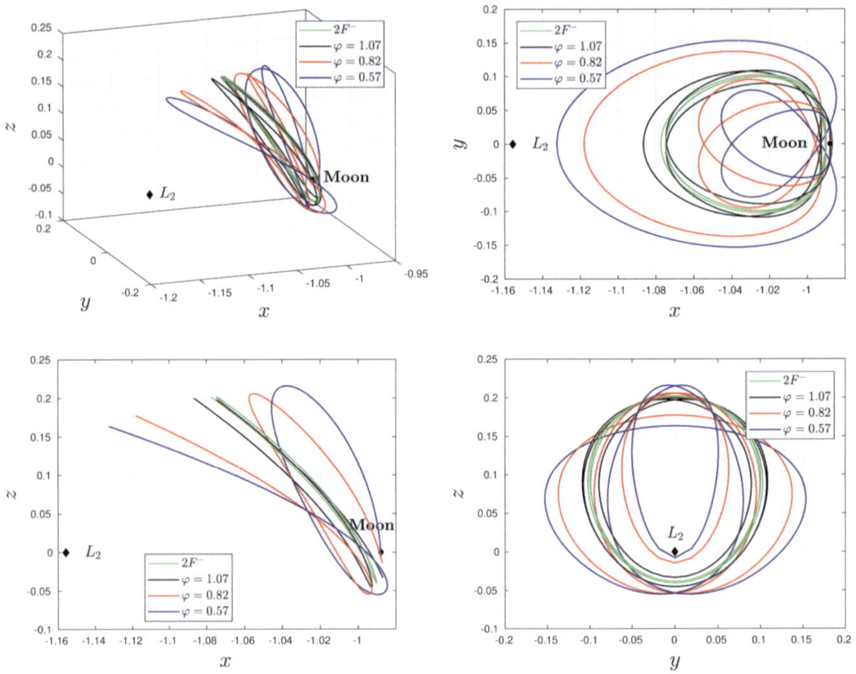

Fig. 4.25 3D representation and coordinate projections of two orbits of the families, computed varying φ with $\beta = 0.01$, departing from $\boldsymbol{2F^-}$ which is near rectilinear halo orbits associated to a 3:1

periodic orbits. The results show an intricated web of families of orbits from which it follows that one can move from orbits of one family to others, of a completely different one, just varying continuously the values of the parameters. Although the SRP perturbation on the QBCP is small, it produces and enlarges the possibilities of using a solar sail spacecraft under the current technology. The SRP not only can greatly change the size and shape of the QBCP orbits but it can also stabilize some orbits, with the clear advantages of this fact with respect to station keeping.

The computed families of periodic orbits contain the dynamical substitutes of the two equilibria and also low order resonant periodic orbits nearby. Among them, there are quasi-colliding orbits whose traces are close to collision with one of the primaries. Since no regularization of the equations of motion has been implemented, the continuation of some families has been finished once reaching close encounters. A maximum value of $\beta_{\max} = 0.04$, which corresponds to the near-future maximum practical value, is considered for the quasi-colliding orbits. For the other periodic orbits, $\beta_{\max} = 0.2$ has been set for the β-continuation, while $\beta_{\max} = 0.1$ has been set for the φ-continuation. Moreover, for the β-continuation of planar Lyapunov-type orbits, the pitch angle φ is set to zero which means that there is no out-of-plane acceleration component due to the effect of the SRP and the continued families for this case remain planar.

Although the motion around the collinear libration points is, in general, unstable, it must be remarked that in some cases the SRP can stabilize the motion, as it has been shown for some of the orbits in the families $\mathbf{1C}^{\pm}$ and $\mathbf{1N}^{\pm}$ around the L_1 point, and $\mathbf{2E}^{\pm}$ and $\mathbf{2F}^{\pm}$ around the L_2 point. Moreover, all the turning points of the characteristic curves (except in one case corresponding to the **13–14** connection) are associated to stability transitions.

It has also been found that some of resonant periodic orbits are connected for suitable values of the parameters, such as **11** and **12** by β-continuation, **11** and **12** by φ-continuation for $\beta \geq 0.096403$, **21** and **24** by β-continuation, and **21** and $\mathbf{2C}^{\pm}$ by φ-continuation for $\beta \geq 0.0109704$, etc. Moreover, some new resonant periodic orbits of the QBCP have been discovered ($\mathbf{1H}^{\pm}$, $\mathbf{1N}^{\pm}$, **210**, $\mathbf{2I}^{\pm}$, $\mathbf{2K}^{\pm}$, $\mathbf{2L}^{\pm}$) and the corresponding families have been incorporated into the web of connections.

The initial conditions given in the Appendix 4 can be used to obtain numerical seeds for the computation of quasi-periodic orbits in more accurate Sun-(Earth+Moon) system models (including, for instance, the perturbations of the remaining bodies of the Solar System) by means of a parallel shooting algorithm. The tables provide the initial data of the characteristic curves, and first, a continuation procedure in the QBCP, with respect to the solar sail parameters and up to their desired values, must be done starting at the given orbits.

Appendix A

See Table 4.4.

Appendix B

See Tables 4.5, 4.6, 4.7 and 4.8.

Table 4.4 Coefficients of the functions α_i ($i = 1, \ldots, 8$)

α_1		α_2		α_3	
k	a_k	k	b_k	k	a_k
0	1.001841608924835E+00	0	0.0E+00	0	9.999999999999983E-01
1	5.767517726198399E-04	1	-2.644376028499938E-04	1	5.634125997553694E-04
2	1.438777025507630E-02	2	-1.328686903400173E-02	2	1.889687440172882E-02
3	-2.630362974972015E-06	3	9.386093208089751E-06	3	-9.911758802567132E-06
4	1.176278356118933E-04	4	-1.218509057517414E-04	4	1.568708136031134E-04
5	-8.068581391005552E-08	5	1.522127598557008E-07	5	-1.707762576173484E-07
6	9.843249766501285E-07	6	-1.072102664277996E-06	6	1.319613679707437E-06
7	-1.172054394418197E-09	7	1.889371261374048E-09	7	-2.136550041985646E-09
8	8.311905970879588E-09	8	-9.324985038927486E-09	8	1.117168916673893E-08
9	-1.408584238695393E-11	9	2.114490981280258E-11	9	-2.387253631031108E-11
10	7.050713786466840E-11	10	-8.071111743144353E-11	10	9.490879622095902E-11
11	-1.494259634910463E-13	11	2.218118050420168E-13	11	-2.462732581558427E-13
12	5.982418979451232E-13	12	-7.036155161882012E-13	12	8.101067708009743E-13
		13	0.0E+00	13	0.0E+00
		14	-3.967229896122177E-15	14	1.354231998469834E-15

α_4		α_5		α_6	
k	a_k	k	b_k	k	a_k
0	-9.755242327484885E-04	0	0.0E+00	0	1.000907457708158E+00
1	2.154764362707107E+00	1	-2.192570751040067E+00	1	2.870921750053134E-04
2	3.657484468968697E-04	2	-3.337210485472868E-04	2	7.187177998612875E-03
3	3.295673376166588E-03	3	-3.295001430200974E-03	3	-2.351183147213254E-06
4	3.301031400812427E-07	4	-3.100635053052634E-07	4	4.585758971122060E-05
5	1.278840687376320E-05	5	-1.277777336854128E-05	5	-3.848683620107037E-08
6	-2.623797952127926E-09	6	2.652806405498111E-09	6	3.270677504935666E-07
7	6.533805514561511E-08	7	-6.528479245085066E-08	7	-4.406966481041876E-10
8	-3.891720707783511E-11	8	3.891720707783511E-11	8	2.452600662570259E-09
9	3.812275838944432E-10	9	-3.812275838944432E-10	9	-4.542938800673444E-12
10	-3.907906049834876E-13	10	3.907906049834876E-13	10	1.892348855112616E-11
11	2.407471187576443E-12	11	-2.407471187576443E-12	11	-4.178420101480123E-14
				12	1.480048946961583E-13

α_7		α_8	
k	a_k	k	b_k
0	-6.314069568006227E-02	0	0.0E+00
1	3.885638623098048E+02	1	-3.897437256237654E+02
2	1.736910203345558E-01	2	-1.734279166322518E-01
3	3.382908071669699E+00	3	-3.385696486642120E+00
4	1.574837565380491E-04	4	-1.555886632413398E-04
5	2.936360489004438E-02	5	-2.937582671967532E-02
6	-1.224434550116014E-05	6	1.225851213107933E-05
7	2.538935434262443E-04	7	-2.539596887692642E-04
8	-2.278929040007574E-07	8	2.280029220202363E-07
9	2.190432706181655E-06	9	-2.190834624429040E-06
10	-3.033311961234353E-09	10	3.036109035120856E-09
11	1.886971545290216E-08	11	-1.887457647579322E-08
12	-3.432375106898453E-11	12	3.432375106898453E-11
13	1.611513703999101E-10	13	-1.631723641506449E-10

Table 4.5 Coordinate values of the Q_i points used in the multiple shooting for the Lyapunov-type bifurcated symmetric periodic orbits around L_1 of the QBCP

Label	Points	x	y	z	$\dot{x}$	$\dot{y}$	$\dot{z}$
11	Q_0	−0.8369141677649318	0.0	0.0	0.0	−6.247467360987357e-06	0.0
	Q_1	−0.8369143549059166	2.709573745419631e-07	0.0	−5.082161707329508e-08	−6.480648443085597e-06	0.0
	Q_2	−0.8369143549059166	−2.709573743429859e-07	0.0	5.082161640371648e-08	−6.480648442412153e-06	0.0
12	Q_0	−0.9104743857262561	0.0	0.0	0.0	0.4739377810674483	0.0
	Q_1	−0.8459929511151905	−0.1655628977875012	0.0	−0.1400166994560041	−0.1684982087531857	0.0
	Q_2	−0.8459929511145754	0.1655628977867639	0.0	0.1400166994556543	−0.1684982087555812	0.0
13	Q_0	−0.7974817205388037	0.0	0.0	0.0	−0.3777122446326868	0.0
	Q_1	−0.9192373592373829	0.1704424715748812	0.0	0.07011144243197555	0.1465655587030538	0.0
	Q_2	−0.9192373592373768	−0.170442471574892	0.0	−0.07011144243200068	0.1465655587030152	0.0
14	Q_0	−0.9707903423961286	0.0	0.0	0.0	1.227264255095871	0.0
	Q_1	−0.723421302212652	0.6027270456931847	0.0	0.2355858114170835	−0.1869649782472948	0.0
	Q_2	−0.7234213022127389	−0.6027270456932223	0.0	−0.2355858114171684	−0.1869649782471134	0.0
16	Q_0	−0.6319707096253417	0.0	0.0	0.0	−0.8048438775487698	0.0
	Q_1	−1.037184304091327	−0.4908901530078815	0.0	−0.215879039779001	0.3343739615235425	0.0
	Q_2	−1.037184304091485	0.4908901530075798	0.0	0.2158790397787566	0.3343739615237423	0.0
18	Q_0	−0.9833707656190481	0.0	0.0	0.0	2.567074732637494	0.0
	Q_1	0.3854770398263357	1.191899113228741	0.0	0.8265479569412737	−0.700945456951042	0.0
	Q_2	0.3854770398256071	−1.191899113229555	0.0	−0.8265479569426034	−0.7009454569499792	0.0
110	Q_0	−0.03157060844895042	0.0	0.0	0.0	−6.660529956097482	0.0
	Q_1	−0.8318469392081019	−1.156379494307932	0.0	−0.9223185175510227	0.7660233481063474	0.0
	Q_2	−0.8318469392061989	1.156379494309332	0.0	0.9223185175529419	0.7660233481040829	0.0

Table 4.6 Coordinate values of the Q_i points used in the multiple shooting for the Lyapunov-type bifurcated symmetric periodic orbits around L_2 of the QBCP

Label	Points	x	y	z	$\dot{x}$	$\dot{y}$	$\dot{z}$
21	Q_0	−1.155683607833269	0.0	0.0	0.0	6.017986197378593e-06	0.0
	Q_1	−1.155683080445066	−1.35724654307585e-06	0.0	−2.43602628718126e-07	5.607671283536216e-06	0.0
	Q_2	−1.155683080445076	1.357246559802415e-06	0.0	2.436026377669816e-07	5.607671335107336e-06	0.0
22	Q_0	−1.176464998903009	0.0	0.0	0.0	0.1258366678951945	0.0
	Q_1	−1.135865746791441	−0.06064397292310485	0.0	−0.02912786246749816	−0.06027198298464091	0.0
	Q_2	−1.135865746791306	0.06064397292282585	0.0	0.02912786246717376	−0.06027198298563409	0.0
24	Q_0	−1.129103186619581	0.0	0.0	0.0	−0.1360498544279882	0.0
	Q_1	−1.160974341506647	0.05721410538987393	0.0	0.04527386517353094	0.06508560473874013	0.0
	Q_2	−1.160974341618912	−0.05721410522911606	0.0	−0.04527386509575504	0.06508560518575468	0.0
26	Q_0	−1.327907131342701	0.0	0.0	0.0	0.6273575150937968	0.0
	Q_1	−0.7899355396205242	0.4323467196724335	0.0	0.0601083431113279	−0.3012283778570424	0.0
	Q_2	−0.7899355396204807	−0.4323467196749719	0.0	−0.06010834311355049	−0.3012283778552309	0.0
28	Q_0	−0.9968163858146138	0.0	0.0	0.0	−1.672727233760122	0.0
	Q_1	−1.054911715001196	−0.5177326140480463	0.0	−0.3494404140645731	0.1921511446135178	0.0
	Q_2	−1.05491171500148	0.5177326140479851	0.0	0.3494404140646158	0.1921511446140117	0.0
210	Q_0	−0.8782074171644279	0.0	0.0	0.0	−0.4770285175327934	0.0
	Q_1	−1.042353740805561	−0.1061932496039309	0.0	−0.3522244951518779	0.2172495277945835	0.0
	Q_2	−1.042353740768837	0.1061932496267413	0.0	0.3522244952195452	0.2172495276450336	0.0

Table 4.7 Coordinate values of the Q_i points used in the multiple shooting for the halo-type bifurcated symmetric periodic orbits around L_1 of the QBCP

Label	Points	x	y	z	$\dot{x}$	$\dot{y}$	$\dot{z}$
$1C^+$	Q_0	−0.9879629720523689	0.0	0.05108383968268531	0.0	0.662266081824031	0.0
	Q_1	−0.9866409311202474	−3.448909680810834e-05	0.05358638966998008	0.001026780002226913	0.6323868971803503	−0.0004766709804533378
	Q_2	−0.9866409311201429	3.44890616636967e-05	0.05358638967006777	−0.001026780066583258	0.6323868971797344	0.0004766712163103312
$1D^+$	Q_0	−0.8681380897120423	0.0	0.1888325357210265	0.0	−0.2473565712365277	0.0
	Q_1	−0.8696919662741807	0.004555119133436505	0.1881666842836814	0.004303223205110978	−0.2394573747008942	0.007383293591091827
	Q_2	−0.8696919662740482	−0.004555119121853662	0.1881666842840072	−0.004303223191235746	−0.2394573747015072	−0.007383293569076564
$1E^+$	Q_0	−0.9879722904635280	0.0	0.0024620950605023	0.0	3.184909363998671	0.0
	Q_1	−0.9870867890410873	−0.01093418466976307	−0.01001480267949663	−0.08606982701096791	0.5231026409004002	1.171218780305022
	Q_2	−0.9870867870195466	0.01093419695386825	−0.01001483016693704	0.0860697854225894	0.5231016923764815	−1.171217914492649
$1F^+$	Q_0	−0.9236861280052134	0.0	0.3180943609744754	0.0	−0.08454341835457363	0.0
	Q_1	−0.9235522643168672	−0.001370189810478059	0.317487021014671	−2.410724962884926e-05	−0.08239232250483541	0.002366524002208882
	Q_2	−0.9235522643154103	0.001370190567395954	0.317487021032232	2.410866911956809e-05	−0.08239232249695087	−0.002366519927220817
$1H^+$	Q_0	−0.9025027848630731	0.0	0.1374532435264076	0.0	−0.3076016202874576	0.0
	Q_1	−0.8872262130386557	−0.1859065894055493	0.1444901405224834	−0.1213385259952692	−0.1177891228979688	−0.04830894534488246
	Q_2	−0.8872262130386072	0.1859065894055026	0.1444901405225003	0.1213385259952341	−0.117789122898134	0.04830894534477374
$1N^+$	Q_0	−0.9958329358116138	0.0	0.04170135653064389	0.0	0.7324812203758224	0.0
	Q_1	−0.9200005341968871	0.1131884237448591	−0.08888187174491974	0.1391866211115352	−0.02890181625177237	−0.2960744534631971
	Q_2	−0.8892697884037004	0.1026091623869626	−0.1552823859531374	0.0991948104028349	−0.1738210354771334	−0.145066411702705

Table 4.8 Coordinate values of the Q_i points used in the multiple shooting for the halo-type bifurcated symmetric periodic orbits around L_2 of the QBCP

Label	Points	x	y	z	$\dot{x}$	$\dot{y}$	$\dot{z}$
$2A^+$	Q_0	−1.17888192842499	0.0	0.04366399706256908	0.0	0.1671602937502363	0.0
	Q_1	−1.122569204897854	−0.08366573209195621	−0.007181302683639001	−0.03982485454000831	−0.07696311205741586	0.06665606627067616
	Q_2	−1.122569204897855	0.08366573209196024	−0.00718130268363609	0.0398248545400124	−0.07696311205740111	−0.06665606627067716
$2C^+$	Q_0	−1.113829762948272	0.0	0.03103557789877939	0.0	−0.196202368818838	0.0
	Q_1	−1.157502933098094	0.07734562728625115	−0.0293917006188531	0.06341045907250445	0.09035281076041549	0.04933522606923745
	Q_2	−1.157502933039179	−0.07734562736981195	−0.02939170057342491	−0.06341045911543587	0.09035281054384216	−0.04933522613396623
$2E^+$	Q_0	−1.077049895739678	0.0	0.2015253011952968	0.0	0.1997859497018886	0.0
	Q_1	−1.074889833747149	−0.000657497532414305	0.2017442892796588	−0.0002614593225596462	0.1900810832779645	0.00514010115137467
	Q_2	−1.07488983374662	0.0006574974897381545	−0.2017442892809748	0.0002614592783244464	0.1900810832767272	−0.005140101066242974
$2F^+$	Q_0	−0.9904316673796592	0.0	0.03844291276661251	0.0	−0.7619049294735506	0.0
	Q_1	−0.9902080912896167	−0.000263333721796341	0.0404864428607886	0.01166416007331324	−0.726748146537556	0.001210145915641109
	Q_2	−0.9902080912866439	0.0002633339060171025	0.04048644286161254	−0.01166415981139307	−0.7267481465269423	−0.001210144051574125
$2G^+$	Q_0	−1.036008068335598	0.0	0.1910211094926181	0.0	0.1350819667635153	0.0
	Q_1	−1.012878007587121	0.05766940227754755	0.1121734194179288	0.08288964423005191	0.03163283443528857	−0.2952379198874565
	Q_2	−1.012878007584373	−0.05766940227855085	0.1121734194081135	−0.08288964423378739	0.0316328344192571	0.2952379199119221
$2H^+$	Q_0	−0.9871245890063371	0.0	0.01536478051466306	0.0	−1.239507689039373	0.0
	Q_1	−1.030053662230724	−0.03534870378554685	−0.1706682804010934	−0.04559954419625001	0.1108030377277615	−0.1356858379224256
	Q_2	−1.030053662229623	0.03534870378813639	−0.1706682803979521	0.04559954419979169	0.1108030377240398	0.1356858379340769
$2I^+$	Q_0	−1.196699300549432	0.0	0.1303077741018548	0.0	0.435912306912807	0.0
	Q_1	−1.034409017595348	−0.03212874837549233	−0.009380147596847093	−0.222124063432269	−0.1585835447515381	0.5739227753388807
	Q_2	−1.034409017589272	0.03212874837114243	−0.009380147612191221	0.2221240635001763	−0.1585835448191122	−0.5739227753232441
$2K^+$	Q_0	−0.9947914993132853	0.0	0.01053549475987681	0.0	−1.404401469653503	0.0
	Q_1	−1.140334534100885	0.017379476425884	−0.08947483908914515	0.03578766323584347	0.3271351565923236	0.2801103323436752
	Q_2	−1.140334534099683	−0.01737947643737417	−0.08947483907938625	−0.03578766325606404	0.3271351565888527	−0.2801103323521913
$2L^+$	Q_0	−0.9965789824580257	0.0	0.1812794590816482	0.0	0.2706205945796978	0.0
	Q_1	−1.092276580227493	−0.004764847248825923	−0.1054694527552328	−0.1122929523219291	0.274398642525683	0.2523017718724777
	Q_2	−1.092276580227238	0.00476484724944888	−0.1054694527558054	0.1122929523233752	0.2743986425251427	−0.2523017718714267

References

1. Jorba-CuscÃ M. Periodic time dependent Hamiltonian systems and applications[D]. Ph.D dissertation, University of Barcelona, 2018.
2. Rosales de Cáceres J J. On the effect of the Sun's gravity around the Earth-Moon L_1 and L_2 libration points[D]. Ph.D dissertation, University of Barcelona, 2020.
3. Andreu M A. The quasi-bicircular problem[D]. Ph.D dissertation, University of Barcelona, 1998.
4. Heiligers J, Diedrich B, Derbes W, et al. Sunjammer: Preliminary end-to-end mission design[C]. AIAA/AAS Astrodynamics Specialist Conference, San Diego, USA, 2014.
5. Farrés A. Transfer orbits to L_4 with a solar sail in the Earth-Sun system[J]. Acta Astronautica, 2017, 137: 78-90.
6. Stoer J, Bulirsch R, Bartels R, et al. Introduction to numerical analysis[M]. New York: Springer, 1980.

Chapter 5
Station-Keeping for Earth-Moon Solar-Sail Resonant Libration Point Orbits

This chapter investigates a dynamical reshaping control method for the station keeping of unstable resonant periodic orbits using the solar radiation pressure in the solar-sail-augmented Earth-Moon Quasi-Bicircular Problem. The control strategy aims to reshape the geometrical structures defined by the unstable Floquet modes, furthermore, asymptotic stability can be achieved by incorporating the central Floquet modes into the procedures. The control variables considered are the attitude angles of the sail, including two scenarios, one with the ligthness number of the sail being constant and another one being variable. The nominal point definition also plays a role in the stability and convergence rate, but independently of this fact, the Jet Transport technique enables the control laws to be developed in a unified way. In addition, the controllers are obtained as Taylor polynomials in terms of the deviation between the state of the spacecraft and the nominal point. The numerical simulations analyze the stability, dynamical and long-term behaviors, robustness to orbit determination errors and control costs. The results verify the feasibility of considering solar radiation pressure for long-term maintenance of libration point orbits in the Earth-Moon system.

5.1 Introduction

In the Chap. 4, we have investigated the natural dynamics of solar sailing in the Earth-Moon system based on the Quasi-Bicircular Problem (QBCP), which includes the gravitational effect of the Sun into the dynamical model. The study presents a complete picture on the evolution of families of solar-sail resonant libration point orbits (LPOs) in the neighborhood of L_1 and L_2 in terms of shape and linear normal behavior. We have shown an intricate web of connections, and found that some orbits are stable (or less unstable) under the influence of the solar radiation pressure (SRP) acceleration, which opens additional opportunities for long-term maintenance of

© The Author(s), under exclusive license to Springer Nature Singapore Pte Ltd. 2026
C. Gao and W. Wang, *Solar Sailing for Deep-Space Explorations*,
https://doi.org/10.1007/978-981-95-0109-0_5

solar-sail LPOs. However, most of the interesting solar-sail resonant orbits are very unstable. Therefore, it is necessary to achieve effective station-keeping strategies when these unstable solar-sail orbits are required candidates for a space mission.

According to [1], station-keeping strategies can be divided into two categories: the first is based on control theory techniques, such as the backstepping controller proposed by Tamakoshi et al. [2] and the optimal periodic orbital controller in Chap. 3; the second is based on the characteristics of the dynamical system, such as the dynamical systems approaches proposed by Farrés et al. [3, 4] and Hamiltonian structure-preserving (HSP) approach proposed by Scheeres et al. [5]. Compared to control theory-based approaches, strategies using tools from Dynamical Systems theory are more dynamically insightful and fuel effective. In this context, Gao et al. [6, 7] recently have investigated a dynamical reshaping method which aims to reshape unstable and/or central Floquet modes into stable ones by means of low-thrust propulsion. In fact, the formalism provides a control framework where control laws with specific characteristics can be designed. Compared to the HSP method, the dynamical reshaping method is more dynamically intuitive and adaptable to particular adjustments and performance by selecting the parameters in the algorithm. Additionally, by means of the Jet Transport (JT) technique [8], the dynamical reshaping control laws can be explicitly written as polynomials in terms of the deviation between the state of the spacecraft and a nominal point on the nominal orbit. Therefore, the control acceleration can be trivially computed by evaluation of these polynomials, which makes the control laws to be easily implemented and autonomously executed on-board. This is of significant importance for solar-sail missions due to the fact that the onboard computation capacities are quite limited.

The aim of this chapter is to exploit the dynamical reshaping method to derive and study some SRP assisted station-keeping control laws for unstable resonant periodic orbits in the solar-sail-augmented QBCP (QBCP-SRP). The control variables of the procedures are the lightness number and the attitude angles of the sail. Due to the fact that the lightness number is difficult to change, two scenarios are considered, one with this parameter time-constant and another one time-varying. In addition, several criteria for selecting the nominal point are discussed in the reshaping algorithms. By means of the JT technique, the control laws are derived and obtained semi-analytically in a polynomial formalism despite of the definition of the nominal point.

The structure of this chapter is as follows: Section 5.2 is devoted to set the equations of motion of the QBCP-SRP, to recall some known results about solar-sail resonant LPOs, and to introduce the notation used for the JT procedure; Sect. 5.3 contains the main contributions of the study, several reshaping station-keeping algorithms are developed in a semi-analytical way, considering the control parameters of the sail within the framework of JT methodology; Sect. 5.4 discusses the stability, dynamical and long-term behaviors, robustness to orbit determination errors and control costs for the proposed station-keeping approaches; Sect. 5.5 finally concludes and summarizes the work.

5.2 Mathematical Preliminaries

5.2.1 Equations of Motion and Solar-Sail Resonant LPOs

According to Eq. (4.4), the equations of motion for the QBCP-SRP system, in the pulsating rotating frame, are

$$\begin{cases} \ddot{x} - \dfrac{\dot{\alpha}_1}{\alpha_1}\dot{x} - 2\alpha_3\dot{y} = \left(\dot{\alpha}_2 + \alpha_2^2 + \alpha_3^2 - \dfrac{\dot{\alpha}_1\alpha_2}{\alpha_1}\right)x + \left(\dot{\alpha}_3 - \dfrac{\dot{\alpha}_1\alpha_3}{\alpha_1}\right)y - \alpha_1\alpha_4 \\ \qquad - \alpha_1\alpha_6\left(\dfrac{1-\mu}{r_{PE}^3}(x-\mu) + \dfrac{\mu}{r_{PM}^3}(x-\mu+1) + \dfrac{m_S}{r_{PS}^3}(x-\alpha_7)\right) + a_x \\ \ddot{y} - \dfrac{\dot{\alpha}_1}{\alpha_1}\dot{y} + 2\alpha_3\dot{x} = \left(\dot{\alpha}_2 + \alpha_2^2 + \alpha_3^2 - \dfrac{\dot{\alpha}_1\alpha_2}{\alpha_1}\right)y - \left(\dot{\alpha}_3 - \dfrac{\dot{\alpha}_1\alpha_3}{\alpha_1}\right)x - \alpha_1\alpha_5 \\ \qquad - \alpha_1\alpha_6\left(\dfrac{1-\mu}{r_{PE}^3}y + \dfrac{\mu}{r_{PM}^3}y + \dfrac{m_S}{r_{PS}^3}(y-\alpha_8)\right) + a_y \\ \ddot{z} - \dfrac{\dot{\alpha}_1}{\alpha_1}\dot{z} = \left(\dot{\alpha}_2 + \alpha_2^2 - \dfrac{\dot{\alpha}_1\alpha_2}{\alpha_1}\right)z - \alpha_1\alpha_6\left(\dfrac{1-\mu}{r_{PE}^3}z + \dfrac{\mu}{r_{PM}^3}z + \dfrac{m_S}{r_{PS}^3}z\right) + a_z \end{cases} \tag{5.1}$$

with $[\boldsymbol{a}]_{\mathcal{T}_R} = [a_x, a_y, a_z]^{\mathrm{T}}$ the SRP acceleration.

To describe the SRP acceleration, it is convenient to introduce the sunlight reference frame $\mathcal{T}_S(O;\ \hat{\boldsymbol{x}}_s, \hat{\boldsymbol{y}}_s, \hat{\boldsymbol{z}}_s)$ (see Fig. 5.1). The origin of this frame is at the center of the sail, the $\hat{\boldsymbol{x}}_s$ axis is directed along the Sun-sail direction, the $\hat{\boldsymbol{z}}_s$ axis is perpendicular to the orbital plane of the primaries and parallel to $\hat{\boldsymbol{z}}_r$, and the $\hat{\boldsymbol{y}}_s$ axis completes the right-handed unitary reference frame. Therefore, $\hat{\boldsymbol{n}}_S$ and $\hat{\boldsymbol{n}}$ in the sunlight frame can be written as

$$[\hat{\boldsymbol{n}}_S]_{\mathcal{T}_S} = [1,\ 0,\ 0]^{\mathrm{T}}, \qquad [\hat{\boldsymbol{n}}]_{\mathcal{T}_S} = [\cos\varphi\cos\gamma,\ \cos\varphi\sin\gamma,\ -\sin\varphi]^{\mathrm{T}}$$

where γ and φ are known as the yaw and pitch angles, respectively (see Fig. 5.1). In this chapter, we consider an ideal solar sail model. Therefore, according to Eq. (4.6), the SRP acceleration is given by

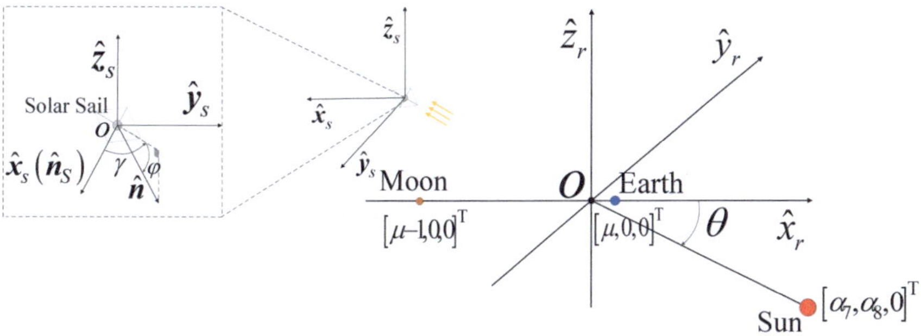

Fig. 5.1 Schematic representation of a solar-sail spacecraft in the Earth-Moon QBCP

$$[\boldsymbol{a}]_{\mathcal{T}_S} = \alpha_1 \alpha_6 \, \beta \frac{m_S}{r_{PS}^2} \begin{bmatrix} \cos^3\varphi \, \cos^3\gamma \\ \cos^3\varphi \, \cos^2\gamma \, \sin\gamma \\ -\cos^2\varphi \, \cos^2\gamma \, \sin\varphi \end{bmatrix} \tag{5.2}$$

Taking into account that the position of the Sun is given by the functions α_7 and α_8, the rotation matrix between the sunlight and synodic QBCP frames is

$$\boldsymbol{L} = \begin{bmatrix} -\frac{\alpha_7}{\sqrt{\alpha_7^2+\alpha_8^2}} & \frac{\alpha_8}{\sqrt{\alpha_7^2+\alpha_8^2}} & 0 \\ -\frac{\alpha_8}{\sqrt{\alpha_7^2+\alpha_8^2}} & -\frac{\alpha_7}{\sqrt{\alpha_7^2+\alpha_8^2}} & 0 \\ 0 & 0 & 1 \end{bmatrix} \tag{5.3}$$

Because the spacecraft moves in the neighborhood of the Earth-Moon barycenter, the Sun-sail distance r_{PS} can be further simplified considering

$$r_{PS} \approx r_S = \sqrt{\alpha_7^2 + \alpha_8^2}$$

Finally, rotating the acceleration of Eq. (5.2) by $\boldsymbol{L}$, the three components of the SRP acceleration in the synodic QBCP frame become:

$$\begin{aligned} a_x &= -\alpha_1\alpha_6 \, \beta \frac{m_S}{\alpha_7^2 + \alpha_8^2} \frac{\alpha_7 \cos^3\varphi \cos^3\gamma - \alpha_8 \cos^3\varphi \cos^2\gamma \sin\gamma}{\sqrt{\alpha_7^2 + \alpha_8^2}} \\ a_y &= -\alpha_1\alpha_6 \, \beta \frac{m_S}{\alpha_7^2 + \alpha_8^2} \frac{\alpha_8 \cos^3\varphi \cos^3\gamma + \alpha_7 \cos^3\varphi \cos^2\gamma \sin\gamma}{\sqrt{\alpha_7^2 + \alpha_8^2}} \\ a_z &= -\alpha_1\alpha_6 \, \beta \frac{m_S}{\alpha_7^2 + \alpha_8^2} \cos^2\varphi \cos^2\gamma \sin\varphi \end{aligned} \tag{5.4}$$

Equations (5.1) and (5.4) define the QBCP-SRP model.

Table 4.3 in Chap. 4 lists the QBCP periodic orbits that have been identified. We see that, when the SRP acceleration is incorporated into the QBCP model, many new periodic orbits bifurcate from the resonant orbits of the QBCP forming an intricate web of families and connections. Starting from the symmetric resonant orbits in Table 4.3 and varying the lightness number β and the pitch angle φ of the sail (while keeping $\gamma = 0°$) one can find the evolution of the families, including characteristic curves, the maximal Floquet exponent and the linear normal behavior of QBCP-SRP periodic orbits. Considering the halo-type orbit $\boldsymbol{2C^-}$ as example, Fig. 5.2 presents the characteristic and stability curves for the β-continuation of solar-sail resonant orbits for several fixed values of φ. The continuation procedure ends at $\beta = 0.04$, which is the maximum near-future technological feasible value of this parameter [9]. The characteristic curves show the value of the adimensional z-coordinate of the initial condition (first orthogonal intersection of the orbit with the $x - z$ plane) as a function of the continuation parameter β. The figure shows that, for positive values of φ (i.e., $\varphi = 30°$ and $60°$), the families departing from orbit $\boldsymbol{2C^-}$ end in the Lyapunov-

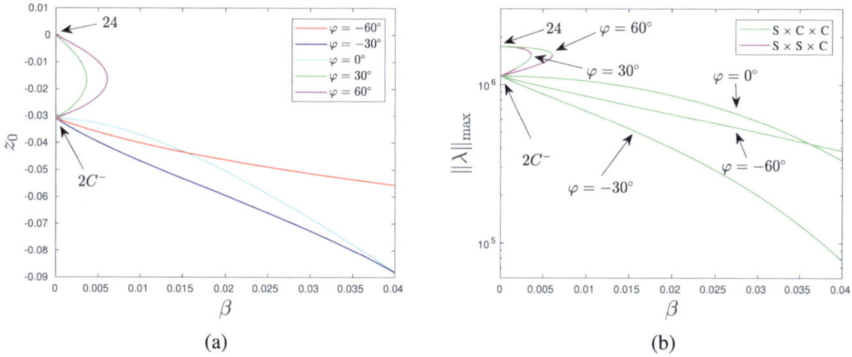

Fig. 5.2 Characteristic (**a**) and stability (**b**) curves of the families departing from the halo-type orbit $\mathbf{2C}^-$

type resonant orbit **24** after turning points. Furthermore, the instability of solar-sail orbits increases along the continuation direction and the turning points correspond to transitions from $(S \times C \times C)$ to $(S \times S \times C)$ in the linear normal behavior. For $\varphi = 0°$ and negative values, no turning points nor bifurcations are found in the characteristic curves up to $\beta \leq 0.04$. Moreover, the stability curves show that the instability of the orbits decreases monotonically with increasing β. This means that some resonant orbits in the QBCP can be made less unstable under the influence of the SRP acceleration.

Figure 5.3 shows four solar-sail resonant orbits of the QBCP-SRP with different values of β in the family of $\varphi = 0°$. These four solar-sail orbits are labeled as $\mathbf{2C}_1^-$, $\mathbf{2C}_2^-$, $\mathbf{2C}_3^-$ and $\mathbf{2C}_4^-$, and their corresponding values of β are 0.02, 0.025, 0.03 and 0.035, respectively. The cross symbol $\times$ denotes the position of the spacecraft at the initial time, located in the $x - z$ plane and below the $x - y$ plane. Though these solar-sail orbits are still unstable, they are more stable than $\mathbf{2C}^-$ which means they are more useful in space missions. Since these orbits are always behind the Moon and are free of eclipse, they are useful as translunar nominal orbits to support, for instance, lunar far-side communications and observations. In addition, these orbits are good candidates to be employed for bounded formation flying as they have the same period T_S. The non-zero initial adimensional coordinates for these four orbits are given in Table 5.1. Note that because of the symmetry of these orbits with respect to the $x - z$ plane, the initial y_0, $\dot{x}_0$ and $\dot{z}_0$ coordinates are zero.

5.2.2 *Jet Transport Technique*

The Jet Transport (JT) is an automatic differentiation technique suitable to compute high-order Taylor expansions of nonlinear functions by means of computers. It implements algebras based on truncated Taylor polynomials at any order, instead of

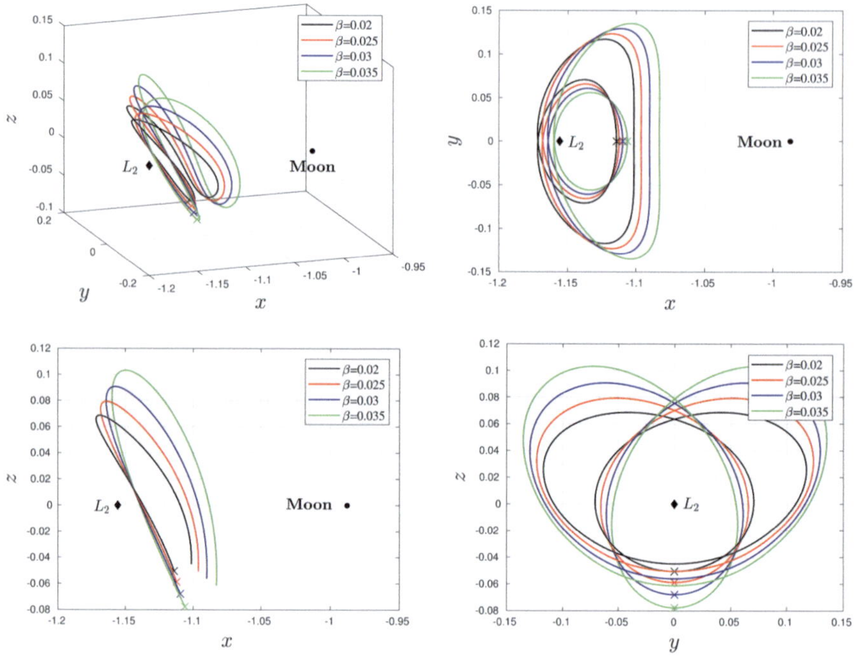

Fig. 5.3 Four solar-sail periodic orbits emanating from the halo-type orbit $\mathbf{2C^-}$. Top-Left: 3D projection of the orbits in the $\mathcal{T}_R(O;\ \hat{\boldsymbol{x}}_r, \hat{\boldsymbol{y}}_r, \hat{\boldsymbol{z}}_r)$ reference frame; Top-Right: $\hat{\boldsymbol{x}}_r - \hat{\boldsymbol{y}}_r$ projection of the orbits; Bottom-Left: $\hat{\boldsymbol{x}}_r - \hat{\boldsymbol{z}}_r$ projection of the orbits; Bottom-Right: $\hat{\boldsymbol{y}}_r - \hat{\boldsymbol{z}}_r$ projection of the orbits

Table 5.1 Non-zero initial adimensional coordinates and sail parameters for the solar-sail resonant orbits in Fig. 5.3

Orbit	x_0	z_0	$\dot{y}_0$	β	φ	γ	Period
$\mathbf{2C_1^-}$	−1.114070009317700	−0.05053380434419807	−0.1647491768627307	0.02	0.0	0.0	T_S
$\mathbf{2C_2^-}$	−1.112278384417229	−0.05883078336815999	−0.1583173195180186	0.025	0.0	0.0	T_S
$\mathbf{2C_3^-}$	−1.109684099454931	−0.06795212316285419	−0.1520924704866176	0.03	0.0	0.0	T_S
$\mathbf{2C_4^-}$	−1.106238510293822	−0.07780133739303541	−0.1458457242078510	0.035	0.0	0.0	T_S

using real numbers. In this chapter, we consider this technique to semi-analytically develop station-keeping control laws and to obtain the stroboscopic map of the controlled system for the stability analysis.

Given a nonlinear function f, we denote by $\boldsymbol{x} \in \mathbb{R}^n$ a selected central base point and by $\delta\boldsymbol{x}$ a symbolic variation with respect to it. The initial jet is then parameterized as $[\boldsymbol{x}] = \boldsymbol{x} + \delta\boldsymbol{x}$. Starting from $[\boldsymbol{x}] = \boldsymbol{x} + \delta\boldsymbol{x}$ and implementing the polynomial algebra defined in the JT technique, we can obtain the Taylor expansion of function f around the central base point $\boldsymbol{x}$. This is:

$$[f([\boldsymbol{x}])] = [f(\boldsymbol{x} + \delta\boldsymbol{x})] = f(\boldsymbol{x}) + \mathcal{P}_f(\delta\boldsymbol{x}) = f(\boldsymbol{x}) + \sum_{1\le k_1+\cdots+k_n\le k} c_{k_1\cdots k_n}\delta x_1^{k_1}\cdots\delta x_n^{k_n} \tag{5.5}$$

where $c_{k_1\cdots k_n}$ are the coefficients of the expansion provided by the procedure, $\delta\boldsymbol{x} = [\delta x_1, \cdots, \delta x_n]^{\mathrm{T}}$, k is the expansion order, and $[k_1, \cdots, k_n]$ is the multi-index exponent of the corresponding monomial. Note that the symbol $[\cdot]$ means JT variable throughout this chapter.

In particular, for an ordinary differential equation $\dot{\boldsymbol{x}} = \boldsymbol{f}(\boldsymbol{x}, t)$, JT enables to automatically compute high-order Taylor expansions of its associating flow map $\boldsymbol{\Phi}(t; \boldsymbol{x}_0, t_0)$: denote by $\boldsymbol{x}(t_0) = \boldsymbol{x}_0 \in \mathbb{R}^n$ the initial condition of the dynamical system, the initial jet is then expressed as $[\boldsymbol{x}_0] = \boldsymbol{x}_0 + \delta\boldsymbol{x}_0$ and represents the neighborhood of the initial state. By means of any numerical integration method (Runge-Kutta, Taylor, symplectic, etc) with its code implemented in the JT polynomial algebra, we can propagate this initial jet $[\boldsymbol{x}_0] = \boldsymbol{x}_0 + \delta\boldsymbol{x}_0$ from time t_0 to any final time t_f and automatically we obtain the final jet $[\boldsymbol{x}_f]$ as the Taylor expansion around the nominal final state $\boldsymbol{x}_f = \boldsymbol{\Phi}(t_f; \boldsymbol{x}_0, t_0)$. This is:

$$[\boldsymbol{x}_f] = \boldsymbol{x}_f + \mathcal{P}_{\boldsymbol{x}_f}(\delta\boldsymbol{x}_0) = \boldsymbol{x}_f + \begin{bmatrix} \sum\limits_{1\le k_1+\cdots+k_n\le k} c^1_{k_1\cdots k_n}\delta x_{0,1}^{k_1}\cdots\delta x_{0,n}^{k_n} \\ \vdots \\ \sum\limits_{1\le k_1+\cdots+k_n\le k} c^n_{k_1\cdots k_n}\delta x_{0,1}^{k_1}\cdots\delta x_{0,n}^{k_n} \end{bmatrix} \tag{5.6}$$

where $c^j_{k_1\cdots k_n}$ are the coefficients of the expansion in the jth component of the state, and $\delta\boldsymbol{x}_0 = [\delta x_{0,1}, \cdots, \delta x_{0,n}]^{\mathrm{T}}$. In this chapter, we use the symbol $\mathcal{P}_{\boldsymbol{y}}(z)$ to denote the polynomial expansion of $\boldsymbol{y}$ in the variables $z = (z_1, \cdots, z_q)$.

5.3 Dynamical Reshaping Station-Keeping Approaches

5.3.1 Floquet Modes and Projections Factors

The geometrical structure around the nominal orbit is the basis to develop efficient station-keeping strategies. As it has been mentioned, the solar-sail resonant orbits of Fig. 5.3 have a $(S \times C \times C)$ normal behavior. Due to the Hamiltonian structure of the dynamical system, the eigenvalues of their associated monodromy matrices can be divided into three pairs with particular geometrical meanings:

- The first pair (λ_1, λ_2) with $\lambda_1 \cdot \lambda_2 = 1$ is associated with the normal hyperbolic behavior of the orbit. The unstable eigenvalue λ_1 is the largest one and its associated eigenvector $\boldsymbol{e}_1(0)$ gives the most expanding direction. They are related with the unstable manifold. In the same way, the stable eigenvalue λ_2 and the corresponding eigenvalue $\boldsymbol{e}_2(0)$ are related with the stable manifold.

- The second pair (λ_3, λ_4) consists in complex conjugated eigenvalues of modulus one, i.e., $\lambda_3 = \bar{\lambda}_4$ and $\|\lambda_3\| = \|\lambda_4\| = 1$. In this case, $\boldsymbol{e}_3(0)$ and $\boldsymbol{e}_4(0)$ refer to the real and imaginary parts of their complex eigenvectors. They are related with a central behavior (i.e., a central manifold), around the periodic orbit, giving rise to quasi-periodic orbits nearby.
- The third pair (λ_5, λ_6) is similar to the second one. The only difference lies in the value of the frequency of the rotation.

To analyze the state of the spacecraft in the neighborhood of the nominal orbit, it is convenient to introduce the Floquet frame. Following Gómez et al. [10], the Floquet modes for the periodic orbits in this case are defined as

$$
\begin{aligned}
\tilde{\boldsymbol{e}}_1(\tau) &= \boldsymbol{e}_1(\tau)\exp\left(-\frac{\tau}{T_S}\ln\lambda_1\right)\\
\tilde{\boldsymbol{e}}_2(\tau) &= \boldsymbol{e}_2(\tau)\exp\left(-\frac{\tau}{T_S}\ln\lambda_2\right)\\
\tilde{\boldsymbol{e}}_3(\tau) &= \cos\left(-\frac{\Gamma_3\tau}{T_S}\right)\boldsymbol{e}_3(\tau) - \sin\left(-\frac{\Gamma_3\tau}{T_S}\right)\boldsymbol{e}_4(\tau)\\
\tilde{\boldsymbol{e}}_4(\tau) &= \sin\left(-\frac{\Gamma_3\tau}{T_S}\right)\boldsymbol{e}_3(\tau) + \cos\left(-\frac{\Gamma_3\tau}{T_S}\right)\boldsymbol{e}_4(\tau)\\
\tilde{\boldsymbol{e}}_5(\tau) &= \cos\left(-\frac{\Gamma_5\tau}{T_S}\right)\boldsymbol{e}_5(\tau) - \sin\left(-\frac{\Gamma_5\tau}{T_S}\right)\boldsymbol{e}_6(\tau)\\
\tilde{\boldsymbol{e}}_6(\tau) &= \sin\left(-\frac{\Gamma_5\tau}{T_S}\right)\boldsymbol{e}_5(\tau) + \cos\left(-\frac{\Gamma_5\tau}{T_S}\right)\boldsymbol{e}_6(\tau)
\end{aligned}
\tag{5.7}
$$

where $\boldsymbol{e}_i(\tau) = A(\tau)\boldsymbol{e}_i(0)$ with $A(\tau)$ being the state transition matrix, and $\tau \in [0, T_S]$ is used to parameterize the resonant orbit considered. Γ_3 and Γ_5 are the angles corresponding to λ_3 and λ_5, respectively.

Looking for an efficient computation of a given state deviation in terms of the Floquet basis, the so-called *projection factors* are used. Assume that, at a given time t, the spacecraft is close to a nominal state on the periodic nominal orbit under consideration. This nominal state is characterized by a certain value of the parameter τ. The state deviation $\delta\boldsymbol{x}(t) = [\delta x(t), \delta y(t), \delta z(t), \delta\dot{x}(t), \delta\dot{y}(t), \delta\dot{z}(t)]^{\mathrm{T}}$, is then defined as the difference between the state of the spacecraft and the associated nominal state at epoch t. Such difference can be written in terms of the Floquet modes as

$$
\delta\boldsymbol{x}(t) = \sum_{i=1}^{6} c_i(t)\tilde{\boldsymbol{e}}_i(\tau(t))
\tag{5.8}
$$

where $c_i(t)$ denote the components of $\delta\boldsymbol{x}(t)$ in the Floquet basis at time t. From now on, unless necessary, the explicit time dependence will be omitted. Inverting the linear system (5.8) yields

$$c_i = \pi_1^i(\tau)\delta x + \pi_2^i(\tau)\delta y + \pi_3^i(\tau)\delta z + \pi_4^i(\tau)\delta \dot{x} + \pi_5^i(\tau)\delta \dot{y} + \pi_6^i(\tau)\delta \dot{z}, \quad i = 1 \ldots 6 \tag{5.9}$$

where, given the Floquet matrix $[\tilde{\boldsymbol{e}}_1(\tau), \cdots, \tilde{\boldsymbol{e}}_6(\tau)]$, the value of $\pi_j^i(\tau)$ is equal to the adjoint of the jth element of the ith column divided by the determinant of the matrix.

To this end, c_i can be computed just by the dot product of the state deviation $\delta \boldsymbol{x}$ and the ith projection factor $\boldsymbol{\pi}^i = [\pi_1^i, \cdots, \pi_6^i]^{\mathrm{T}}$. In practice, it is convenient to compute the projection factors $\boldsymbol{\pi}^i(\tau)$, $\tau \in [0, T_s]$, only once, and to store them together with the nominal resonant orbit and associated Floquet modes, by means of their Fourier coefficients. In this way, a fast on-board evaluation is feasible when required.

5.3.2 Dynamical Reshaping Algorithms

The dynamical reshaping methodology provides a low-thrust station-keeping control framework making use of the geometric structures discussed in Sect. 5.3.1. The idea is presented in [6] for low-thrust station keeping of Sun-Earth L_1 halo orbits in the autonomous RTBP and, as we will see, is a good candidate to develop SRP assisted station-keeping controllers in the non-autonomous Earth-Moon QBCP-SRP. Moreover, as it has been mentioned, by means of the JT procedure, the control laws can be explicitly expressed in a high-order polynomial formalism.

In solar-sail missions, the SRP acceleration is determined by the lightness number and the attitude angles of the sail (see Eq. 5.4). Denote by β_n, φ_n and γ_n the nominal parameters considered for a solar-sail mission in a resonant orbit, and $\delta\beta$, $\delta\varphi$ and $\delta\gamma$ the deviations with respect to these nominal values. According to Eqs. (5.1) and (5.4), the equations of motion for the controlled solar sail becomes

$$\begin{cases} \ddot{x} - \dfrac{\dot{\alpha}_1}{\alpha_1}\dot{x} - 2\alpha_3\dot{y} = \left(\dot{\alpha}_2 + \alpha_2^2 + \alpha_3^3 - \dfrac{\dot{\alpha}_1\alpha_2}{\alpha_1}\right)x + \left(\dot{\alpha}_3 - \dfrac{\dot{\alpha}_1\alpha_3}{\alpha_1}\right)y - \alpha_1\alpha_4 - \alpha_1\alpha_6\left(\dfrac{1-\mu}{r_{PE}^3}(x-\mu) + \dfrac{\mu}{r_{PM}^3}(x-\mu+1) + \dfrac{m_S}{r_{PS}^3}(x-\alpha_7)\right) \\ \qquad - \alpha_1\alpha_6(\beta_n + \delta\beta)\dfrac{m_S}{\alpha_7^2+\alpha_8^2}\dfrac{\alpha_7\cos^3(\varphi_n+\delta\varphi)\cos^3(\gamma_n+\delta\gamma) - \alpha_8\cos^3(\varphi_n+\delta\varphi)\cos^2(\gamma_n+\delta\gamma)\sin(\gamma_n+\delta\gamma)}{\sqrt{\alpha_7^2+\alpha_8^2}} \\ \ddot{y} - \dfrac{\dot{\alpha}_1}{\alpha_1}\dot{y} + 2\alpha_3\dot{x} = \left(\dot{\alpha}_2 + \alpha_2^2 + \alpha_3^3 - \dfrac{\dot{\alpha}_1\alpha_2}{\alpha_1}\right)y - \left(\dot{\alpha}_3 - \dfrac{\dot{\alpha}_1\alpha_3}{\alpha_1}\right)x - \alpha_1\alpha_5 - \alpha_1\alpha_6\left(\dfrac{1-\mu}{r_{PE}^3}y + \dfrac{\mu}{r_{PM}^3}y + \dfrac{m_S}{r_{PS}^3}(y-\alpha_8)\right) \\ \qquad - \alpha_1\alpha_6(\beta_n + \delta\beta)\dfrac{m_S}{\alpha_7^2+\alpha_8^2}\dfrac{\alpha_8\cos^3(\varphi_n+\delta\varphi)\cos^3(\gamma_n+\delta\gamma) + \alpha_7\cos^3(\varphi_n+\delta\varphi)\cos^2(\gamma_n+\delta\gamma)\sin(\gamma_n+\delta\gamma)}{\sqrt{\alpha_7^2+\alpha_8^2}} \\ \ddot{z} - \dfrac{\dot{\alpha}_1}{\alpha_1}\dot{z} = \left(\dot{\alpha}_2 + \alpha_2^2 - \dfrac{\dot{\alpha}_1\alpha_2}{\alpha_1}\right)z - \alpha_1\alpha_6\left(\dfrac{1-\mu}{r_{PE}^3}z + \dfrac{\mu}{r_{PM}^3}z + \dfrac{m_S}{r_{PS}^3}z\right) \\ \qquad - \alpha_1\alpha_6(\beta_n + \delta\beta)\dfrac{m_S}{\alpha_7^2+\alpha_8^2}\cos^2(\varphi_n+\delta\varphi)\cos^2(\gamma_n+\delta\gamma)\sin(\varphi_n+\delta\varphi) \end{cases} \tag{5.10}$$

where $\delta\beta$, $\delta\varphi$ and $\delta\gamma$ are the control variables for the station keeping.

In the following, two nominal point definitions are used. The isochronous nominal point is the state in the selected nominal orbit moving in an isochronous way with respect to the state of the spacecraft. The nominal point at minimum distance is the state on the nominal orbit at closest distance to the spacecraft measured in

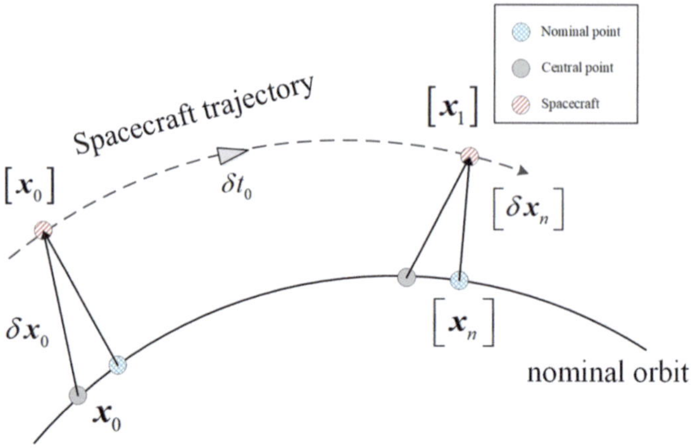

Fig. 5.4 Schematic representation of the central point, spacecraft and nominal point states

space components. Other suitable nominal points definitions could be implemented in practice. Fortunately, the JT technique provides a unified manner to develop the reshaping algorithms regardless of the nominal point definition adopted.

Consider the isochronous nominal point $\boldsymbol{x}_0$ as the central base point at epoch t_0, and $\delta\boldsymbol{x}_0$ a symbolic arbitrary state deviation at this epoch. With these notations, the state of the spacecraft can be written as $[\boldsymbol{x}_0] = \boldsymbol{x}_0 + \delta\boldsymbol{x}_0$ (it will be determined by a particular value of $\delta\boldsymbol{x}_0$). In addition, let δt_0 be a symbolic time deviation, and consider the control values $\{\delta\beta_0, \delta\varphi_0, \delta\gamma_0\}$ also as JT variables at epoch t_0. Using Eqs. (5.10), the Taylor expansion of the solar-sail spacecraft state at epoch $[t_0] = t_0 + \delta t_0$ is

$$[\boldsymbol{x}_1] = \boldsymbol{x}_0 + \mathcal{P}_{x_1}(\delta\boldsymbol{x}_0, \delta t_0, \delta\beta_0, \delta\varphi_0, \delta\gamma_0)$$

Implementing the computation of the selected nominal point definition for $[\boldsymbol{x}_1]$ within the JT framework, the nominal point and its associated τ parameter are obtained. Denoting by $[\boldsymbol{x}_n]$ this nominal point and by $[\tau_n]$ the associated τ parameter, their polynomial expressions in terms of the JT variables $\{\delta\boldsymbol{x}_0, \delta t_0, \delta\beta_0, \delta\varphi_0, \delta\gamma_0\}$ can be written as

$$[\boldsymbol{x}_n] = \boldsymbol{x}_0 + \mathcal{P}_{x_n}(\delta\boldsymbol{x}_0, \delta t_0, \delta\beta_0, \delta\varphi_0, \delta\gamma_0), \quad [\tau_n] = \boldsymbol{t}_0 + \mathcal{P}_{\tau_n}(\delta\boldsymbol{x}_0, \delta t_0, \delta\beta_0, \delta\varphi_0, \delta\gamma_0) \tag{5.11}$$

In particular, note that when the isochronous point is chosen as the nominal point, then $\mathcal{P}_{\tau_n}(\delta\boldsymbol{x}_0, \delta t_0, \delta\beta_0, \delta\varphi_0, \delta\gamma_0) = \delta t_0$ and $[\tau_n] = t_0 + \delta t_0$. See Fig. 5.4 for the relationships among the central base point, nominal point and spacecraft states.

Then, the state deviation of the solar-sail spacecraft with respect to the nominal point at epoch $[t_0]$ is

$$[\delta \boldsymbol{x}_n] = [\boldsymbol{x}_1] - [\boldsymbol{x}_n] = \mathcal{P}_{\boldsymbol{x}_n}(\delta \boldsymbol{x}_0, \delta t_0, \delta \beta_0, \delta \varphi_0, \delta \gamma_0) \tag{5.12}$$

Moreover, the ith projection factor associated with the nominal point can be computed by means of a JT-based evaluation of its corresponding Fourier expansion at time parameter $[\tau_n]$, so

$$[\boldsymbol{\pi}_n^i] = \boldsymbol{\pi}_0^i + \mathcal{P}_{\boldsymbol{\pi}_n^i}(\delta \boldsymbol{x}_0, \delta t_0, \delta \beta_0, \delta \varphi_0, \delta \gamma_0) \tag{5.13}$$

where $\boldsymbol{\pi}_0^i$ is the ith projection factor associated with the isochronous nominal point $\boldsymbol{x}_0$. The dot product of Eqs. (5.12) and (5.13) then gives the component along the ith Floquet mode, i.e.,

$$[c_i] = \langle [\boldsymbol{\pi}_n^i], [\delta \boldsymbol{x}_n] \rangle = \mathcal{P}_{c_i}(\delta \boldsymbol{x}_0, \delta t_0, \delta \beta_0, \delta \varphi_0, \delta \gamma_0) \tag{5.14}$$

Furthermore, one can compute the time derivative of $[c_i]$,

$$[\dot{c}_i] = \frac{\mathrm{d}[c_i]}{\mathrm{d}t} = \mathcal{P}_{\dot{c}_i}(\delta \boldsymbol{x}_0, \delta t_0, \delta \beta_0, \delta \varphi_0, \delta \gamma_0) \tag{5.15}$$

Evaluating Eqs. (5.14) and (5.15) with $\delta t_0 = 0$, the ith Floquet component and its first derivative at epoch t_0 are obtained. Their polynomial expressions are

$$[c_i] = \mathcal{P}_{c_i}(\delta \boldsymbol{x}_0, \delta \beta_0, \delta \varphi_0, \delta \gamma_0), \quad [\dot{c}_i] = \mathcal{P}_{\dot{c}_i}(\delta \boldsymbol{x}_0, \delta \beta_0, \delta \varphi_0, \delta \gamma_0) \tag{5.16}$$

The dynamical reshaping methodology consists in forcing the decayment of selected Floquet modes requiring that

$$[f_i] := [\dot{c}_i] + G_i[c_i] = \mathcal{P}_{f_i}(\delta \boldsymbol{x}_0, \delta \beta_0, \delta \varphi_0, \delta \gamma_0) = 0, \tag{5.17}$$

where $G_i > 0$ is a selected value to modulate the decayment rate versus the magnitude of the control variables. So, the problem of deriving control laws turns into solving parametric-implicit equations like Eq. (5.17) to obtain the required control variables at epoch t_0 in terms of $\delta \boldsymbol{x}_0$:

$$[\delta \beta_0] = \mathcal{P}_{\delta \beta_0}(\delta \boldsymbol{x}_0), \qquad [\delta \varphi_0] = \mathcal{P}_{\delta \varphi_0}(\delta \boldsymbol{x}_0), \qquad [\delta \gamma_0] = \mathcal{P}_{\delta \gamma_0}(\delta \boldsymbol{x}_0)$$

In solar-sail missions, the control variables $\delta \varphi$ and $\delta \gamma$ can be easily achieved by attitude maneuvers, while $\delta \beta$ implies an adjustment of the sail membrane area. Because the difficulty of changing β requires a more complex mechanism, two scenarios are considered: time-constant and time-varying β.

5.3.2.1 Scenario I: Time-Constant β

According to the geometrical analysis of Sect. 5.3.1, the Floquet component along the mode $\bar{\boldsymbol{e}}_1$ is unstable. Therefore, to achieve stability, the control of c_1 is mandatory. Because β is time-constant, Eq. (5.17) for the c_1 component becomes

$$[f_1] = [\dot{c}_1] + G_1[c_1] = \mathcal{P}_{f_1}(\delta\boldsymbol{x}_0, \delta\varphi_0, \delta\gamma_0) = 0 \tag{5.18}$$

which is under-determined since the equation has two unknowns: $\delta\varphi_0$ and $\delta\gamma_0$. If this mode is the only one reshaped, the chosen option is to impose a minimum norm constraint on the vector $[\delta\varphi_0, \delta\gamma_0]^{\mathrm{T}}$, thus, the resulting control law can be considered as the optimal one to keep the motion bounded. However, other options could be also possible, for instance, one could consider only $\delta\varphi_0$, or $\delta\gamma_0$, as the control variable. But in this case, at some epochs, Eq. (5.18) is singular, this is, has no solutions, then the control procedure should include some additional conditions.

In addition to the unstable c_1 component, the central components c_3, c_4, c_5 and c_6 can be also included into the reshaping procedure (evidently, the stable component c_2 can be left alone as it is already associated with an exponential decay). Although these four components are already neutral stable, to incorporate them into the reshaping algorithms opens the possibility toward asymptotic stability. In this case weighted sums are used. For instance, if a pair of central components are included, the component to be reshaped is defined as $c_{1ij} = w_1c_1 + w_ic_i + w_jc_j$ with associated reshaping equation

$$[f_{1ij}] = [\dot{c}_{1ij}] + G_{1ij}[c_{1ij}] = \mathcal{P}_{f_{1ij}}(\delta\boldsymbol{x}_0, \delta\varphi_0, \delta\gamma_0) = 0 \tag{5.19}$$

As usual, only positive weights, such that $w_1 + w_i + w_j = 1$, are used,. The reshaped two central components can be chosen from the same pair or different pairs, therefore, for the nominal orbit of $(S \times C \times C)$ type, the possible $\{i, j\}$ combinations are $\{3, 4\}$, $\{3, 5\}$, $\{3, 6\}$, $\{4, 5\}$, $\{4, 6\}$ and $\{5, 6\}$. In addition, since the unstable component c_1 is the main concern in the reshaping procedure, it is reasonable to set $w_1 \geq (w_i + w_j)$, i.e., $w_1 \geq 0.5$.

In a similar way, the four central components can be included into the reshaping algorithms by considering $c_{13456} = w_1c_1 + w_3c_3 + w_4c_4 + w_5c_5 + w_6c_6$. In this case

$$[f_{13456}] = [\dot{c}_{13456}] + G_{13456}[c_{13456}] = \mathcal{P}_{f_{13456}}(\delta\boldsymbol{x}_0, \delta\varphi_0, \delta\gamma_0) = 0 \tag{5.20}$$

with $w_1 + w_3 + w_4 + w_5 + w_6 = 1$ and $w_1 \geq 0.5$, as mentioned before.

Again, one can consider the minimum norm constraint on the vector $[\delta\varphi_0, \delta\gamma_0]^{\mathrm{T}}$ for Eqs. (5.19) or (5.20). Many other possibilities on constraints could be used following the ideas introduced in this section, however, further computations have not been done.

5.3.2.2 Scenario II: Time-Varying β

Including $\delta\beta$, the number of control variables is three, and the possibility to reshape more Floquet modes simultaneously can be studied. In this way, and using Eq. (5.17), one can reshape up to three components considering

$$\begin{cases} [f_1] = \mathcal{P}_{f_1}(\delta \boldsymbol{x}_0, \delta\beta_0, \delta\varphi_0, \delta\gamma_0) = 0 \\ [f_i] = \mathcal{P}_{f_i}(\delta \boldsymbol{x}_0, \delta\beta_0, \delta\varphi_0, \delta\gamma_0) = 0 \\ [f_j] = \mathcal{P}_{f_j}(\delta \boldsymbol{x}_0, \delta\beta_0, \delta\varphi_0, \delta\gamma_0) = 0 \end{cases} \tag{5.21}$$

where possible choices for $\{i, j\}$ are $\{3, 4\}$ and $\{5, 6\}$. Mixed choices ($\{3, 5\}$, $\{3, 6\}$, $\{4, 5\}$ and $\{4, 6\}$) could be also possible, but in these cases, there appear incompatibility issues in Eqs. (5.21) at some epochs. i.e., the control variables $\delta\beta$, $\delta\varphi$ and $\delta\gamma$ may not be determined for all values of $t \in [0; T_S]$ (mod T_S), then, an extension of the strategy should be considered in a neighborhood of these epochs.

Other procedures come from using weighted sums like the ones introduced in the previous scenario. For instance, one can consider $c_{35} = w_3 c_3 + w_5 c_5$ and $c_{46} = w_4 c_4 + w_5 c_5$ with a reshaping procedure of the form

$$\begin{cases} [f_1] = \mathcal{P}_{f_1}(\delta \boldsymbol{x}_0, \delta\beta_0, \delta\varphi_0, \delta\gamma_0) = 0 \\ [f_{35}] = \mathcal{P}_{f_{35}}(\delta \boldsymbol{x}_0, \delta\beta_0, \delta\varphi_0, \delta\gamma_0) = 0 \\ [f_{46}] = \mathcal{P}_{f_{46}}(\delta \boldsymbol{x}_0, \delta\beta_0, \delta\varphi_0, \delta\gamma_0) = 0 \end{cases} \tag{5.22}$$

where $w_3 + w_5 = 1$ and $w_4 + w_6 = 1$. In a similar way, $c_{36} = w_3 c_3 + w_6 c_6$ and $c_{45} = w_4 c_4 + w_5 c_5$ with $w_3 + w_6 = w_4 + w_5 = 1$. can also be considered. Other options, such as $c_{34} = w_3 c_3 + w_4 c_4$ and $c_{56} = w_5 c_5 + w_6 c_6$, have incompatibility issues at some epochs and their implementation has not been considered.

Note that the controllers obtained depend on the definition of the nominal point, but the procedure and the expression of the results is general regardless of how the nominal point is defined. The control variables $\delta\beta$, $\delta\varphi$ and $\delta\gamma$ are derived semi-analytically and explicitly expressed as Taylor polynomials in terms of the state deviation with respect to the isochronous (central) point. Because the polynomial coefficients of the control variables are periodic with the same period as the nominal orbit, they are computed just once and stored by means of their Fourier coefficients.

5.4 Numerical Simulations and Discussions

5.4.1 Stability of the Reshaping Algorithms

This subsection shows how the reshaping approaches stabilize the solar-sail spacecraft in a neighborhood of the nominal orbit. As a case example, the solar-sail orbit with $\beta = 0.03$ displayed in Fig. 5.3 is considered as the nominal orbit for the numerical simulations. Similar results and conclusions are expected for other solar-sail

Table 5.2 Eigenvalues of the monodromy matrix M_0 (natural system)

Monodromy matrix	λ_1	λ_2	λ_3, λ_4	λ_5, λ_6
M_0	588463.8	1.6993×10^{-6}	$0.4523 \pm 0.8919\,\mathrm{i}$	$0.9649 \pm 0.2626\,\mathrm{i}$

Table 5.3 Eigenvalues of M_{c_1} for different gain factors G_1. Isochronous nominal point

Gain factor G_1	λ_1	λ_2	λ_3, λ_4	λ_5, λ_6
0.0	1.0	1.6993×10^{-6}	$0.4523 \pm 0.8919\,\mathrm{i}$	$0.9649 \pm 0.2626\,\mathrm{i}$
1.0	1.1237×10^{-3}	1.6993×10^{-6}	$0.4523 \pm 0.8919\,\mathrm{i}$	$0.9649 \pm 0.2626\,\mathrm{i}$
3.0	1.4174×10^{-9}	1.6983×10^{-6}	$0.4523 \pm 0.8919\,\mathrm{i}$	$0.9649 \pm 0.2626\,\mathrm{i}$
5.0	2.0522×10^{-15}	1.6983×10^{-6}	$0.4523 \pm 0.8919\,\mathrm{i}$	$0.9649 \pm 0.2626\,\mathrm{i}$

orbits in the neighborhood because all of them are of $(S \times C \times C)$ type and the maximum Floquet multiplier is of the same order of magnitude. In addition, the expansion order in the JT procedure is set to 2 taking into account the trade-off between computational burden and accuracy.

The first analysis considers the eigenvalues of the monodromy matrix of the associated controlled system. For convenience, this matrix, associated to a particular algorithm, will be denoted by means of a subscript: M_{c_1}, $M_{c_{1ij}}$, and $M_{c_{13456}}$ will be the monodromy matrices respectively associated to the $\{c_1\}$, $\{c_{1ij}\}$, and $\{c_{13456}\}$ reshaping control laws of scenario I, and $M_{c_{1,3,4}}$, $M_{c_{1,5,6}}$, $M_{c_{1,35,46}}$ and $M_{c_{1,36,45}}$ will be the ones that reshape $\{c_1, c_3, c_4\}$, $\{c_1, c_5, c_6\}$, $\{c_1, c_{35}, c_{46}\}$, and $\{c_1, c_{36}, c_{45}\}$ in scenario II. In addition, the monodromy matrix for the natural system with no control law applied will be denoted by M_0.

Table 5.2 lists the eigenvalues of M_0 showing the $(S \times C \times C)$ normal behavior of the nominal orbit. The maximum norm eigenvalue $\lambda_1 = 588463.8$ indicates a highly unstable component along the Floquet mode $\tilde{e}_1$, which is the reason of the instability of the natural system. Table 5.3 gives the eigenvalues of M_1, i.e., the normal behavior when applying the c_1 component reshaping law of Eq. (5.18), for several different values of the gain factor G_1. The isochronous nominal point has been considered for tracking the deviations in this example. One can see the way the orbit has been stabilized: the unstable eigenvalue λ_1 becomes less than 1 for positive gain factors, and decreases for lager values of G_1. This is reasonable because the reshaping algorithm requires the c_1 component to decay exponentially, the value of λ_1 should be in the order of $\mathrm{e}^{-G_i T_s}$. Moreover, note that this reshaping control law has no effect on the other stable or central eigenvalues, therefore, the controlled solar-sail spacecraft oscillates and remains inside the center manifold of the nominal orbit once the motion is in the steady regime.

Tables 5.4, 5.5, 5.6 and 5.7 summarize weighted cases of the reshaping algorithms of scenario I including central pairs c_{1ij} (for $\{i, j\}$ are $\{3, 4\}$, $\{5, 6\}$, $\{3, 6\}$ and $\{4, 6\}$) as in Eq. (5.19). For the sake of simplicity, the weights w_i and w_j take the same value,

Table 5.4 Eigenvalues of $M_{c_{134}}$ for different values of w_1. Gain factor $G_{134} = 1$. Isochronous nominal point

Weight w_1	λ_1	λ_2	λ_3, λ_4	λ_5, λ_6
0.55	1.1251×10^{-3}	1.6970×10^{-6}	$-0.1752 \pm 1.2084\,i$	$0.9649 \pm 0.2626\,i$
0.60	1.1247×10^{-3}	1.6976×10^{-6}	$-0.0643 \pm 1.0065\,i$	$0.9649 \pm 0.2626\,i$
0.65	1.1245×10^{-3}	1.6980×10^{-6}	$0.0084 \pm 0.9090\,i$	$0.9649 \pm 0.2626\,i$
0.70	1.1243×10^{-3}	1.6983×10^{-6}	$0.0751 \pm 0.8685\,i$	$0.9649 \pm 0.2626\,i$
0.75	1.1242×10^{-3}	1.6985×10^{-6}	$0.1414 \pm 0.8573\,i$	$0.9649 \pm 0.2626\,i$
0.80	1.1240×10^{-3}	1.6987×10^{-6}	$0.2077 \pm 0.8601\,i$	$0.9649 \pm 0.2626\,i$
0.85	1.1239×10^{-3}	1.6989×10^{-6}	$0.2732 \pm 0.8683\,i$	$0.9649 \pm 0.2626\,i$
0.90	1.1238×10^{-3}	1.6991×10^{-6}	$0.3366 \pm 0.8774\,i$	$0.9649 \pm 0.2626\,i$
0.95	1.1237×10^{-3}	1.6992×10^{-6}	$0.3966 \pm 0.8855\,i$	$0.9649 \pm 0.2626\,i$

Table 5.5 Eigenvalues of $M_{c_{156}}$ for different values of w_1. Gain factor $G_{156} = 1$. Isochronous nominal point

Weight w_1	λ_1	λ_2	λ_3, λ_4	λ_5, λ_6
0.50	1.1245×10^{-3}	1.6974×10^{-6}	$0.4523 \pm 0.8919\,i$	0.8614, 29460.2
0.55	1.1247×10^{-3}	1.6970×10^{-6}	$0.4523 \pm 0.8919\,i$	0.8919, 12195.5
0.60	1.1250×10^{-3}	1.6965×10^{-6}	$0.4523 \pm 0.8919\,i$	0.9470, 3108.3
0.65	1.1253×10^{-3}	1.6958×10^{-6}	$0.4523 \pm 0.8919\,i$	1.0736, 326.8
0.70	1.1258×10^{-3}	1.6952×10^{-6}	$0.4523 \pm 0.8919\,i$	1.4425, 14.4
0.75	1.1265×10^{-3}	1.6948×10^{-6}	$0.4523 \pm 0.8919\,i$	$0.7166 \pm 1.0202\,i$
0.80	1.1266×10^{-3}	1.6954×10^{-6}	$0.4523 \pm 0.8919\,i$	$0.4179 \pm 0.3994\,i$
0.85	1.1254×10^{-3}	1.6968×10^{-6}	$0.4523 \pm 0.8919\,i$	$0.4470 \pm 0.2017\,i$
0.90	1.1244×10^{-3}	1.6981×10^{-6}	$0.4523 \pm 0.8919\,i$	$0.5320 \pm 0.2186\,i$
0.95	1.1239×10^{-3}	1.6989×10^{-6}	$0.4523 \pm 0.8919\,i$	$0.6897 \pm 0.2991\,i$

Table 5.6 Eigenvalues of $M_{c_{136}}$ for different values of w_1. Gain factor $G_{136} = 1$. Isochronous nominal point

Weight w_1	λ_1	λ_2	λ_3, λ_4	λ_5, λ_6
0.75	1.1373×10^{-3}	1.6830×10^{-6}	$0.6958 \pm 0.6021\,i$	$0.3119 \pm 0.1177\,i$
0.80	1.1286×10^{-3}	1.6931×10^{-6}	$0.4379 \pm 0.7970\,i$	0.1161, 0.8864
0.85	1.1254×10^{-3}	1.6969×10^{-6}	$0.4094 \pm 0.8467\,i$	0.1799, 0.9213
0.90	1.1242×10^{-3}	1.6984×10^{-6}	$0.4040 \pm 0.8752\,i$	0.3361, 0.9136
0.95	1.1238×10^{-3}	1.6990×10^{-6}	$0.4191 \pm 0.8963\,i$	0.6993, 0.8004

Table 5.7 Eigenvalues of $M_{c_{146}}$ for different values of w_1. Gain factor $G_{146} = 1$. Isochronous nominal point

Weight w_1	λ_1	λ_2	λ_3, λ_4	λ_5, λ_6
0.50	1.1246×10^{-3}	1.6971×10^{-6}	$0.4752 \pm 1.0454\,\mathrm{i}$	1.0253, 41184.8
0.55	1.1249×10^{-3}	1.6965×10^{-6}	$0.4857 \pm 1.0319\,\mathrm{i}$	1.0439, 15322.3
0.60	1.1253×10^{-3}	1.6956×10^{-6}	$0.5110 \pm 1.0124\,\mathrm{i}$	1.0824, 2454.3
0.65	1.1257×10^{-3}	1.6946×10^{-6}	$0.5649 \pm 0.9297\,\mathrm{i}$	1.2288, 90.6
0.70	1.1263×10^{-3}	1.6937×10^{-6}	$0.5562 \pm 0.7301\,\mathrm{i}$	$1.4946 \pm 1.2992\,\mathrm{i}$
0.75	1.1263×10^{-3}	1.6937×10^{-6}	$0.5562 \pm 0.7301\,\mathrm{i}$	$1.4946 \pm 1.2992\,\mathrm{i}$
0.80	1.1270×10^{-3}	1.6952×10^{-6}	$0.4594 \pm 0.8871\,\mathrm{i}$	0.0913034, 0.892656
0.85	1.1253×10^{-3}	1.6971×10^{-6}	$0.4610 \pm 0.8666\,\mathrm{i}$	0.122838, 0.920338
0.90	1.1242×10^{-3}	1.6983×10^{-6}	$0.4573 \pm 0.8567\,\mathrm{i}$	0.262837, 0.91629
0.95	1.1238×10^{-3}	1.6989×10^{-6}	$0.4468 \pm 0.8626\,\mathrm{i}$	0.627981, 0.832721

i.e., $w_i = w_j = (1 - w_1)/2$ with gain factor $G_{1ij} = 1.0$ in all cases and considering the isochronous nominal point. Apparently, the stability of the control laws depends on the chosen weights. For the c_{134} reshaping algorithm in Table 5.4, w_1 starts at the value of 0.55 and this is because of incompatibility issues occur for $w_1 = 0.50$. For $w_1 \leq 0.60$, the controlled system is slightly unstable because the norm of the pairs of complex eigenvalues $\{\lambda_3, \lambda_4\}$ is larger than one. This phenomenon shows that appropriate weights should be chosen to ensure the stability of the system when considering this approach. This same fact is also displayed in Tables 5.5 and 5.7, where stability is achieved for $w_1 \geq 0.80$. In addition, for the c_{134} and c_{156} reshaping algorithms, the two central components are from the same pair and, as expected, Tables 5.4 and 5.5 show that only the reshaped eigenvalues change, whereas the others remain the same as in the natural system. So, in these cases the system can be stabilized, but asymptotic stability can not be achieved. However, if the central components are taken from different pairs, it is possible to achieve asymptotic stability for appropriately chosen weights. See Table 5.6 for $w_1 \geq 0.75$ and Table 5.7 for $w_1 \geq 0.80$.

Tables 5.8 and 5.9 present the results for the c_{13456} component reshaping algorithm of Eq. (5.20) that includes the four central components in Scenario I considering the isochronous nominal point. The gain factors in Tables 5.8 and 5.9 are 1.0 and 5.0, respectively, whereas $w_3 = w_4 = w_5 = w_6 = (1 - w_1)/4$. From the tables it follows that all the unstable and central eigenvalues depend on the chosen weights and asymptotic stability is achieved for $w_1 \geq 0.70$.

For Scenario II, where the variations in the lightness number β are included, Tables 5.10 and 5.11 present the results for the $\{c_1, c_3, c_4\}$ and $\{c_1, c_5, c_6\}$ reshaping algorithms of Eqs. (5.21). For the sake of simplicity, the values of the gain factors of the central components are taken equal to the one of the c_1 component. Notice that the reshaped eigenvalues decrease when the gain factors increase, being of the order of $e^{-G_i T_s}$, just as it happened with the case related with Table 5.3. Again, one

Table 5.8 Eigenvalues of $M_{c_{13456}}$ for different values of w_1. Gain factor $G_{13456} = 1$. Isochronous nominal point

Weight w_1	λ_1	λ_2	λ_3, λ_4	λ_5, λ_6
0.50	1.1256×10^{-3}	1.6953×10^{-6}	$0.4802 \pm 0.8041\,i$	$1.0862,\ 155.7$
0.55	1.1260×10^{-3}	1.6949×10^{-6}	$0.4535 \pm 0.7854\,i$	$1.2967,\ 13.3$
0.60	1.1262×10^{-3}	1.6948×10^{-6}	$0.4521 \pm 0.7693\,i$	$1.2639 \pm 1.1219\,i$
0.65	1.1262×10^{-3}	1.6953×10^{-6}	$0.5038 \pm 0.8773\,i$	$0.5248 \pm 0.6384\,i$
0.70	1.1258×10^{-3}	1.6961×10^{-6}	$0.4374 \pm 0.8414\,i$	$0.4905 \pm 0.4269\,i$
0.75	1.1258×10^{-3}	1.6961×10^{-6}	$0.4263 \pm 0.8307\,i$	$0.5016 \pm 0.3253\,i$
0.80	1.1247×10^{-3}	1.6977×10^{-6}	$0.4179 \pm 0.8317\,i$	$0.5438 \pm 0.2932\,i$
0.85	1.1243×10^{-3}	1.6983×10^{-6}	$0.4114 \pm 0.8425\,i$	$0.6104 \pm 0.2993\,i$
0.90	1.1241×10^{-3}	1.6987×10^{-6}	$0.4127 \pm 0.8618\,i$	$0.6997 \pm 0.3144\,i$
0.95	1.1238×10^{-3}	1.6991×10^{-6}	$0.4273 \pm 0.8818\,i$	$0.8139 \pm 0.3150\,i$

Table 5.9 Eigenvalues of $M_{c_{13456}}$ for different values of w_1. Gain factor $G_{13456} = 5$. Isochronous nominal point

Weight w_1	λ_1	λ_2	λ_3, λ_4	λ_5, λ_6
0.50	4.8229×10^{-15}	1.7001×10^{-6}	$0.4802 \pm 0.8040\,i$	$1.0863,\ 155.7$
0.55	3.3984×10^{-15}	1.7001×10^{-6}	$0.4535 \pm 0.7853\,i$	$1.2968,\ 13.3$
0.60	1.4801×10^{-15}	1.6998×10^{-6}	$0.4521 \pm 0.7692\,i$	$1.2639 \pm 1.1220\,i$
0.65	4.7999×10^{-15}	1.6993×10^{-6}	$0.5039 \pm 0.8773\,i$	$0.5248 \pm 0.6383\,i$
0.70	2.6062×10^{-15}	1.6987×10^{-6}	$0.4374 \pm 0.8414\,i$	$0.4905 \pm 0.4269\,i$
0.75	1.5847×10^{-15}	1.6982×10^{-6}	$0.4263 \pm 0.8307\,i$	$0.5016 \pm 0.3253\,i$
0.80	1.3306×10^{-15}	1.6977×10^{-6}	$0.4179 \pm 0.8316\,i$	$0.5438 \pm 0.2932\,i$
0.85	1.1888×10^{-15}	1.6975×10^{-6}	$0.4114 \pm 0.8425\,i$	$0.6104 \pm 0.2993\,i$
0.90	5.5590×10^{-15}	1.6973×10^{-6}	$0.4127 \pm 0.8618\,i$	$0.6997 \pm 0.3144\,i$
0.95	8.1905×10^{-15}	1.6973×10^{-6}	$0.4273 \pm 0.8818\,i$	$0.8139 \pm 0.3150\,i$

Table 5.10 Eigenvalues of $M_{c_{1,3,4}}$ for different gain factors $G_1 = G_3 = G_4$. Isochronous nominal point

Gain factor $G_{1,3,4}$	λ_1	λ_2	λ_3, λ_4	λ_5, λ_6
0.0	1.0	1.6977×10^{-6}	1.0, 1.0	$0.9649 \pm 0.2626\,i$
1.0	1.1270×10^{-3}	1.6943×10^{-6}	$1.1236 \times 10^{-3},\ 1.1236 \times 10^{-3}$	$0.9649 \pm 0.2626\,i$
3.0	1.4079×10^{-9}	1.7122×10^{-6}	$1.4187 \times 10^{-9},\ 1.4187 \times 10^{-9}$	$0.9649 \pm 0.2626\,i$
5.0	1.7453×10^{-15}	1.7043×10^{-6}	$1.9095 \times 10^{-15},\ 1.9095 \times 10^{-15}$	$0.9649 \pm 0.2626\,i$

Table 5.11 Eigenvalues of $M_{c_{1,5,6}}$ for different gain factors $G_1 = G_5 = G_6$. Isochronous nominal point

Gain factor G	λ_1	λ_2	λ_3, λ_4	λ_5, λ_6
0.0	1.0	1.6977×10^{-6}	$0.4523 \pm 0.8919\,\mathrm{i}$	1.0, 1.0
1.0	1.1270×10^{-3}	1.6944×10^{-6}	$0.4523 \pm 0.8919\,\mathrm{i}$	1.1236×10^{-3}, 1.1236×10^{-3}
3.0	1.4084×10^{-9}	1.7115×10^{-6}	$0.4523 \pm 0.8919\,\mathrm{i}$	1.4187×10^{-9}, 1.4187×10^{-9}
5.0	1.7564×10^{-15}	1.7040×10^{-6}	$0.4523 \pm 0.8919\,\mathrm{i}$	1.9865×10^{-15}, 1.9865×10^{-15}

Table 5.12 Eigenvalues of $M_{c_{1,35,46}}$ for different gain factors $G_1 = G_{35} = G_{46} = 1.0$. Isochronous nominal point

Weights w_3 and w_4	λ_1	λ_2	λ_3, λ_4	λ_5, λ_6
0.05	1.1270×10^{-3}	1.6943×10^{-6}	$0.4476 \pm 0.8909\,\mathrm{i}$	1.1236×10^{-3}, 1.1236×10^{-3}
0.10	1.1271×10^{-3}	1.6943×10^{-6}	$0.4416 \pm 0.8903\,\mathrm{i}$	1.1236×10^{-3}, 1.1236×10^{-3}
0.15	1.1272×10^{-3}	1.6942×10^{-6}	$0.4335 \pm 0.8906\,\mathrm{i}$	1.1236×10^{-3}, 1.1236×10^{-3}
0.20	1.1273×10^{-3}	1.6940×10^{-6}	$0.4216 \pm 0.8929\,\mathrm{i}$	1.1236×10^{-3}, 1.1236×10^{-3}
0.25	1.1274×10^{-3}	1.6938×10^{-6}	$0.4011 \pm 0.9002\,\mathrm{i}$	1.1236×10^{-3}, 1.1236×10^{-3}
0.30	1.1278×10^{-3}	1.6932×10^{-6}	$0.3399 \pm 0.9328\,\mathrm{i}$	1.1236×10^{-3}, 1.1236×10^{-3}
0.90	1.1276×10^{-3}	1.6935×10^{-6}	1.1236×10^{-3}, 1.1236×10^{-3}	$0.9751 \pm 0.1493\,\mathrm{i}$
0.95	1.1272×10^{-3}	1.6935×10^{-6}	1.1236×10^{-3}, 1.1234×10^{-3}	$0.9604 \pm 0.2418\,\mathrm{i}$

can see how the methodology can address the decay of specific modes while leaving some others of the original system unvaried, and this property that can be of interest not only for the particular problem under consideration in this chapter, but for other general control problems.

Table 5.12 shows the results of mixing pairs in Scenario II using Eqs. (5.22). For this case, the $\{c_1, c_{35}, c_{46}\}$ reshaping algorithm with gain factors $G_1 = G_{35} = G_{46} = 1.0$ is considered and, again for convenience: $w_3 = w_4$, $w_5 = w_6$, and $w_3 + w_5 = w_4 + w_6 = 1.0$. From the results, it follows that, except for the cases $\{w_3, w_4\} \in [0.30, 0.90]$, where the equations present singularities at some epochs and the implementation has not been analyzed, the controlled system becomes asymptotically stable under the control laws because the norms of λ_3 and λ_4 are slightly smaller than 1 for $0.05 \leq \{w_3, w_4\} \leq 0.30$, and so do λ_5 and λ_6 for $0.90 \leq \{w_3, w_4\} \leq 0.95$.

Table 5.13 gives the results for the mixed pairs controller $\{c_1, c_{36}, c_{45}\}$ with $w_3 + w_6 = w_4 + w_5 = 1.0$ and $w_3 = w_4$, $w_5 = w_6$. In this case, the incompatibility issues occur for $0.40 \leq \{w_3, w_4\} \leq 0.80$. For $\{w_3, w_4\} \leq 0.40$, the controlled system is asymptotically stable and the convergence rate increases when increasing $\{w_3, w_4\}$. For $0.80 \leq \{w_3, w_4\}$, the controlled system becomes slightly unstable. Again we notice that the technique of mixing components by means of weights can be helpful to achieve asymptotic stability, but the weights have to be selected carefully.

The previous tables show the characteristics of the controllers when considering the isochronous state as the nominal point. In what follows, the effect of the choice of the nominal point on the performance of the controller is briefly

Table 5.13 Eigenvalues of $M_{c_{1,36,45}}$ for different gain factors $G_1 = G_{36} = G_{45} = 1.0$. Isochronous nominal point

Weights w_3 and w_4	λ_1	λ_2	λ_3, λ_4	λ_5, λ_6
0.05	1.1270×10^{-3}	1.6943×10^{-6}	$0.4448 \pm 0.8826\,i$	1.1236×10^{-3}, 1.1236×10^{-3}
0.10	1.1271×10^{-3}	1.6943×10^{-6}	$0.4332 \pm 0.8737\,i$	1.1236×10^{-3}, 1.1236×10^{-3}
0.15	1.1272×10^{-3}	1.6942×10^{-6}	$0.4156 \pm 0.8651\,i$	1.1236×10^{-3}, 1.1236×10^{-3}
0.20	1.1273×10^{-3}	1.6940×10^{-6}	$0.3893 \pm 0.8569\,i$	1.1236×10^{-3}, 1.1236×10^{-3}
0.25	1.1274×10^{-3}	1.6938×10^{-6}	$0.3496 \pm 0.8487\,i$	1.1236×10^{-3}, 1.1236×10^{-3}
0.30	1.1276×10^{-3}	1.6935×10^{-6}	$0.2869 \pm 0.8384\,i$	1.1236×10^{-3}, 1.1236×10^{-3}
0.35	1.1281×10^{-3}	1.6928×10^{-6}	$0.1772 \pm 0.8174\,i$	1.1236×10^{-3}, 1.1236×10^{-3}
0.40	1.1299×10^{-3}	1.6900×10^{-6}	$-0.0834 \pm 0.7176\,i$	1.1235×10^{-3}, 1.1235×10^{-3}
0.80	1.1267×10^{-3}	1.6950×10^{-6}	1.1235×10^{-3}, 1.1236×10^{-3}	$1.0015 \pm 0.8345\,i$
0.85	1.1266×10^{-3}	1.6948×10^{-6}	1.1236×10^{-3}, 1.1236×10^{-3}	$1.0485 \pm 0.4863\,i$
0.90	1.1268×10^{-3}	1.6947×10^{-6}	1.1236×10^{-3}, 1.1236×10^{-3}	$1.0300 \pm 0.3473\,i$
0.95	1.1269×10^{-3}	1.6945×10^{-6}	1.1236×10^{-3}, 1.1236×10^{-3}	$0.9997 \pm 0.2847\,i$

Table 5.14 Eigenvalues of M_{c_1} with $G_1 = 1.0$ for several minimum-distance nominal points

$\{w_p, w_v\}$	λ_1	λ_2	λ_3, λ_4	λ_5, λ_6
{0.95, 0.05}	1.1232×10^{-3}	1.7016×10^{-6}	$0.4560 \pm 0.8915\,i$	$0.9684 \pm 0.0503\,i$
{0.90, 0.10}	1.1230×10^{-3}	1.6853×10^{-6}	$0.4264 \pm 0.8901\,i$	$0.9763 \pm 0.0339\,i$
{0.85, 0.15}	1.1246×10^{-3}	1.6689×10^{-6}	$0.4213 \pm 0.8915\,i$	$0.9822 \pm 0.0112\,i$

Table 5.15 Eigenvalues of $M_{c_{1,3,4}}$ with $G_1 = G_3 = G_4 = 1.0$ for several minimum-distance nominal points

$\{w_p, w_v\}$	λ_1	λ_2	λ_3, λ_4	λ_5, λ_6
{1.00, 0.00}	1.1244×10^{-3}	1.3121×10^{-6}	1.1244×10^{-3}, 1.1244×10^{-3}	0.9094, 1.0
{0.90, 0.10}	1.1237×10^{-3}	1.3589×10^{-6}	1.1237×10^{-3}, 1.1237×10^{-3}	0.9455, 1.0
{0.80, 0.20}	1.1237×10^{-3}	1.3923×10^{-6}	1.1237×10^{-3}, 1.1237×10^{-3}	0.9675, 1.0
{0.70, 0.30}	1.1246×10^{-3}	1.4176×10^{-6}	1.1237×10^{-3}, 1.1237×10^{-3}	0.9826, 1.0
{0.60, 0.40}	1.1246×10^{-3}	1.4538×10^{-6}	1.1237×10^{-3}, 1.1237×10^{-3}	0.9936, 1.0

summarized. In general, one can define minimum-distance nominal points considering the minimization of a weighted sum of position and velocity deviations, $d_n = w_p\sqrt{\delta x_n^2 + \delta y_n^2 + \delta z_n^2} + w_v\sqrt{\delta \dot{x}_n^2 + \delta \dot{y}_n^2 + \delta \dot{z}_n^2}$ with $w_p + w_v = 1.0$ and being $\delta \boldsymbol{x}_n = [\delta x_n, \delta y_n, \delta z_n, \delta \dot{x}_n, \delta \dot{y}_n, \delta \dot{z}_n]^{\mathrm{T}}$ the state deviation between spacecraft and the nominal point. In particular, the nominal point is named as minimum position deviation point for $w_p = 1.0$ and $w_v = 0$, which will play an important role in assessing the distance between the controlled trajectory and the nominal orbit in the following. Using the c_1 component reshaping law from Scenario I and $\{c_1, c_3, c_4\}$ components reshaping law from Scenario II as example, Tables 5.14 and 5.15 give the eigenvalues of their monodromy matrices, respectively, considering several different minimum-distance nominal point definitions. Table 5.14 presents the results for $w_p \in [0.85, 0.95]$ ($w_v \in [0.05, 0.15]$) since the control law is slightly unstable for other points. Similarly, Table 5.15 presents the results for $w_p \in [0.60, 1.00]$ ($w_v \in [0.00, 0.40]$). According to Table 5.14, the control law is asymptotically stable for points with $w_p \in [0.85, 0.95]$ ($w_v \in [0.05, 0.15]$) showing that a neutrally stable system can become asymptotically stable by choosing appropriate nominal points. For the $\{c_1, c_3, c_4\}$ reshaping law in Table 5.15, although the stability property does not change (i.e., neutrally stable in the linear sense), the pair (λ_5, λ_6) is no longer the same to the ones of the natural system for the minimum-distance nominal points. Therefore, although both reshaping laws in Tables 5.15 and 5.10 are neutrally stable in the linear sense, the eigenstructures of the associated monodromy matrices are different and determined by the definition of the nominal point. Moreover, it is interesting to see that $\lambda_6 = 1.0$ always exits regardless of the weight selected. However, the way in which the eigenvalues persist or change according to different definitions of nominal points is out of the scope of this book.

5.4.2 *Dynamical and Long-Term Behavior in the Presence of Insertion Errors*

Using the c_1 and c_{13456} cases from Scenario I, and the $\{c_1, c_{35}, c_{46}\}$ case from Scenario II as examples, this section shows their geometrical behaviors in the presence of orbit insertion errors of 5 km, 5 km, 5 km, 5 cm/s, 5 cm/s, and 5 cm/s.

Figure 5.5 gives the time evolution of the Floquet coefficients of the controlled trajectory for the c_1 component reshaping algorithm over 120 orbital periods (about 10 years), using the isochronous and minimum-distance nominal points. For both cases, the c_1 component decreases exponentially according to the requirement. Moreover, the c_2 component decreases to zero because it is already associated with the stable motion. For the isochronous nominal case, the c_1 component reshaping algorithm has no effect on the central eigenvalues of the system, and the central components c_i $(i = 3, 4, 5, 6)$ present cosine/sine type curves, meaning the spacecraft evolves inside the center manifold of the nominal orbit. However, using the minimum-distance nominal point with $w_p = 0.9$ and $w_v = 0.1$, the controlled system is asymptotically stable (see Table 5.14) and the coefficients c_i $(i = 3, 4, 5, 6)$ decay toward zero. Considering the isochronous nominal point and incorporating the central components into the reshaping algorithm, Figs. 5.6 and 5.7 show the results of the evolution of the Floquet weighted components, $c_{13456} = 0.8c_1 + 0.05(c_3 + c_4 + c_5 + c_6)$ and $\{c_1,\ c_{36} = 0.4\,c_3 + 0.6\,c_6,\ c_{45} = 0.4\,c_4 + 0.6\,c_5\}$, respectively. According to the design, the modes c_{13456} in Fig. 5.6, and c_1, c_{36} and c_{45} in Fig. 5.7 decay exponentially, as well as all the c_i $(i = 1, \ldots, 6)$ components, showing that asymptotic stability has been achieved for these two control laws. Similar results can be expected for the other reshaping approaches.

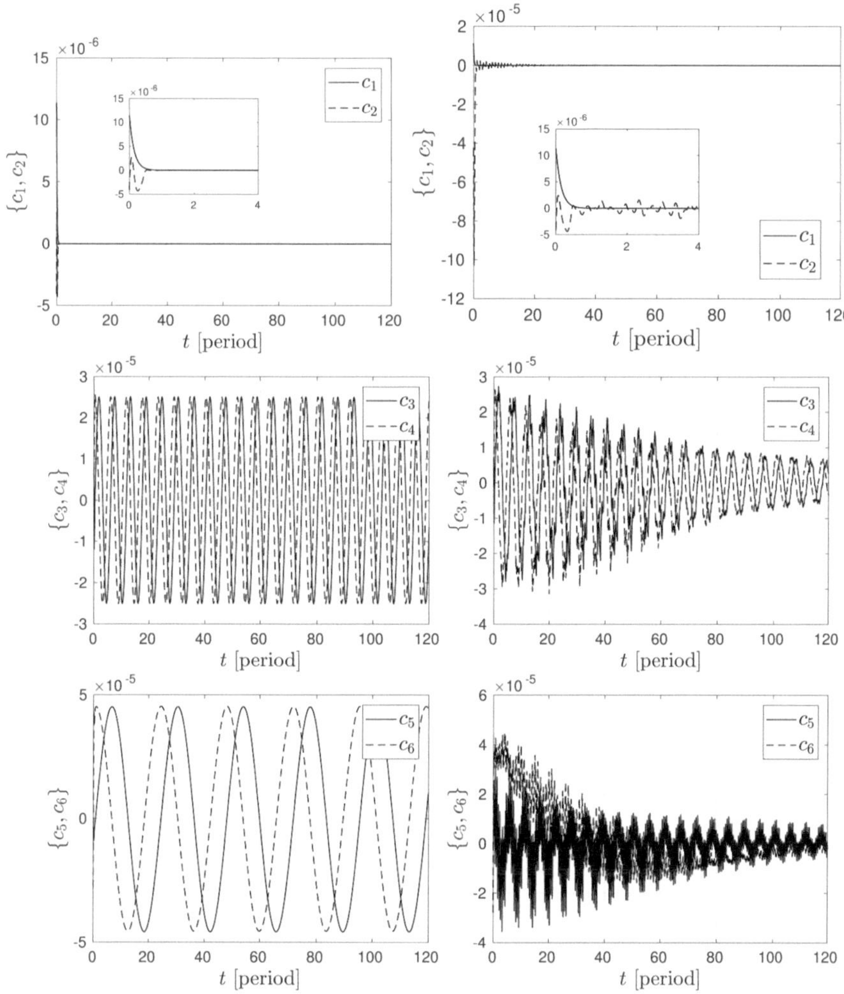

Fig. 5.5 Time evolution of the Floquet coefficients over 120 orbital periods for the c_1 component reshaping algorithm with $G_1 = 1.0$. Left column: isochronous nominal point; Right column: minimum-distance nominal point with $w_p = 0.9$ and $w_v = 0.1$

The long-term behavior of the dynamical reshaping algorithms has also been studied iterating their associated Poincaré maps 12, 36, 60, and 72 times. Taking into account that the period of the resonant orbit is approximately one month, these iterates roughly correspond to 1, 3, 5, and 7 years, respectively. The initial state deviations are taken in a circle inside the hypersurface $\mathcal{N} = \{y = 0, z > 0\}$, which for convenience is denoted by C_R:

$$C_R = \left\{\delta \boldsymbol{x}_0 = [R\cos\theta,\ 0.0,\ R\sin\theta,\ 0.0,\ 0.0,\ 0.0]^{\mathrm{T}},\quad \theta \in [0, 2\pi]\right\}.$$

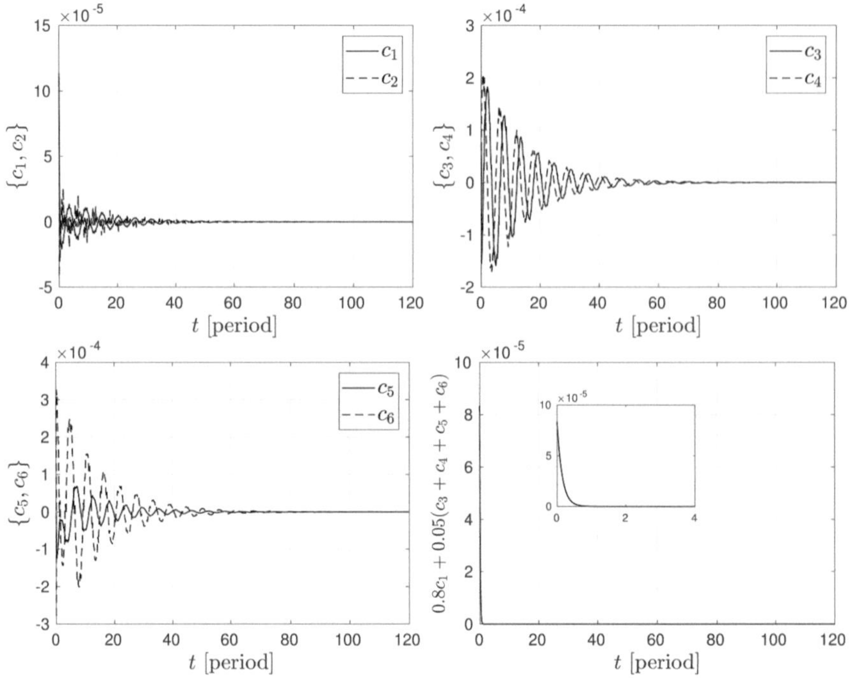

Fig. 5.6 Time evolution of the Floquet coefficients over 120 orbital periods for $c_{13456} = 0.8c_1 + 0.05(c_3 + c_4 + c_5 + c_6)$ component reshaping algorithm with $G_{13456} = 1.0$ and considering the isochronous nominal point

Figure 5.8 shows the iterates of C_R, with $R = 5$ km, under the reshaping algorithms corresponding to Figs. 5.5, 5.6, and 5.7, respectively. The long-term behavior of these station-keeping procedures are stable. More specifically, the algorithms in Figs. 5.5(right column), 5.6, and 5.7 are asymptotically stable whereas the one in Fig. 5.5(left column) is neutrally stable. Besides, the algorithm of Fig. 5.7 displays a faster asymptotic stability when compared to the others.

5.4.3 Robustness of the Reshaping Laws to Orbit Determination Errors and Cost Estimations

In the following, the behavior of the controllers in the presence of orbit determination (OD) errors is tested. Following [7], three levels of OD errors are considered. Error level 1: mean 0 and standard deviation (3σ) of 1 km in each position component and 1 cm/s in each velocity component; error level 2: mean 0 and standard deviation (3σ) of 3 km in each position component and 3 cm/s in each velocity component; and error level 3: mean 0 and standard deviation (3σ) of 5 km in each position component

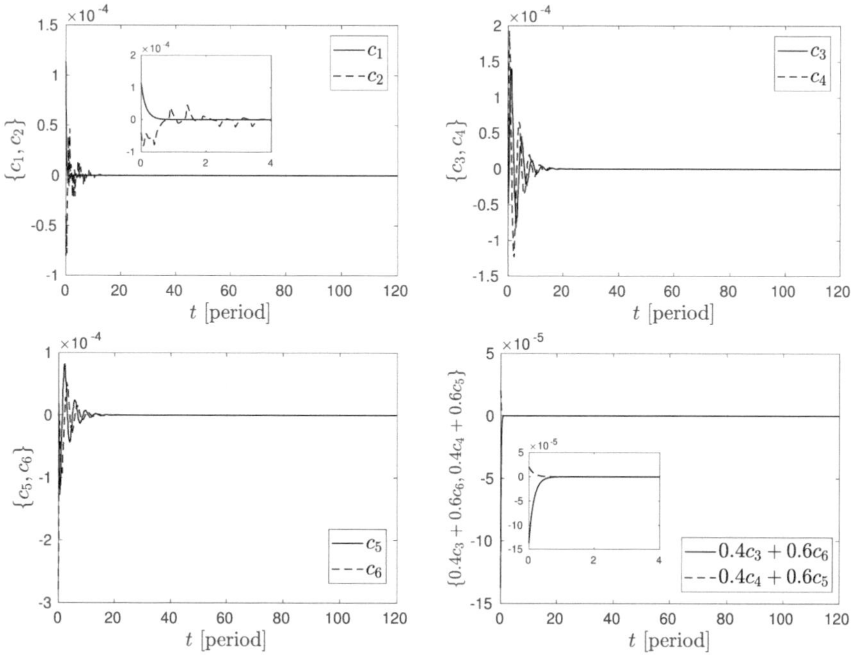

Fig. 5.7 Time evolution of the Floquet coefficients over 120 orbital periods for $\{c_1,\ c_{36} = 0.4c_3 + 0.6c_6,\ c_{45} = 0.4c_4 + 0.6c_5\}$ components reshaping algorithm with $G_1 = G_{36} = G_{45} = 1.0$ and considering the isochronous nominal point

and 5 cm/s in each velocity component. The observational frequency is set to 1 hour in the numerical simulations.

Figure 5.9 shows the distance, denoted by δp, between the solar-sail spacecraft and the nominal orbit under the c_1 reshaping laws. For the isochronous nominal point, the controller is neutrally stable in a linear sense, therefore, OD errors make the distance between the spacecraft and the isochronous nominal point quite large, reaching about 300 km for level 3, when the motion is into steady regime. However, the controlled trajectory is still close to the nominal orbit, according to the measure of the distance between the spacecraft and the minimum position deviation point in the orbit (i.e., the distance to the nominal orbit). For instance, the minimum-distance with respect to the minimum position deviation point is about 150 km for level 3, which is half of the distance with respect to the isochronous point. When considering a minimum-distance nominal point with asymptotic stability, for instance when $w_p = 0.9$ and $w_v = 0.1$, the controlled trajectory is closer to the nominal point, in terms of both the nominal point or the minimum position deviation point. From Fig. 5.9b, it follows that for error level 3 the minimum-distance with respect to the minimum position deviation point is about 50 km, one third of the one in the isochronous point case.

Figures 5.10 and 5.11 show the results for the c_{13456} and $\{c_1, c_{36}, c_{45}\}$ reshaping algorithms, respectively. Here, the isochronous nominal point is used for the first two

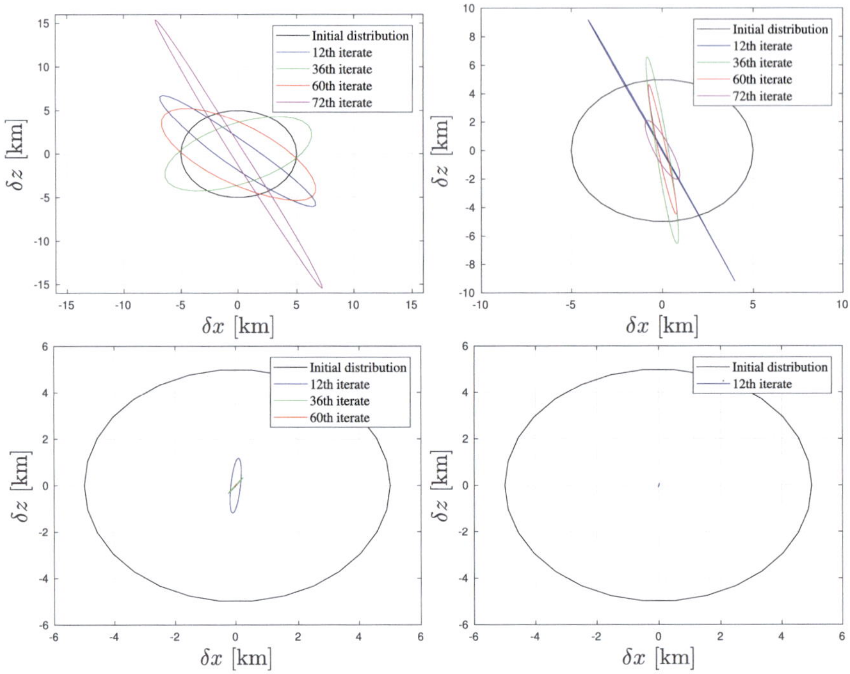

Fig. 5.8 Poincaré map iterates of initial state deviations using the set C_R for $R = 5\,\text{km}$. Top-Left: Associated to the reshaping algorithm (left column) in Fig. 5.5; Top-Right: Associated to the reshaping algorithm (right column) in Fig. 5.5; Bottom-Left: Associated to the reshaping algorithm in Fig. 5.6; Bottom-Right: Associated to the reshaping algorithm in Fig. 5.7

cases, and the minimum position deviation nominal point for the third case, because they are already asymptotically stable (see Figs. 5.6 and 5.7). From these figures, one can realize that, even for the largest error level 3, the spacecraft is maintained in a small neighborhood of the nominal orbit under any of the reshaping algorithms, showing their robustness in the presence of OD errors. Comparing the above results, one can also conclude that station-keeping strategies with asymptotic stability are more robust to OD errors than the ones with neutral stability.

Finally, some cost estimations of the control variables to be applied for the proposed station-keeping strategies, are given. Table 5.16 gives the maximum and minimum values of control variables corresponding to the station-keeping controllers of Figs. 5.9, 5.10 and 5.11. In addition, taking the controller of Fig. 5.11 as example, Fig. 5.12 displays the time history of the control variables (which is qualitatively similar to the other cases). It follows that the cost increases with larger OD errors, however, the required variations of the control variables are small, even for error level 3. For the c_1 and c_{13456} reshaping algorithms, where the two attitude angles are control variables, $\delta\varphi$ essentially varies between $-0.7°$ and $0.6°$, and $\delta\gamma$ between $-0.6°$ and $0.4°$ for error 3. Therefore, the sail needs to perform small oscillations around its nominal

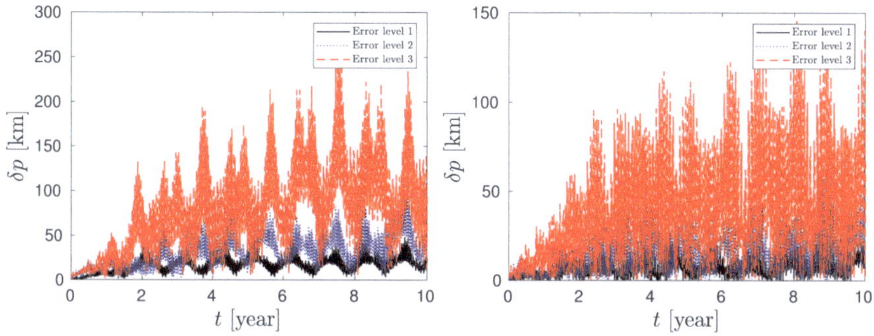

(a) Isochronous nominal point.

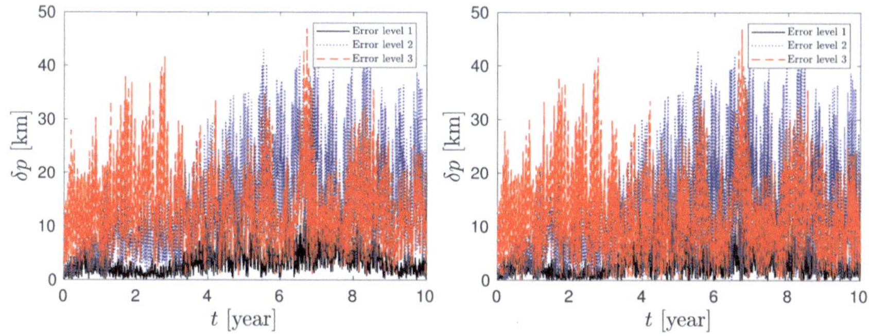

(b) Minimum-distance nominal point with $w_p = 0.9$ and $w_v = 0.1$.

Fig. 5.9 Distance between the solar-sail spacecraft and the nominal orbit over 10 years for c_1 reshaping algorithm with $G_1 = 1.0$. Left column: measured with respect to the chosen nominal point; Right column: measured with respect to the nominal orbit (minimum position deviation point)

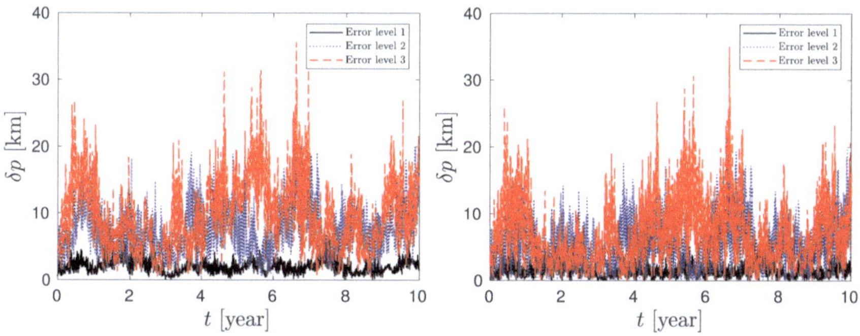

Fig. 5.10 Distance between the solar-sail spacecraft and the nominal orbit over 10 years for the $c_{13456} = 0.8c_1 + 0.05(c_3 + c_4 + c_5 + c_6)$ reshaping algorithm with $G_{13456} = 1.0$ considering the isochronous nominal point. Left column: measured with respect to the nominal point; Right column: measured with respect to the nominal orbit (minimum position deviation point)

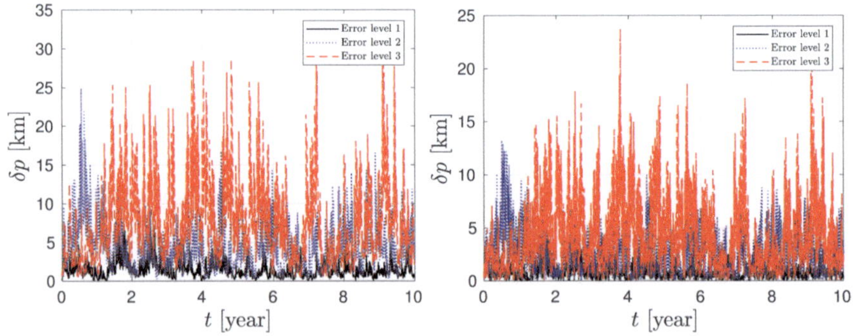

Fig. 5.11 Distance between the solar-sail spacecraft and the nominal orbit over 10 years for the $\{c_1, 0.4c_3 + 0.6c_6, 0.4c_4 + 0.6c_5\}$ reshaping algorithm with $G_1 = G_{36} = G_{45} = 1.0$ considering the isochronous nominal point. Left column: measured with respect to the nominal point; Right column: measured with respect to the nominal orbit (minimum position deviation point)

Table 5.16 Maximum and minimum values of control variables corresponding to the station-keeping controllers of Figs. 5.9, 5.10 and 5.11. The units of angular variables are in degrees

Station-keeping controller	Error level	$\delta\varphi_{max}$	$\delta\varphi_{min}$	$\delta\gamma_{max}$	$\delta\gamma_{min}$	$\delta\beta_{max}$	$\delta\beta_{min}$
Controller of Fig. 5.9a	1	0.1039	−0.1073	0.0592	−0.0673	–	–
	2	0.2903	−0.2592	0.1841	−0.1799	–	–
	3	0.5055	−0.4655	0.3142	−0.3344	–	–
Controller of Fig. 5.9b	1	0.0988	−0.1015	0.0636	−0.0624	–	–
	2	0.2617	−0.2811	0.1861	−0.1779	–	–
	3	0.5059	−0.5125	0.3277	−0.2925	–	–
Controller of Fig. 5.10	1	0.1510	−0.1494	0.0883	−0.0789	–	–
	2	0.3720	−0.3495	0.2488	−0.3178	–	–
	3	0.5886	−0.6906	0.3807	−0.5684	–	–
Controller of Fig. 5.11	1	0.0828	−0.0619	0.2926	−0.3117	5.8553×10^{-5}	-6.5859×10^{-5}
	2	0.2087	−0.1774	1.0177	−0.8848	2.1399×10^{-4}	-1.6510×10^{-4}
	3	0.4168	−0.3664	1.4795	−1.6802	3.2118×10^{-4}	-3.2285×10^{-4}

orientation. In addition, the lightness number varies slightly (around 4×10^{-4}), as is shown in Table 5.16 and Fig. 5.12 (case $\{c_1, c_{36}, c_{45}\}$) for the largest error level. Therefore, the proposed station-keeping strategies are very affordable for solar-sail missions, even when the lightness number is taken as a control variable (Fig. 5.12).

Of course, the accuracy and cost of the station-keeping depends on the observational frequency of the OD. Using the c_{13456} reshaping algorithm of Fig. 5.10 as an example, some results using the Monte Carlo method for different observational frequencies are shown. The following indices are used to compute the average needs for the attitude maneuvers in one period for the ith Monte Carlo simulation:

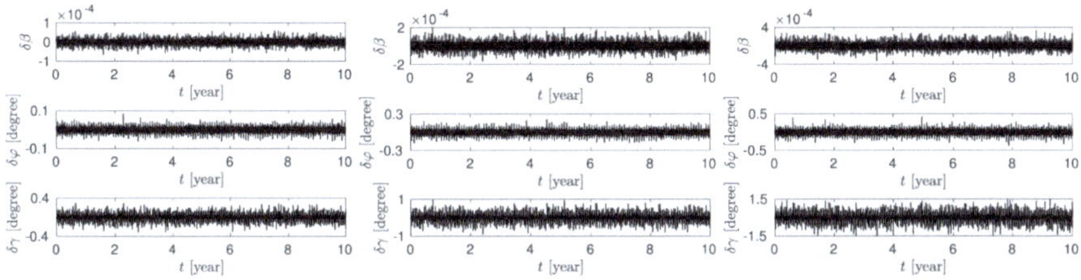

Fig. 5.12 History of the control variables corresponding to the station-keeping controller of Fig. 5.11. Left column: error level 1; Middle column: error level 2; Right column: error level 3

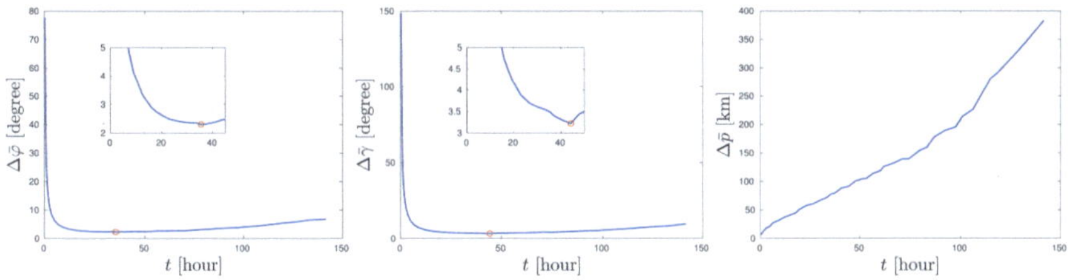

Fig. 5.13 Average cost per period and mean distance of the spacecraft to the nominal orbit as a function of the OD observational frequency. Results obtained from the Monte Carlo simulation using the $c_{13456} = 0.8c_1 + 0.05(c_3 + c_4 + c_5 + c_6)$ reshaping algorithm with $G_{13456} = 1.0$. Error level 3 and the isochronous nominal point have been considered in the simulations

$$\Delta\varphi_i = \frac{1}{120}\int_0^{120T_s} \|\delta\dot{\varphi}_i(t)\|dt, \qquad \Delta\gamma_i = \frac{1}{120}\int_0^{120T_s} \|\delta\dot{\gamma}_i(t)\|dt$$

where T_s is the period of the nominal orbit and the simulation time is set to $120T_s$ (about 10 years). Note that the physical meaning of the two indices are the average angular lengths sweept by φ and γ within one period. The averaged values $\Delta\bar{\varphi} = \text{mean}(\Delta\varphi_1, \cdots, \Delta\varphi_N)$ and $\Delta\bar{\gamma} = \text{mean}(\Delta\gamma_1, \cdots, \Delta\gamma_N)$, being $N = 100$ the number of samples in the Monte Carlo simulations to get the final estimates, have also been computed. Moreover, the mean distance from the spacecraft to the nominal point in a sampled simulation is defined as:

$$\Delta p_i = \frac{1}{120T_s}\int_0^{120T_s} \|\delta p_i(t)\|dt,$$

with again, a final value of the Monte Carlo simulation given by, $\Delta\bar{p} = \text{mean}(\Delta p_1, \cdots, \Delta p_N)$.

Figure 5.13 gives $\Delta\bar{\varphi}$, $\Delta\bar{\gamma}$ and $\Delta\bar{p}$ in terms of the observational time span in OD, considering the error level 3. More specifically, the OD error is included into the measurement state at each observational time and control variables are recomputed and updated at these epochs. The asymptotic behaviour in the cost shown in the figure, when the time step goes to zero, is due to the independence of the OD measurements errors, because no filter has been implemented to smooth this procedure. The small red circle in the left and middle subfigures indicate the minimum values of $\Delta\bar{\varphi}$ and $\Delta\bar{\gamma}$. The procedure reaches a minimum value in $\Delta\bar{\varphi}$ for 35.4 h and in $\Delta\bar{\gamma}$ for 44.25 h. For larger time intervals the costs start increasing, as well as the distance to the nominal orbit. Similar results are expected for other reshaping laws. To sum it up, again the proposed station-keeping approaches are affordable, because $\Delta\bar{\varphi}$ is less than 80° and $\Delta\bar{\gamma}$ is less than 150° within one period (a synodic month).

5.5 Conclusions

This chapter investigates the SRP assisted station-keeping of unstable resonant LPOs in the Earth-Moon system. The dynamical model adopted is the high-fidelity QBCP, and the station-keeping approaches are designed within the framework of a dynamical reshaping methodology. Several reshaping procedures have been proposed in accordance to the normal behavior of the nominal orbit and with the use of different definitions of nominal points. By means of the JT technique, the reshaping algorithms have been formulated in a semi-analytic way, and explicitly expressed in a polynomial formalism in terms of the deviation between the state of spacecraft and the associated nominal point. Moreover, both scenarios, one with the lightness number being constant, and in another one adjustable, have been discussed for the implementation of the control laws.

The analysis of the eigenvalues of the monodromy matrix has shown that the proposed reshaping algorithms can stabilize the motion by canceling the unstable Floquet modes. For the isochronous nominal point, it is shown that incorporating the central modes from different pairs one can achieve asymptotic stability, whereas incorporating the ones from the same pair one achieves neutral stability. It is also shown that the performance of the control laws depends on the nominal point associated. For instance, stable control laws can become asymptotically stable by choosing appropriate minimum-distance nominal points. Through the analyses of the geometrical behaviors, it is seen that the reshaped components decrease according to the requirements and follow the dynamical intuitions underlying in their design.

To test the robustness of the station-keeping strategies, the behavior for three different error levels in OD simulations has been considered. The results show that, in the presence of these errors, the solar-sail spacecraft can be maintained in a small neighborhood of the nominal orbit under the proposed station-keeping approaches. In addition, the control laws with asymptotic stability perform better than the ones with neutral stability in terms of maintenance accuracy. Additionally, it is shown that the variations of control variables, including attitude angles and lightness number, are small, even for the largest OD errors considered. Therefore, the sail just needs to perform small oscillations around its nominal attitude, and to slightly change its lightness number in case this parameter is considered variable. Based on these results, the feasibility of using SRP for long-term maintenance in Earth-Moon resonant LPOs has been verified.

References

1. Shirobokov M, Trofimov S, Ovchinnikov M. Survey of station-keeping techniques for libration point orbits[J]. Journal of Guidance, Control, and Dynamics, 2017, 40(5): 1085-1105.
2. Tamakoshi D, Kojima H. Solar sail orbital control using reflectivity variations near the Earth–Moon L_2 point[J]. Journal of Guidance, Control, and Dynamics, 2018, 41(2): 417-430.
3. Farrés A, Jorba, Á. On the station keeping of a solar sail in the elliptic Sun-Earth system[J]. Advances in Space Research, 2011, 48(11): 1785-1796.
4. Farrés A, Jorba Á. Station keeping of a solar sail around a halo orbit[J]. Acta Astronautica, 2014, 94(1): 527-539.
5. Scheeres D J, Hsiao F Y, Vinh N X. Stabilizing motion relative to an unstable orbit: Applications to spacecraft formation flight[J]. Journal of Guidance, Control, and Dynamics, 2003, 26(1): 62-73.
6. Gao C, Masdemont J J, Gómez, et al. High order dynamical systems approaches for low-thrust station-keeping of libration point orbits[J]. Acta Astronautica, 2021, 190: 349-364.
7. Gao C, Masdemont J J, Gómez G, et al. Low-thrust station-keeping control for lunar near rectilinear halo orbits[J]. Celestial Mechanics and Dynamical Astronomy, 2023, 135(2): 1-14.
8. Pérez-Palau D, Masdemont J J, Gómez G. Tools to detect structures in dynamical systems using Jet Transport[J]. Celestial mechanics and dynamical astronomy, 2015, 123, 239-262.
9. Heiligers J, Diedrich B, Derbes B, et al. Sunjammer: PreliminaryPreliminary end-to-end mission design[C]. AIAA/AAS Astrodynamics Specialist Conference, San Diego, USA, 2014.
10. Gomez G, Jorba A, Masdemont J J, et al. Study of the transfer between halo orbits[J]. Acta Astronautica, 1998, 43(9-10): 493-520.

Chapter 6
Coupled Orbit-Attitude Control for Solar Sail Formation Flying in the Earth-Moon Circular Restricted Three Body Problem

This chapter investigates the problem of coupled orbit-attitude control for solar sail formation flying where the leader follows a solar-sail resonant halo orbit in the Earth-Moon Circular Restricted Three Body Problem, whereas the follower is controlled to track a predefined reference relative trajectory around the leader. Taking the reflectivity control devices (RCD) as joined orbit and attitude control actuators, their force and torque models are first established for a spinning solar sail. In particular, the saturation and coupling of the control variables for the coupled orbit-attitude control are focused. Then the orbit and attitude dynamics of the spinning solar sail are discussed and the equations of relative orbit-attitude motion are given. Furthermore, constraints on the relative attitude angles between the sails and their rates of change are considered for the attitude control. Together with the optimal switch regulations of RCD, the problem addressed in this chapter is essentially a problem of constrained optimal coupled orbit-attitude control. Therefore, the framework of model predictive control technique is applied to deal with this problem by solving a convex quadratic programming problem subject to linear constraints. Finally, illustrative scenarios are simulated to show the effectiveness and controllability of the proposed coupled orbit-attitude controller.

6.1 Introduction

In the above chapters, we have introduced the methodologies on orbit design and station keeping using solar sailing and focusing on the Earth-Moon system. We see that these solar-sail displaced orbits above the L_2 point [1–3] and solar-sail resonant libration point orbits [4–9] provide additional opportunities for observation or communication capabilities. However, a single solar sail may be insufficient for certain scientific missions due to a limitation in the payload mass. A feasible solution is to distribute the payload among multiple solar-sail spacecraft with each one only

© The Author(s), under exclusive license to Springer Nature Singapore Pte Ltd. 2026
C. Gao and W. Wang, *Solar Sailing for Deep-Space Explorations*,
https://doi.org/10.1007/978-981-95-0109-0_6

carrying a functional module. That way, multiple solar sails work cooperatively to achieve complex missions. In many formation flying missions, it is desirable that multiple spacecraft maintain a geometric configuration and achieve relative attitude alignment [10, 11], which means active control is indispensable for formation keeping, reconfiguration and reorientation. Therefore, the objective of this chapter is to design an active joined orbit and attitude controller for solar sail formation flying in the Earth-Moon system where the leader follows a solar-sail resonant halo orbit and the follower is controlled to track a predefined reference relative trajectory around the leader. Taking the reflectivity control devices (RCD) as joined orbit and attitude control actuators, their force and torque models are first established for a spinning solar sail. In particular, the saturation and coupling of the control variables for the coupled orbit-attitude control are focused. Then the orbit and attitude dynamics of the spinning solar sail are discussed and the equations of relative orbit-attitude motion are given. Furthermore, constraints on the relative attitude angles between the sails and their rates of change are considered for the attitude control. Together with the optimal switch regulations of RCD, the problem addressed in this chapter is essentially a problem of constrained optimal coupled orbit-attitude control. Therefore, the framework of model predictive control (MPC) technique is applied to deal with this problem by solving a convex quadratic programming problem subject to linear constraints. Finally, illustrative scenarios are simulated to show the effectiveness and controllability of the proposed coupled orbit-attitude controller.

This chapter is organized as follows. Section 6.2 presents the force and torque models of the RCD for a spinning solar sail and gives the saturation and coupling of the orbit-attitude control variables. Section 6.3 briefly describes the solar-sail resonant halo orbits in the Earth-Moon system and establishes the equations of relative orbit-attitude motion of the spinning solar sail. MPC procedures are detailed in Sect. 6.4 including the discretization of the state and output equations, constraints reformulation and the definition of an optimization index. The effectiveness and controllability of the proposed MPC-based coupled orbit-attitude controller for solar sail formation flying are illustrated in Sect. 6.5. Finally, some concluding remarks are given in Sect. 6.6.

6.2 Force and Torque Models of RCD

The basic configuration of the spinning solar sail discussed in this chapter is shown in Fig. 6.1. Assuming a high spin rate which is generally equal or greater than 0.5 rmp [12, 13], the dynamics of the solar sail can be approximated as that of a rigid spacecraft. The sail consists of a payload and a flat, Lambertian annular membrane. A thin layer of electrochromic material panels (EMP), which are distributed evenly, is mounted on the edge of the sail membrane. r_a and r_c are the inner and outer radii of the sail annular membrane respectively, whereas r_b is the inner radius of the RCDs ring. A spin-free coordinate frame $\mathcal{T}_N(O_n;\ \hat{\boldsymbol{x}}_n, \hat{\boldsymbol{y}}_n, \hat{\boldsymbol{z}}_n)$ and a body-fixed coordinate frame $\mathcal{T}_B(O_b;\ \hat{\boldsymbol{x}}_b, \hat{\boldsymbol{y}}_b, \hat{\boldsymbol{z}}_b)$ are defined to describe the attitude of the solar

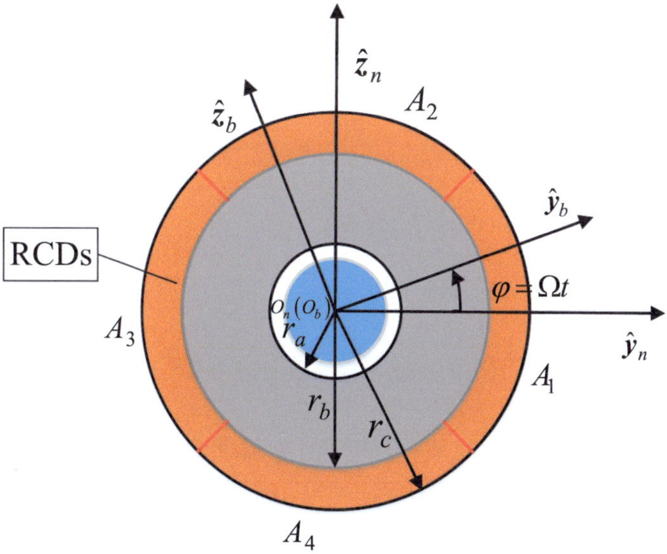

Fig. 6.1 Configuration of the spinning solar sail

sail. The origins of both frames are at the center of the sail. The $\hat{\boldsymbol{y}}_n$ and $\hat{\boldsymbol{z}}_n$ axes as well as the $\hat{\boldsymbol{y}}_b$ and $\hat{\boldsymbol{z}}_b$ axes are in the plane of the sail membrane, whereas the $\hat{\boldsymbol{x}}_n$ and $\hat{\boldsymbol{x}}_b$ axes are along the sail spin axis which is normal to the sail membrane. In addition, the RCDs ring is divided into four equal parts A_1, A_2, A_3 and A_4 in the frame $\mathcal{T}_N$ as shown in Fig. 6.1. For the flat and Lambertian membrane with an area ΔA, the SRP force exerted can be written as

$$\Delta \boldsymbol{F} = P \Delta A \left(\hat{\boldsymbol{n}}_S \cdot \hat{\boldsymbol{n}}\right) \left[\left(c_a + c_d\right) \hat{\boldsymbol{n}}_S + \left(\frac{2}{3} c_d + 2 c_s \left(\hat{\boldsymbol{n}}_S \cdot \hat{\boldsymbol{n}}\right)\right) \hat{\boldsymbol{n}} \right] \tag{6.1}$$

It is assumed that, when the EMP are switched on, their optical properties are the same as those of the sail membrane. Let us denote by A the total area of the sail membrane including the RCD ring and by u the area ratio of the switched-off EMP to A. Then, the total SRP force exerted on the sail is

$$\begin{aligned} \boldsymbol{F} = {} & u A P \left(\hat{\boldsymbol{n}}_S \cdot \hat{\boldsymbol{n}}\right) \left[\left(c_{a_{\text{off}}} + c_{d_{\text{off}}}\right) \hat{\boldsymbol{n}}_S + \left(\frac{2}{3} c_{d_{\text{off}}} + 2 c_{s_{\text{off}}} \left(\hat{\boldsymbol{n}}_S \cdot \hat{\boldsymbol{n}}\right)\right) \hat{\boldsymbol{n}} \right] \\ & + (1-u) A P \left(\hat{\boldsymbol{n}}_S \cdot \hat{\boldsymbol{n}}\right) \left[\left(c_{a_{\text{on}}} + c_{d_{\text{on}}}\right) \hat{\boldsymbol{n}}_S + \left(\frac{2}{3} c_{d_{\text{on}}} + 2 c_{s_{\text{on}}} \left(\hat{\boldsymbol{n}}_S \cdot \hat{\boldsymbol{n}}\right)\right) \hat{\boldsymbol{n}} \right] \end{aligned} \tag{6.2}$$

where $c_{a_{\text{off}}}, c_{d_{\text{off}}}, c_{s_{\text{off}}}$ and $c_{a_{\text{on}}}, c_{d_{\text{on}}}, c_{s_{\text{on}}}$ are coefficients of absorption, diffuse reflection and specular reflection of the switched-off and switched-on EMPs, respectively.

The area ratios of the switched-off EMPs in A_1, A_2, A_3 and A_4 are denoted as u_{y+}, u_{z+}, u_{y-} and u_{z-}, respectively. In frame $\mathcal{T}_N$, the control torque exerted about the $\hat{z}_n$ axis due to the SRP force on part A_1 can be expressed as [13]

$$\begin{aligned} T_{z+} &= -\frac{4u_{y+}}{u_{\max}}\int_{-\frac{\pi}{4}}^{\frac{\pi}{4}}\int_{r_b}^{r_c}\boldsymbol{P}_{\text{off}}\cdot\hat{\boldsymbol{n}}r^2\cos\theta\,\mathrm{d}r\,\mathrm{d}\theta - \frac{u_{\max}-4u_{y+}}{u_{\max}}\int_{-\frac{\pi}{4}}^{\frac{\pi}{4}}\int_{r_b}^{r_c}\boldsymbol{P}_{\text{on}}\cdot\hat{\boldsymbol{n}}r^2\cos\theta\,\mathrm{d}r\,\mathrm{d}\theta \\ &= -\left(\frac{u_{\max}}{4}\boldsymbol{P}_{\text{on}} + u_{y+}\left(\boldsymbol{P}_{\text{off}} - \boldsymbol{P}_{\text{on}}\right)\right)\cdot\hat{\boldsymbol{n}}r_{eq} \end{aligned} \tag{6.3}$$

where $u_{\max}$ is the maximum value of u, and

$$\begin{aligned} \boldsymbol{P}_{\text{off}} &= P\left(\hat{\boldsymbol{n}}_S\cdot\hat{\boldsymbol{n}}\right)\left[\left(c_{a_{\text{off}}}+c_{d_{\text{off}}}\right)\hat{\boldsymbol{n}}_S + \left(\frac{2}{3}c_{d_{\text{off}}} + 2c_{s_{\text{off}}}\left(\hat{\boldsymbol{n}}_S\cdot\hat{\boldsymbol{n}}\right)\right)\hat{\boldsymbol{n}}\right] \\ \boldsymbol{P}_{\text{on}} &= P\left(\hat{\boldsymbol{n}}_S\cdot\hat{\boldsymbol{n}}\right)\left[\left(c_{a_{\text{on}}}+c_{d_{\text{on}}}\right)\hat{\boldsymbol{n}}_S + \left(\frac{2}{3}c_{d_{\text{on}}} + 2c_{s_{\text{on}}}\left(\hat{\boldsymbol{n}}_S\cdot\hat{\boldsymbol{n}}\right)\right)\hat{\boldsymbol{n}}\right] \\ r_{eq} &= \frac{4}{u_{\max}}\int_{-\frac{\pi}{4}}^{\frac{\pi}{4}}\int_{r_b}^{r_c} r^2\cos\theta\,\mathrm{d}r\,\mathrm{d}\theta = \frac{4\sqrt{2}(r_c^3 - r_b^3)}{3u_{\max}} \end{aligned} \tag{6.4}$$

Similarly, the control torque exerted about the $\hat{z}_n$ axis in frame $\mathcal{T}_N$ due to the SRP force on part A_3 is

$$T_{z-} = \left(\frac{u_{\max}}{4}\boldsymbol{P}_{\text{on}} + u_{y-}\left(\boldsymbol{P}_{\text{off}} - \boldsymbol{P}_{\text{on}}\right)\right)\cdot\hat{\boldsymbol{n}}r_{eq} \tag{6.5}$$

Thus, the total control torque exerted about the $\hat{z}_n$ axis in frame $\mathcal{T}_N$ is

$$T_z = T_{z+} + T_{z-} = -u_y\left(\boldsymbol{P}_{\text{off}} - \boldsymbol{P}_{\text{on}}\right)\cdot\hat{\boldsymbol{n}}r_{eq} \tag{6.6}$$

where $u_y = u_{y+} - u_{y-}$.

In the same way, the total control torque exerted about the $\hat{y}_n$ axis in frame $\mathcal{T}_N$ can be expressed as

$$T_y = T_{y+} + T_{y-} = u_z\left(\boldsymbol{P}_{\text{off}} - \boldsymbol{P}_{\text{on}}\right)\cdot\hat{\boldsymbol{n}}r_{eq} \tag{6.7}$$

where $u_z = u_{z+} - u_{z-}$.

Bearing Eqs. (6.6) and (6.7) in mind, the control torque T_z is independent of u_z and the control torque T_y is independent of u_y. This is because A_1 and A_3 are split into half by the $\hat{y}_n$ axis while A_2 and A_4 are split into half by the $\hat{z}_n$ axis and the switched-off EMPs are distributed evenly in these four parts [13].

From Eqs. (6.2), (6.6) and (6.7), it can be seen that u is the orbit control variable whereas u_y and u_z are the attitude control variables. These three parameters are constrained as [13]

$$
\begin{aligned}
&0 \le u \le u_{\max} \\
&-\frac{u_{\max}}{4} \le u_y \le \frac{u_{\max}}{4} \\
&-\frac{u_{\max}}{4} \le u_z \le \frac{u_{\max}}{4} \\
&|u_y| + |u_z| \le \min(u, u_{\max} - u)
\end{aligned}
\tag{6.8}
$$

Equation (6.8) indicates that the value of u_y can vary between the maximum value $u_{\max}/4$ and the minimum value $-u_{\max}/4$. When all EMPs in part A_1 are switched off and all EMPs in part A_3 are switched on, i.e., $u_{y+} = u_{\max}/4$ and $u_{y-} = 0$, the value of u_y equals $u_{\max}/4$. Contrarily, when all EMPs in part A_1 are switched on and all EMPs in part A_3 are switched off, i.e., $u_{y+} = 0$ and $u_{y-} = u_{\max}/4$, the value of u_y equals $-u_{\max}/4$. The same holds for u_z. When $u \le u_{\max}/2$, the sum of the absolute values of u_y and u_z is constrained by u. When $u > u_{\max}/2$, the sum is constrained by $(u_{\max} - u)$ which is the area ratio of the switched-on RCDs to the total area of the sail membrane.

Assuming that the frames $\mathcal{T}_N$ and $\mathcal{T}_B$ coincide with each other at the initial time, the control torques in frame $\mathcal{T}_B$ can be expressed as [13]

$$
\begin{bmatrix} T_{by} \\ T_{bz} \end{bmatrix} = \begin{bmatrix} \cos(\varphi) & \sin(\varphi) \\ -\sin(\varphi) & \cos(\varphi) \end{bmatrix} \begin{bmatrix} T_y \\ T_z \end{bmatrix}
\tag{6.9}
$$

where $\varphi = \omega t$ is the roll angle of the spinning solar sail, ω is the spin rate, and t is the time.

Idealized diffuse reflection and absorption models of EMPs are often studied to control a solar sail [14, 15]. In the diffuse reflection model, EMPs can be switched between specular reflection and diffuse reflection states; whereas in the absorption model, EMPs can be switched between specular reflection and absorption states. According to [16], the absorption model outperforms the diffuse reflection model in both orbit control and attitude control. Hence, only the absorption model is discussed in this chapter, so $c_{d_{\text{off}}} = c_{s_{\text{off}}} = c_{a_{\text{on}}} = c_{d_{\text{on}}} = 0$ and $c_{a_{\text{off}}} = c_{s_{\text{on}}} = 1$.

6.3 Equations of Motion

6.3.1 Orbit Dynamics of a Solar Sail in the Earth-Moon System

The synodic reference frame $\mathcal{T}_R(O_r;\ \hat{\boldsymbol{x}}_r, \hat{\boldsymbol{y}}_r, \hat{\boldsymbol{z}}_r)$ in Fig. 6.2 is used to describe the orbit motion of a solar sail in the Earth-Moon system under the framework of the CRTBP. When the SRP acceleration is included into the CRTBP, the equations of motion (see Eq. 3.16) are

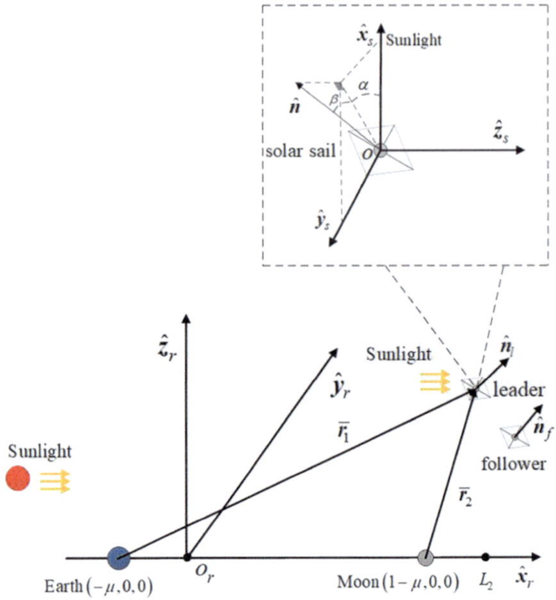

Fig. 6.2 Schematic of solar sail formation flying in the Earth-Moon system

$$
\begin{aligned}
\ddot{x} - 2\dot{y} &= \frac{\partial \Omega}{\partial x} + a_x \\
\ddot{y} + 2\dot{x} &= \frac{\partial \Omega}{\partial y} + a_y \\
\ddot{z} &= \frac{\partial \Omega}{\partial z} + a_z
\end{aligned}
\tag{6.10}
$$

where $[\boldsymbol{a}]_{\mathcal{T}_R} = [a_x, a_y, a_z]^{\mathrm{T}}$ is the adimensional SRP acceleration expressed in frame $\mathcal{T}_R$.

For convenience, a sunlight reference frame $\mathcal{T}_S(O_s;\ \hat{\boldsymbol{x}}_s, \hat{\boldsymbol{y}}_s, \hat{\boldsymbol{z}}_s)$ in Fig. 6.2 is introduced to describe the SRP acceleration exerted on the sail. The origin O_s of the frame is located at the center of the sail, the $\hat{\boldsymbol{x}}_s$ axis is along the Sun-sail direction, the $\hat{\boldsymbol{z}}_s$ axis is in the same direction as the $\hat{\boldsymbol{z}}_r$ axis, and the $\hat{\boldsymbol{y}}_s$ axis satisfies the right-handed rule. Thus, the transformation matrix L_{sr} from frame $\mathcal{T}_R$ to frame $\mathcal{T}_S$ is

$$
L_{sr} = L_z(-\omega_S \tau) = \begin{bmatrix} \cos(\omega_S \tau) & -\sin(\omega_S \tau) & 0 \\ \sin(\omega_S \tau) & \cos(\omega_S \tau) & 0 \\ 0 & 0 & 1 \end{bmatrix}
\tag{6.11}
$$

where $\omega_S = 0.9252$ is the adimensional frequency of the periodic motion of the Sun. Since the motion of the Sun is clockwise in frame $\mathcal{T}_R$, there is a minus-sign "–" in Eq. (6.11).

Denoting by L_x, L_y, and L_z the rotation matrices about the $\hat{\boldsymbol{x}}$, $\hat{\boldsymbol{y}}$, and $\hat{\boldsymbol{z}}$ axes, respectively, we have

$$L_x(\gamma) = \begin{bmatrix} 1 & 0 & 0 \\ 0 & \cos\gamma & \sin\gamma \\ 0 & -\sin\gamma & \cos\gamma \end{bmatrix}, \quad L_y(\gamma) = \begin{bmatrix} \cos\gamma & 0 & -\sin\gamma \\ 0 & 1 & 0 \\ \sin\gamma & 0 & \cos\gamma \end{bmatrix}, \quad L_z(\gamma) = \begin{bmatrix} \cos\gamma & \sin\gamma & 0 \\ -\sin\gamma & \cos\gamma & 0 \\ 0 & 0 & 1 \end{bmatrix} \tag{6.12}$$

where γ means the rotation angle about the axes. Therefore, with the $3-2-1$ Euler angles, the transformation matrix L_{bs} from $\mathcal{T}_S$ to $\mathcal{T}_B$ is

$$\begin{aligned} L_{bs} &= L_x(\varphi)\, L_y(\beta)\, L_z(\alpha) \\ &= \begin{bmatrix} \cos\alpha\cos\beta & \sin\alpha\cos\beta & -\sin\beta \\ a & b & \cos\beta\sin\varphi \\ c & d & \cos\beta\cos\varphi \end{bmatrix} \end{aligned} \tag{6.13}$$

with

$$\begin{aligned} a &= \cos\alpha\sin\beta\sin\varphi - \sin\alpha\cos\varphi \\ b &= \sin\alpha\sin\beta\sin\varphi + \cos\alpha\cos\varphi \\ c &= \cos\alpha\sin\beta\cos\varphi + \sin\alpha\sin\varphi \\ d &= \sin\alpha\sin\beta\cos\varphi - \cos\alpha\sin\varphi \end{aligned} \tag{6.14}$$

where α and β are the yaw and pitch angles of the solar sail as shown in Fig. 6.2, and φ is the roll angle in Eq. (6.9).

The Sun-sail unit vector $\hat{\boldsymbol{n}}_S$ in frame $\mathcal{T}_S$ can be written as

$$\left[\hat{\boldsymbol{n}}_s\right]_{\mathcal{T}_S} = [1, \quad 0, \quad 0]^{\mathrm{T}} \tag{6.15}$$

The sail normal unit vector $\hat{\boldsymbol{n}}$ in frame $\mathcal{T}_B$ is

$$\left[\hat{\boldsymbol{n}}\right]_{\mathcal{T}_B} = [1, \quad 0, \quad 0]^{\mathrm{T}} \tag{6.16}$$

Hence, $\hat{\boldsymbol{n}}$ in frame $\mathcal{T}_S$ can be represented as

$$\left[\hat{\boldsymbol{n}}\right]_{\mathcal{T}_S} = L_{sb} \begin{bmatrix} 1 \\ 0 \\ 0 \end{bmatrix} = L_{bs}^{\mathrm{T}} \begin{bmatrix} 1 \\ 0 \\ 0 \end{bmatrix} = \begin{bmatrix} \cos\alpha\cos\beta \\ \sin\alpha\cos\beta \\ -\sin\beta \end{bmatrix} \tag{6.17}$$

Substituting Eqs. (6.15)–(6.17) into Eq. (6.2) and dividing by the sail's mass, the expression of $\boldsymbol{a}$ in frame $\mathcal{T}_S$ is

$$[\boldsymbol{a}]_{\mathcal{T}_S} = \frac{\bar{\kappa}}{2} \begin{bmatrix} 2(1-u)\cos^3\alpha\cos^3\beta + u\cos\alpha\cos\beta \\ 2(1-u)\cos^2\alpha\sin\alpha\cos^3\beta \\ -2(1-u)\cos^2\alpha\sin\beta\cos^2\beta \end{bmatrix} \tag{6.18}$$

where $\bar{\kappa} = 2\bar{P}\bar{A}/\bar{m}$ is the adimensional characteristic acceleration of the solar sail which is assumed to be a constant in this chapter because of an assumed uniform SRP throughout the Earth-Moon system; $\bar{P}$, $\bar{A}$, and $\bar{m}$ are the adimensional values of P, A, and the sail's mass m, respectively.

Since the orbit motion of the solar sail is formulated in frame $\mathcal{T}_R$, it is useful to express $\boldsymbol{a}$ in that frame as

$$[\boldsymbol{a}]_{\mathcal{T}_R} = L_{rs}\,[\boldsymbol{a}]_{\mathcal{T}_S} \tag{6.19}$$

In [5], families of solar-sail resonant halo orbits have been designed under four sail steering laws and two initial Sun-sail configurations. These orbits, parameterized by the area ratio of the switched-off EMPs, are generated by perturbing resonant libration point orbits using the parallel shooting technique and continuation method. In this chapter, the solar-sail resonant halo orbits under the in-plane Sun-sail line steering law are taken as the nominal orbits for the leader, so

$$\hat{\boldsymbol{n}} = \hat{\boldsymbol{n}}_S \tag{6.20}$$

which means $\alpha_l = 0$ and $\beta_l = 0$, and the subscript l denotes the leader.

Figure 6.3 shows three solar-sail resonant halo orbits around the L_2 point in case of $\bar{\kappa} = 0.05$. The corresponding non-zero initial adimensional coordinates are given in Table 6.1. These orbits have the same period $T = 2\pi/\omega_S$ and it is equal to that of the orbital motion of the Sun around the Earth-Moon barycenter. The symbol "×" in Fig. 6.3 denotes the position of the solar sail at the initial time which is located in the $\hat{\boldsymbol{x}}_r - \hat{\boldsymbol{z}}_r$ plane and above the L_2 point. Because these orbits are always behind the Moon, they can be used for lunar far-side communication and observation [4, 5].

Figure 6.4 gives the value of the largest eigenvalue, $\|\lambda\|_{\max}$, of the monodromy matrix as the function of u for the family of solar-sail resonant halo orbits in Fig. 6.3. It can be seen that all of the orbits in the family are unstable and the instability increases with increasing u.

6.3.2 Attitude Dynamics of a Solar Sail in the Earth-Moon System

To describe the attitude motion of a solar-sail spacecraft in the Earth-Moon CRTBP, an inertial coordinate frame $\mathcal{T}_I(O_i;\ \hat{\boldsymbol{x}}_i, \hat{\boldsymbol{y}}_i, \hat{\boldsymbol{z}}_i)$ is defined. The origin O_i and the $\hat{\boldsymbol{z}}_i$ axis of frame $\mathcal{T}_I$ coincide with those of frame $\mathcal{T}_R$ and the $\hat{\boldsymbol{x}}_i$ axis is initially aligned with the $\hat{\boldsymbol{x}}_r$ axis. Hence, the frames $\mathcal{T}_I$ and $\mathcal{T}_R$ coincide with each other at the initial time. Denoting $\boldsymbol{\omega}$ as the adimensional angular velocity vector of the solar sail relative to frame $\mathcal{T}_I$, the following relationship holds

$$\boldsymbol{\omega} = \boldsymbol{\omega}_{bs} + \boldsymbol{\omega}_{sr} + \boldsymbol{\omega}_{ri} \tag{6.21}$$

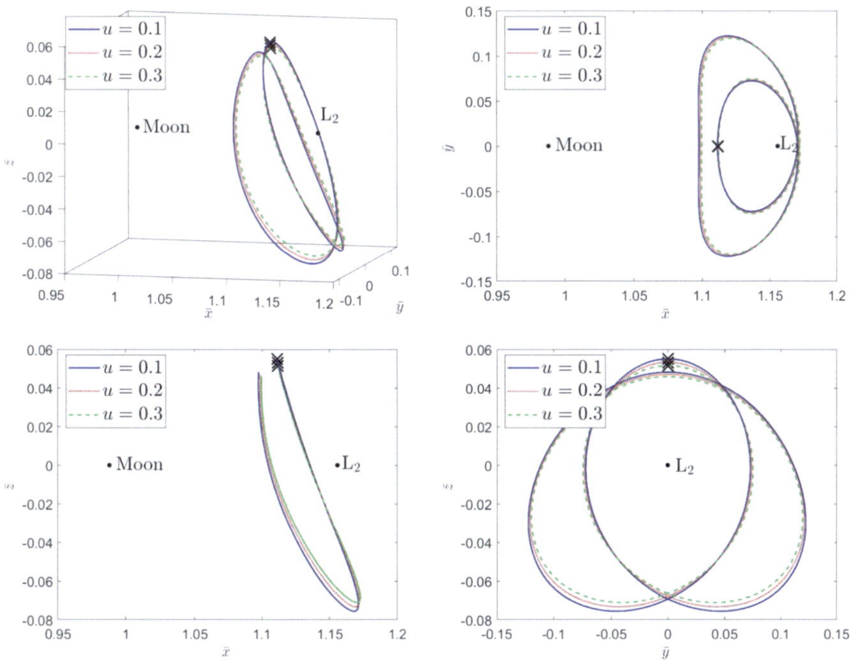

Fig. 6.3 Three solar-sail resonant halo orbits around the L_2 point with $\bar{\kappa} = 0.05$ in the Earth-Moon CRTBP. Top-Left: 3D projection of the orbits in the $\mathcal{T}_R(O_r;\ \hat{\boldsymbol{x}}_r, \hat{\boldsymbol{y}}_r, \hat{\boldsymbol{z}}_r)$ frame; Top-Right: $\hat{\boldsymbol{x}}_r - \hat{\boldsymbol{y}}_r$ projection of the orbits; Bottom-Left: $\hat{\boldsymbol{x}}_r - \hat{\boldsymbol{z}}_r$ projection of the orbits; Bottom-Right: $\hat{\boldsymbol{y}}_r - \hat{\boldsymbol{z}}_r$ projection of the orbits

Table 6.1 Non-zero initial adimensional coordinates and parameter u for the resonant halo orbits in Fig. 6.3

u	x_0	z_0	y'_0
0.1	1.11121696522881	0.0550781348018197	0.168348903548823
0.2	1.11157902884427	0.0533115657668458	0.169842125229864
0.3	1.11190112558243	0.0515960853275869	0.171356234075843

where $\boldsymbol{\omega}_{bs}$ is the adimensional angular velocity vector of frame $\mathcal{T}_B$ relative to frame $\mathcal{T}_S$; $\boldsymbol{\omega}_{sr}$ is the adimensional angular velocity vector of frame $\mathcal{T}_S$ relative to frame $\mathcal{T}_R$; $\boldsymbol{\omega}_{ri}$ is the adimensional angular velocity vector of frame $\mathcal{T}_R$ relative to frame $\mathcal{T}_I$.

With the $3-2-1$ rotational sequence from frame $\mathcal{T}_S$ to frame $\mathcal{T}_B$, Eq. (6.21) can be expressed as

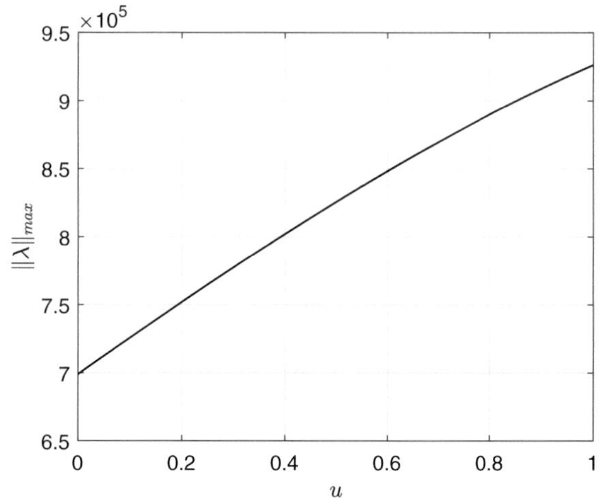

Fig. 6.4 The norm value of the largest eigenvalue of the monodromy matrix

$$\begin{bmatrix} \omega_x \\ \omega_y \\ \omega_z \end{bmatrix} = L_x(\varphi) L_y(\beta) \begin{bmatrix} 0 \\ 0 \\ \dot{\alpha} \end{bmatrix} + L_x(\varphi) \begin{bmatrix} 0 \\ \dot{\beta} \\ 0 \end{bmatrix} + \begin{bmatrix} \dot{\varphi} \\ 0 \\ 0 \end{bmatrix} + L_{bs} \begin{bmatrix} 0 \\ 0 \\ -\omega_S \end{bmatrix} + L_{br} \begin{bmatrix} 0 \\ 0 \\ 1 \end{bmatrix} \tag{6.22}$$

$$= \begin{bmatrix} \dot{\varphi} - \dot{\alpha} \sin\beta - \sin\beta(1-\omega_S) \\ \dot{\beta}\cos\varphi + \dot{\alpha}\sin\varphi\cos\beta + \sin\varphi\cos\beta(1-\omega_S) \\ -\dot{\beta}\sin\varphi + \dot{\alpha}\cos\varphi\cos\beta + \cos\varphi\cos\beta(1-\omega_S) \end{bmatrix} \tag{6.23}$$

where $[\omega]_{\mathcal{T}_B} = [\omega_x, \omega_y, \omega_z]^{\mathrm{T}}$ is the expression of $\boldsymbol{\omega}$ in frame $\mathcal{T}_B$ and $L_{br} = L_{bs} L_{sr}$ is the transformation matrix from frame $\mathcal{T}_R$ to frame $\mathcal{T}_B$.

Hence, the adimensional attitude kinematic equations can be written as

$$\begin{aligned} \dot{\varphi} &= \omega_x + \tan\beta\sin\varphi\omega_y + \tan\beta\cos\varphi\omega_z \\ \dot{\beta} &= \cos\varphi\omega_y - \sin\varphi\omega_z \\ \dot{\alpha} &= \sec\beta\sin\varphi\omega_y + \sec\beta\cos\varphi\omega_z - (1-\omega_S) \end{aligned} \tag{6.24}$$

With the in-plane Sun-sail line steering law, $\alpha_l = 0$ and $\beta_l = 0$ for the leader. Assuming that the formation flying missions require the follower to point accurately to the leader, the attitude angles of the follower α_f and β_f should be small quantities where the subscript f denotes the follower. Thus, for both the leader and the follower Eq. (6.24) can be simplified as

$$\begin{aligned} \dot{\varphi} &= \omega_x = \omega \\ \dot{\beta} &= \cos\varphi\omega_y - \sin\varphi\omega_z \\ \dot{\alpha} &= \sin\varphi\omega_y + \cos\varphi\omega_z - (1-\omega_S) \end{aligned} \tag{6.25}$$

Hence, the solutions for ω_y and ω_z are

$$\begin{aligned}\omega_y &= \dot{\beta}\cos\varphi + \sin\varphi\,(1-\omega_S+\dot{\alpha})\\ \omega_z &= -\dot{\beta}\sin\varphi + \cos\varphi\,(1-\omega_S+\dot{\alpha})\end{aligned} \tag{6.26}$$

For spacecraft moving in the neighborhood of the Earth-Moon libration points, the gravity gradient torques from the Earth and the Moon are much smaller than the control torques [11]. Thus, taking only the control torques into consideration, the attitude dynamics for the axisymmetric spinning solar sail are

$$\begin{aligned}I\dot{\omega}_y - (I - I_x)\,\omega_z\omega &= \tau_{by}\\ I\dot{\omega}_z - (I_x - I)\,\omega_y\omega &= \tau_{bz}\end{aligned} \tag{6.27}$$

where I_x and I are the adimensional moments of inertia about the spin axis and the other two axes of the sail, respectively; τ_{by} and τ_{bz} are the adimensional alternatives of T_{by} and T_{bz}, respectively.

Because α_l and β_l are zero for the leader and α_f and β_f are small quantities for the follower, Eq. (6.6) can be reduced to

$$\begin{aligned}T_z &= -u_y\,(\boldsymbol{P}_{\mathrm{off}} - \boldsymbol{P}_{\mathrm{on}})\cdot\hat{\boldsymbol{n}}r_{eq}\\ &= -u_y\left(P(\hat{\boldsymbol{n}}_S\cdot\hat{\boldsymbol{n}})\hat{\boldsymbol{n}}_S - 2P(\hat{\boldsymbol{n}}_S\cdot\hat{\boldsymbol{n}})^2\hat{\boldsymbol{n}}\right)\cdot\hat{\boldsymbol{n}}r_{eq}\\ &= u_yP\left(\hat{\boldsymbol{n}}_S\cdot\hat{\boldsymbol{n}}\right)^2 r_{eq}\\ &= u_yP\cos^2\alpha\cos^2\beta r_{eq}\\ &= u_yPr_{eq}\end{aligned} \tag{6.28}$$

Similarly, Eq. (6.7) can be written as

$$T_y = -u_zPr_{eq} \tag{6.29}$$

Hence, the adimensional values of the control torques T_z and T_y are

$$\begin{aligned}\tau_z &= u_y\bar{P}\bar{r}_{eq}\\ \tau_y &= -u_z\bar{P}\bar{r}_{eq}\end{aligned} \tag{6.30}$$

Substituting Eqs. (6.26) into (6.27) and using Eqs. (6.9) and (6.30), the attitude equations of motion becomes

$$\begin{bmatrix}\ddot{\alpha}\\ \ddot{\beta}\end{bmatrix} + \frac{I_x\omega}{I}\begin{bmatrix}0 & -1\\ 1 & 0\end{bmatrix}\begin{bmatrix}\dot{\alpha}+1-\omega_S\\ \dot{\beta}\end{bmatrix} = \frac{\bar{P}\bar{r}_{eq}}{I}\begin{bmatrix}1 & 0\\ 0 & -1\end{bmatrix}\begin{bmatrix}u_y\\ u_z\end{bmatrix} \tag{6.31}$$

6.3.3 Coupled Orbit-Attitude Dynamics of a Solar Sail in the Earth-Moon System

In solar sail formation flying, denote by $[\boldsymbol{r}_l]_{\mathcal{T}_R} = [x_l, y_l, z_l]^{\mathrm{T}}$ the adimensional position vector of the leader and by $[\delta\boldsymbol{r}]_{\mathcal{T}_R} = [\delta x, \delta y, \delta z]^{\mathrm{T}}$ the adimensional relative position vector between the follower and the leader. According to Eq. (6.10), the equations of orbit relative motion in frame $\mathcal{T}_R$ are

$$\begin{aligned}
\delta\ddot{x} &= \delta x + 2\delta\dot{y} - (1-\mu)\left(\frac{x_l + \delta x + \mu}{r_{1f}^3} - \frac{x_l + \mu}{r_{1l}^3}\right) - \mu\left(\frac{x_l + \delta x - 1 + \mu}{r_{2f}^3} - \frac{x_l - 1 + \mu}{r_{2l}^3}\right) + a_{fx} - a_{lx} \\
\delta\ddot{y} &= \delta x - 2\delta\dot{y} - (1-\mu)\left(\frac{y_l + \delta y}{r_{1f}^3} - \frac{y_l}{r_{1l}^3}\right) - \mu\left(\frac{y_l + \delta y}{r_{2f}^3} - \frac{y_l}{r_{2l}^3}\right) + a_{fy} - a_{ly} \\
\delta\ddot{z} &= -(1-\mu)\left(\frac{z_l + \delta z}{r_{1f}^3} - \frac{z_l}{r_{1l}^3}\right) - \mu\left(\frac{z_l + \delta z}{r_{2f}^3} - \frac{z_l}{r_{2l}^3}\right) + a_{fz} - a_{lz}
\end{aligned} \tag{6.32}$$

where $\left[\boldsymbol{a}_{sf}\right]_{\mathcal{T}_R} = [a_{fx}, a_{fy}, a_{fz}]^{\mathrm{T}}$ and $[\boldsymbol{a}_{sl}]_{\mathcal{T}_R} = [a_{lx}, a_{ly}, a_{lz}]^{\mathrm{T}}$ are the adimensional SRP accelerations of the follower and the leader, respectively; the expressions for r_{1l}, r_{2l} and r_{1f}, r_{2f} are

$$\begin{aligned}
r_{1l} &= \sqrt{(x_l + \mu)^2 + y_l^2 + z_l^2} \\
r_{2l} &= \sqrt{(x_l - 1 + \mu)^2 + y_l^2 + z_l^2} \\
r_{1f} &= \sqrt{(x_l + \delta x + \mu)^2 + (y_l + \delta y)^2 + (z_l + \delta z)^2} \\
r_{2f} &= \sqrt{(x_l + \delta x - 1 + \mu)^2 + (y_l + \delta y)^2 + (z_l + \delta z)^2}
\end{aligned} \tag{6.33}$$

In particular, the SRP accelerations exerted on both sails are almost equal. Bearing Eq. (6.18) in mind, the adimensional SRP acceleration of the follower can be linearized around that of the leader as

$$\boldsymbol{a}_f = \boldsymbol{a}_l + \left.\frac{\partial \boldsymbol{a}}{\partial u}\right|_{u=u_l} \delta u + \left.\frac{\partial \boldsymbol{a}}{\partial \alpha}\right|_{\alpha=\alpha_l} \delta\alpha + \left.\frac{\partial \boldsymbol{a}}{\partial \beta}\right|_{\beta=\beta_l} \delta\beta \tag{6.34}$$

where $\delta u = u_f - u_l$, $\delta\alpha = \alpha_f - \alpha_l$ and $\delta\beta = \beta_f - \beta_l$.

Thus, the relative adimensional SRP acceleration in frame $\mathcal{T}_R$ can be approximated as

$$[\delta\boldsymbol{a}]_{\mathcal{T}_R} = L_{rs}\left[\left.\frac{\partial\,[\boldsymbol{a}]_{\mathcal{T}_S}}{\partial u}\right|_{u=u_l} \delta u + \left.\frac{\partial\,[\boldsymbol{a}]_{\mathcal{T}_S}}{\partial \alpha}\right|_{\alpha=\alpha_l} \delta\alpha + \left.\frac{\partial\,[\boldsymbol{a}]_{\mathcal{T}_S}}{\partial \beta}\right|_{\beta=\beta_l} \delta\beta\right] \tag{6.35}$$

where $[\delta\boldsymbol{a}]_{\mathcal{T}_R} = \left[\boldsymbol{a}_f\right]_{\mathcal{T}_R} - [\boldsymbol{a}_l]_{\mathcal{T}_R}$ and

$$\frac{\partial\,[\boldsymbol{a}]_{\mathcal{T}_S}}{\partial u} = \bar{\kappa}\begin{bmatrix} -\cos^3\alpha\cos^3\beta + \cos\alpha\cos\beta \\ -\cos^2\alpha\sin\alpha\cos^3\beta \\ \cos^2\alpha\sin\beta\cos^2\beta \end{bmatrix} \tag{6.36}$$

$$\frac{\partial\,[\boldsymbol{a}]_{\mathcal{T}_S}}{\partial \alpha} = \frac{\bar{\kappa}}{2}\begin{bmatrix} -6\,(1-u)\cos^2\alpha\sin\alpha\cos^3\beta - u\sin\alpha\cos\beta \\ 2\,(1-u)\left(-2\cos\alpha\sin^2\alpha + \cos^3\alpha\right)\cos^3\beta \\ 4\,(1-u)\sin\alpha\cos\alpha\sin\beta\cos^2\beta \end{bmatrix} \tag{6.37}$$

$$\frac{\partial\,[\boldsymbol{a}_s]_{\mathcal{T}_S}}{\partial \beta} = \frac{\bar{\kappa}}{2}\begin{bmatrix} -6\,(1-u)\cos^3\alpha\sin\beta\cos^2\beta - u\cos\alpha\sin\beta \\ -6\,(1-u)\cos^2\alpha\sin\alpha\cos^2\beta\sin\beta \\ -2\,(1-u)\cos^2\alpha\left(\cos^3\beta - 2\cos\beta\sin^2\beta\right) \end{bmatrix} \tag{6.38}$$

Along the nominal trajectory of $[\boldsymbol{r}_l]_{\mathcal{T}_R}$, Eq. (6.32) can be linearized as

$$[\delta\ddot{\boldsymbol{r}}]_{\mathcal{T}_R} = \left.\frac{\partial\nabla\Omega}{\partial\boldsymbol{r}}\right|_{\boldsymbol{r}=\boldsymbol{r}_l}[\delta\boldsymbol{r}]_{\mathcal{T}_R} + E\,[\delta\dot{\boldsymbol{r}}]_{\mathcal{T}_R} + L_{rs}\left[\left.\frac{\partial\,[\boldsymbol{a}]_{\mathcal{T}_S}}{\partial u}\right|_{u=u_l}\delta u + \left.\frac{\partial\,[\boldsymbol{a}]_{\mathcal{T}_S}}{\partial \alpha}\right|_{\alpha=\alpha_l}\delta\alpha + \left.\frac{\partial\,[\boldsymbol{a}]_{\mathcal{T}_S}}{\partial \beta}\right|_{\beta=\beta_l}\delta\beta\right] \tag{6.39}$$

where $\nabla\Omega$ means the gradient of the pseudo-potential Ω, and

$$\boldsymbol{E} = \begin{bmatrix} 0 & 2 & 0 \\ -2 & 0 & 0 \\ 0 & 0 & 0 \end{bmatrix} \tag{6.40}$$

According to Eq. (6.31), the equations of relative attitude motion are

$$\begin{bmatrix} \delta\ddot{\alpha} \\ \delta\ddot{\beta} \end{bmatrix} + \frac{I_x\omega}{I}\begin{bmatrix} 0 & -1 \\ 1 & 0 \end{bmatrix}\begin{bmatrix} \delta\dot{\alpha} \\ \delta\dot{\beta} \end{bmatrix} = \frac{\bar{P}\bar{r}_{eq}}{I}\begin{bmatrix} 1 & 0 \\ 0 & -1 \end{bmatrix}\begin{bmatrix} \delta u_y \\ \delta u_z \end{bmatrix} \tag{6.41}$$

where $\delta\alpha = \alpha_f - \alpha_l$, $\delta\beta = \beta_f - \beta_l$, $\delta u_y = u_{yf} - u_{yl}$, and $\delta u_z = u_{zf} - u_{zl}$.

Combining Eqs. (6.39) and (6.41), the linear state and output equations of relative orbit-attitude motion can be written as

$$\begin{aligned} \dot{\boldsymbol{X}} &= \boldsymbol{A}\boldsymbol{X} + \boldsymbol{B}\boldsymbol{u} \\ \dot{\boldsymbol{Y}} &= \boldsymbol{C}\boldsymbol{X} \end{aligned} \tag{6.42}$$

where $\boldsymbol{X} = \left[\delta x, \delta y, \delta z, \delta\dot{x}, \delta\dot{y}, \delta\dot{z}, \delta\alpha, \delta\beta, \delta\dot{\alpha}, \delta\dot{\beta}\right]^{\mathrm{T}}$, $\boldsymbol{u} = \left[\delta u, \delta u_y, \delta u_z\right]^{\mathrm{T}}$ and $\boldsymbol{Y} = [\delta x, \delta y, \delta z, \delta\dot{x}, \delta\dot{y}, \delta\dot{z}]^{\mathrm{T}}$. The matrices $\boldsymbol{A}$, $\boldsymbol{B}$ and $\boldsymbol{C}$ are expressed as

$$\boldsymbol{A} = \begin{bmatrix} \boldsymbol{O}_{3\times3} & \boldsymbol{I}_{3\times3} & \boldsymbol{O}_{3\times2} & \boldsymbol{O}_{3\times2} \\ \boldsymbol{A}_1 & E & \boldsymbol{A}_2 & \boldsymbol{O}_{3\times2} \\ \boldsymbol{O}_{2\times3} & \boldsymbol{O}_{2\times3} & \boldsymbol{O}_{2\times2} & \boldsymbol{I}_{2\times2} \\ \boldsymbol{O}_{2\times3} & \boldsymbol{O}_{2\times3} & \boldsymbol{O}_{2\times2} & \boldsymbol{A}_3 \end{bmatrix},\quad \boldsymbol{B} = \begin{bmatrix} \boldsymbol{O}_{3\times1} & \boldsymbol{O}_{3\times2} \\ \boldsymbol{B}_1 & \boldsymbol{O}_{3\times2} \\ \boldsymbol{O}_{2\times1} & \boldsymbol{O}_{2\times2} \\ \boldsymbol{O}_{2\times1} & \boldsymbol{B}_2 \end{bmatrix},\quad \boldsymbol{C} = \left[\boldsymbol{I}_{6\times6}\ \boldsymbol{O}_{6\times4}\right] \tag{6.43}$$

where $\boldsymbol{O}$ and $\boldsymbol{I}$ represent the zero matrix and the identity matrix, and

$$\boldsymbol{A}_1 = \left.\frac{\partial \nabla \Omega}{\partial \boldsymbol{r}}\right|_{\boldsymbol{r}=\boldsymbol{r}_l}, \quad \boldsymbol{A}_2 = L_{rs}\left[\left.\frac{[\partial \boldsymbol{a}]_{T_S}}{\partial \alpha}\right|_{\alpha=\alpha_l} \quad \left.\frac{[\partial \boldsymbol{a}]_{T_S}}{\partial \beta}\right|_{\beta=\beta_l}\right], \quad \boldsymbol{B}_3 = -\frac{I_x \omega}{I}\begin{bmatrix} 0 & -1 \\ 1 & 0 \end{bmatrix}, \quad \boldsymbol{B}_1 = \left.\frac{\partial\, [\boldsymbol{a}]_{T_S}}{\partial u}\right|_{u=u_l}, \quad \boldsymbol{B}_2 = \frac{\bar{P}\bar{r}_{eq}}{I}\begin{bmatrix} 1 & 0 \\ 0 & -1 \end{bmatrix} \tag{6.44}$$

As mentioned before, the leader is assumed to follow a nominal solar-sail resonant halo orbit with the in-plane Sun-sail line steering law Eq. (6.20), i.e., $\alpha_l = \beta_l = 0$ and $\dot{\alpha}_l = \dot{\beta}_l = 0$. In addition, the constraints on the relative attitude angles between the follower and the leader and the constraints on the rates of change of follower's attitude angles are defined as

$$\begin{aligned} \alpha_{f_{\min}} &= \delta\alpha_{\min} \le \delta\alpha \le \delta\alpha_{\max} = \alpha_{f_{\max}} \\ \beta_{f_{\min}} &= \delta\beta_{\min} \le \delta\beta \le \delta\beta_{\max} = \beta_{f_{\max}} \\ \dot{\alpha}_{f_{\min}} &= \delta\dot{\alpha}_{\min} \le \delta\dot{\alpha} \le \delta\dot{\alpha}_{\max} = \dot{\alpha}_{f_{\max}} \\ \dot{\beta}_{f_{\min}} &= \delta\dot{\beta}_{\min} \le \delta\dot{\beta} \le \delta\dot{\beta}_{\max} = \dot{\beta}_{f_{\max}} \end{aligned} \tag{6.45}$$

where the subscripts max and min represent maximum and minimum values, respectively.

To reduce electric energy consumption, it is expected to switch on EMP as few as possible because switching on EMP could consume the electric energy [17]. Therefore, the value of $(u_{\max} - u_f)$, which is the area ratio of the switched-on RCD to the total area of the sail membrane, should be minimized as much as possible for the follower. Since $u_f = u_l + \delta u$, the control variable δu should be maximized in the control process.

To this end, bearing Eqs. (6.8), (6.42) and (6.45) in mind, the control objective for the follower is to track a predefined relative trajectory around the leader under the optimal switch sequence for RCD, i.e., u_f, in the presence of constraints on the control variables, the attitude angles and the rates of change of attitude angles. In the next section, the MPC technique will be formulated to solve this constrained optimal control problem.

6.4 MPC-Based Coupled Orbit-Attitude Controller Design

Within the framework of the MPC methodology, zero-order holder is used for discretization in this chapter and the sampling period T_n is set to T_S/N where N means the number of sampling point in one period. In this way, Eq. (6.42) is discretized to

$$\begin{aligned} \boldsymbol{X}_{k+1} &= \boldsymbol{A}_k \boldsymbol{X}_k + \boldsymbol{B}_k \boldsymbol{u}_k \\ \boldsymbol{Y}_k &= \boldsymbol{C}_k \boldsymbol{X}_k \end{aligned} \tag{6.46}$$

where $k \in \mathbb{R}^+$ is the sampling instant; matrices $\boldsymbol{A}_k$, $\boldsymbol{B}_k$ and $\boldsymbol{C}_k$ are constant within each sampling period, but are updated at the beginning of each time interval with the states at the end of previous sampling period.

Denoting N_p and N_c as the prediction horizon and control horizon, the stacks of state variables, control variables, and output variables can be written in a compact form as

$$\begin{aligned} \tilde{\boldsymbol{X}}_k &= \left[\boldsymbol{X}_{k+1,k}^{\mathrm{T}}, \boldsymbol{X}_{k+2,k}^{\mathrm{T}}, \cdots, \boldsymbol{X}_{k+N_p,k}^{\mathrm{T}}\right]^{\mathrm{T}} \\ \tilde{\boldsymbol{u}}_k &= \left[\boldsymbol{u}_{k,k}^{\mathrm{T}}, \boldsymbol{u}_{k+1,k}^{\mathrm{T}}, \cdots, \boldsymbol{u}_{k+N_c-1,k}^{\mathrm{T}}\right]^{\mathrm{T}} \\ \tilde{\boldsymbol{Y}}_k &= \left[\boldsymbol{Y}_{k+1,k}^{\mathrm{T}}, \boldsymbol{Y}_{k+2,k}^{\mathrm{T}}, \cdots, \boldsymbol{Y}_{k+N_p,k}^{\mathrm{T}}\right]^{\mathrm{T}} \end{aligned} \tag{6.47}$$

where the subscript $(k+m, k)$ means the predicted value at the instant $k+m$ with the information at the instant k.

Substituting Eq. (6.47) into Eq. (6.46) and assuming that $\boldsymbol{u}_{k+i,k} = \boldsymbol{u}_{k+N_c-1,k}$, $i \in \left[N_c, N_p - 1\right]$, the following relationships hold

$$\begin{aligned} \tilde{\boldsymbol{X}}_k &= \tilde{\boldsymbol{A}} \boldsymbol{X}_k + \tilde{\boldsymbol{B}} \tilde{\boldsymbol{u}}_k \\ \tilde{\boldsymbol{Y}}_k &= \tilde{\boldsymbol{C}} \tilde{\boldsymbol{X}}_k \end{aligned} \tag{6.48}$$

with

$$\tilde{A} = \begin{bmatrix} \boldsymbol{A}_k \\ \boldsymbol{A}_{k+1}\boldsymbol{A}_k \\ \vdots \\ \prod_{j=0}^{N_p-1} \boldsymbol{A}_{k+j} \end{bmatrix}, \quad \tilde{\boldsymbol{B}} = \begin{bmatrix} \boldsymbol{B}_k & \boldsymbol{O} & \cdots & \boldsymbol{O} \\ \boldsymbol{A}_{k+1}\boldsymbol{B}_k & \boldsymbol{B}_{k+1} & \cdots & \boldsymbol{O} \\ \vdots & \vdots & \ddots & \vdots \\ \prod_{j=1}^{N_c-1} \boldsymbol{A}_{k+j}\boldsymbol{B}_k & \prod_{j=2}^{N_c-1} \boldsymbol{A}_{k+j}\boldsymbol{B}_{k+1} & \vdots & \boldsymbol{B}_{k+N_c-1} \\ \vdots & \vdots & \vdots & \vdots \\ \prod_{j=1}^{N_p-1} \boldsymbol{A}_{k+j}\boldsymbol{B}_k & \prod_{j=2}^{N_p-1} \boldsymbol{A}_{k+j}\boldsymbol{B}_{k+1} & \cdots & \sum_{j=N_c-1}^{N_p-1} \left(\prod_{i=j+1}^{N_p-1} \boldsymbol{A}_{k+i}\right) \boldsymbol{B}_{k+j} \end{bmatrix}, \quad \tilde{\boldsymbol{C}} = \boldsymbol{C} \otimes I_{N_p \times N_p} \tag{6.49}$$

where $\otimes$ is the Kronecker product of two matrices and the zero matrix $\boldsymbol{O}$ is of compatible dimensions.

Furthermore, a cost function is defined as [18, 19]

$$\min J_k = \sum_{i=1}^{N_p} \left(\boldsymbol{Y}_{k+i,k} - \boldsymbol{Y}_{ref_{k+i,k}}\right)^{\mathrm{T}} \boldsymbol{Q} \left(\boldsymbol{Y}_{k+i,k} - \boldsymbol{Y}_{ref_{k+i,k}}\right) - \sum_{i=0}^{N_c-1} \boldsymbol{P}^{\mathrm{T}} \boldsymbol{u}_{k+i,k} \tag{6.50}$$

where $\boldsymbol{Y}_{ref_{k+i,k}}$, $i \in \left[1, N_p\right]$ is the reference relative states at the instant $k+i$ with the information at the instant k; the state weighting matrix $\boldsymbol{Q}$ is a positive-definite state matrix; the control weighting matrix has a form of $\boldsymbol{P} = [p, 0, 0]^{\mathrm{T}}$ with $p \in \mathbb{R}^+$ because only the first element of $\boldsymbol{u}_{k+i,k}$, i.e., $\delta u_{k+i,k}$, is expected to be optimized as mentioned before.

Similar to Eqs. (6.48), (6.50) can be rewritten in a compact form as

$$\min J_k = \left(\tilde{\boldsymbol{Y}}_k - \tilde{\boldsymbol{Y}}_{ref_k}\right)^{\mathrm{T}} \tilde{\boldsymbol{Q}} \left(\tilde{\boldsymbol{Y}}_k - \tilde{\boldsymbol{Y}}_{ref_k}\right) - \tilde{\boldsymbol{P}}^{\mathrm{T}} \tilde{\boldsymbol{u}}_k \tag{6.51}$$

with

$$\tilde{\boldsymbol{Y}}_{ref_k} = \left[\boldsymbol{Y}^{\mathrm{T}}_{ref_{k+1,k}}, \boldsymbol{Y}^{\mathrm{T}}_{ref_{k+2,k}}, \cdots, \boldsymbol{Y}^{\mathrm{T}}_{ref_{k+N_p,k}}\right]^{\mathrm{T}}, \quad \tilde{\boldsymbol{Q}} = \boldsymbol{Q} \otimes I_{N_p \times N_p}, \quad \tilde{\boldsymbol{P}} = \boldsymbol{P} \otimes I_{N_c \times 1} \tag{6.52}$$

Bearing in mind the definitions of the control variables and Eq. (6.8), the constraints on the control variables can be formulated in compact forms as

$$\begin{bmatrix} \tilde{\boldsymbol{I}} \\ -\tilde{\boldsymbol{I}} \end{bmatrix} \tilde{\boldsymbol{u}}_k \le \begin{bmatrix} \tilde{\boldsymbol{u}}_{\max} \\ -\tilde{\boldsymbol{u}}_{\min} \end{bmatrix}, \quad \begin{bmatrix} \tilde{\boldsymbol{L}}_1 \\ -\tilde{\boldsymbol{L}}_2 \end{bmatrix} \tilde{\boldsymbol{u}}_k \le \begin{bmatrix} \tilde{\boldsymbol{u}}_1 \\ -\tilde{\boldsymbol{u}}_2 \end{bmatrix} \tag{6.53}$$

with

$$\tilde{\boldsymbol{I}} = \boldsymbol{I}_{3\times 3} \otimes \boldsymbol{I}_{N_c \times N_c} \tag{6.54}$$

$$\tilde{\boldsymbol{L}}_1 = \boldsymbol{L}_1 \otimes \boldsymbol{I}_{N_c \times N_c}, \quad \tilde{\boldsymbol{L}}_2 = \boldsymbol{L}_2 \otimes \boldsymbol{I}_{N_c \times N_c} \tag{6.55}$$

$$\tilde{\boldsymbol{u}}_{\max} = \boldsymbol{u}_{\max} \otimes \boldsymbol{I}_{N_c \times 1}, \quad \tilde{\boldsymbol{u}}_{\min} = \boldsymbol{u}_{\min} \otimes \boldsymbol{I}_{N_c \times 1} \tag{6.56}$$

$$\boldsymbol{u}_{\max} = \left[u_{\max} - u_l, \frac{u_{\max}}{4} - u_{yl}, \frac{u_{\max}}{4} - u_{zl}\right]^{\mathrm{T}}, \quad \boldsymbol{u}_{\min} = \left[-u_{\min}, -\frac{u_{\min}}{4} - u_{yl}, -\frac{u_{\min}}{4} - u_{zl}\right]^{\mathrm{T}} \tag{6.57}$$

$$\tilde{\boldsymbol{u}}_1 = \boldsymbol{u}_1 \otimes \boldsymbol{I}_{N_c \times 1}, \quad \tilde{\boldsymbol{u}}_2 = \boldsymbol{u}_2 \otimes \boldsymbol{I}_{N_c \times 1} \tag{6.58}$$

$$\boldsymbol{u}_1 = \begin{bmatrix} u_l - u_{yl} - u_{zl} \\ u_l - u_{yl} + u_{zl} \\ u_l + u_{yl} - u_{zl} \\ u_l + u_{yl} + u_{zl} \end{bmatrix}, \quad \boldsymbol{u}_2 = \begin{bmatrix} u_{\max} - u_l - u_{yl} - u_{zl} \\ u_{\max} - u_l - u_{yl} + u_{zl} \\ u_{\max} - u_l + u_{yl} - u_{zl} \\ u_{\max} - u_l + u_{yl} + u_{zl} \end{bmatrix} \tag{6.59}$$

$$\boldsymbol{L}_1 = \begin{bmatrix} -1 & 1 & 1 \\ -1 & 1 & -1 \\ -1 & -1 & 1 \\ -1 & -1 & -1 \end{bmatrix}, \quad \boldsymbol{L}_2 = \begin{bmatrix} 1 & 1 & 1 \\ 1 & 1 & -1 \\ 1 & -1 & 1 \\ 1 & -1 & -1 \end{bmatrix} \tag{6.60}$$

In addition, Eq. (6.45) can be rewritten in a compact form as

$$\begin{bmatrix} \tilde{\boldsymbol{C}}_a \\ -\tilde{\boldsymbol{C}}_a \end{bmatrix} \tilde{\boldsymbol{X}}_k \le \begin{bmatrix} \tilde{\boldsymbol{Y}}_{a_{\max}} \\ -\tilde{\boldsymbol{Y}}_{a_{\min}} \end{bmatrix} \tag{6.61}$$

with

$$\begin{aligned} \tilde{\boldsymbol{C}}_a &= \boldsymbol{C}_a \otimes \boldsymbol{I}_{N_p \times N_p} \\ \boldsymbol{C}_a &= \left[\boldsymbol{O}_{4\times 6}\ \boldsymbol{I}_{4\times 4}\right] \\ \tilde{\boldsymbol{Y}}_{a_{\max}} &= \boldsymbol{Y}_{a_{\max}} \otimes \boldsymbol{I}_{N_p \times 1} \\ \tilde{\boldsymbol{Y}}_{a_{\min}} &= \boldsymbol{Y}_{a_{\min}} \otimes \boldsymbol{I}_{N_p \times 1} \\ \boldsymbol{Y}_{a_{\max}} &= \left[\alpha_{f_{\max}}, \beta_{f_{\max}}, \alpha'_{f_{\max}}, \beta'_{f_{\max}}\right]^{\mathrm{T}} \\ \boldsymbol{Y}_{a_{\min}} &= \left[\alpha_{f_{\min}}, \beta_{f_{\min}}, \alpha'_{f_{\min}}, \beta'_{f_{\min}}\right]^{\mathrm{T}} \end{aligned} \tag{6.62}$$

Furthermore, defining an auxiliary variables as $\tilde{\boldsymbol{E}} = \tilde{\boldsymbol{Y}}_{ref_k} - \tilde{\boldsymbol{C}}\tilde{\boldsymbol{A}}\boldsymbol{X}_k$ and then substituting it and Eq. (6.48) into Eq. (6.51) yields

$$\begin{aligned} \min J_k &= \left(\tilde{\boldsymbol{C}}\tilde{\boldsymbol{A}}\boldsymbol{X}_k + \tilde{\boldsymbol{C}}\tilde{\boldsymbol{B}}\tilde{\boldsymbol{u}}_k - \tilde{\boldsymbol{Y}}_{ref_k}\right)^{\mathrm{T}} \tilde{\boldsymbol{Q}} \left(\tilde{\boldsymbol{C}}\tilde{\boldsymbol{A}}\boldsymbol{X}_k + \tilde{\boldsymbol{C}}\tilde{\boldsymbol{B}}\tilde{\boldsymbol{u}}_k - \tilde{\boldsymbol{Y}}_{ref_k}\right) - \tilde{\boldsymbol{P}}^{\mathrm{T}}\tilde{\boldsymbol{u}}_k \\ &= \left(\tilde{\boldsymbol{C}}\tilde{\boldsymbol{B}}\tilde{\boldsymbol{u}}_k - \tilde{\boldsymbol{E}}\right)^{\mathrm{T}} \tilde{\boldsymbol{Q}} \left(\tilde{\boldsymbol{C}}\tilde{\boldsymbol{B}}\tilde{\boldsymbol{u}}_k - \tilde{\boldsymbol{E}}\right) - \tilde{\boldsymbol{P}}^{\mathrm{T}}\tilde{\boldsymbol{u}}_k \\ &= \tilde{\boldsymbol{u}}_k^{\mathrm{T}}\tilde{\boldsymbol{B}}^{\mathrm{T}}\tilde{\boldsymbol{C}}^{\mathrm{T}}\tilde{\boldsymbol{Q}}\tilde{\boldsymbol{C}}\tilde{\boldsymbol{B}}\tilde{\boldsymbol{u}}_k - \left(2\tilde{\boldsymbol{E}}^{\mathrm{T}}\tilde{\boldsymbol{Q}}\tilde{\boldsymbol{C}}\tilde{\boldsymbol{B}} + \tilde{\boldsymbol{P}}^{\mathrm{T}}\right)\tilde{\boldsymbol{u}}_k + const \end{aligned} \tag{6.63}$$

with $const = \tilde{\boldsymbol{E}}^{\mathrm{T}}\tilde{\boldsymbol{Q}}\tilde{\boldsymbol{E}}$ a constant and $\tilde{\boldsymbol{B}}^{\mathrm{T}}\tilde{\boldsymbol{C}}^{\mathrm{T}}\tilde{\boldsymbol{Q}}\tilde{\boldsymbol{C}}\tilde{\boldsymbol{B}}$ is the Hessian matrix which is positive definite.

To this end, the original constrained optimal control problem has been converted into a standard convex quadratic programming (QP) problem represented by Eq. (6.63) which can be solved through a standard QP algorithm. Once the stack of control variables $\tilde{\boldsymbol{u}}_k$ is obtained at each sampling instant, only the first control input $\boldsymbol{u}_{k,k}$ will be applied to the control system. Moreover, based on the receding horizon strategy of the MPC methodology, the optimization process keeps shifting forward with the QP problem being solved at each sampling instant. In particular, the area ratio of the switched-off EMPs u_f and attitude angles α_f and β_f of the follower will be calculated at each instant. Note that the adimensional SRP acceleration $\boldsymbol{a}_f$ of the follower is calculated by Eq. (6.18) and is applied to the nonlinear orbit equations of motion Eq. (6.10) directly in numerical simulations.

6.5 Numerical Simulations

In this section, two geometric configurations of solar sail formation flying, where the predefined reference relative trajectories between two sails are time-invariant and time-varying, are first tested to verify the proposed MPC-based coupled orbit-attitude controller using the RCDs as actuators. Then, the controllability of the MPC-based controller is analyzed in a range of solar-sail resonant halo orbits that the leader follows, displacements when the reference relative trajectories are time-invariant, and frequencies when the reference relative trajectories are time-varying.

The parameters of the leader and follower are the same and given in Table 6.2 according to [13, 16]. It can be calculated that $\bar{\kappa} = 0.05$ and $u_{\max} = 0.4$. In addition, the parameters used in MPC-based controller and numerical simulations are given in Table 6.3. The simulation time is set to one orbital period T_S.

Table 6.2 Dimensional solar sail parameters

Parameters	Symbol	Value
SRP	P	4.563×10^{-6}[Pa]
Mass	m	$85\left[\text{kg}\right]$
Inner radius of sail membrane	r_a	1 [m]
Inner radius of RCDs ring	r_b	15.5 [m]
Outer radius of sail membrane	r_c	20 [m]
Moment of inertia	$[I_x, I]$	$[20, 5] \times 10^3 \left[\text{kg} \cdot \text{m}^2\right]$
Spin rate	ω	$0.5\left[\text{rpm}\right]$

Table 6.3 Parameters in the MPC-based controller and numerical simulations

Parameters	Symbol	Value
Number of sampling time	N	4000
Prediction horizon	N_p	50
Control horizon	N_c	30
State weighting matrix	Q	$diag\left(\left[10^{20}, 10^{20}, 10^{20}, 10^{15}, 10^{15}, 10^{15}\right]\right)$
Control weighting matrix	P	$diag\left([1, 1, 1]\right)$
Constraints on the yaw and pitch angles of the follower	$\left[\alpha_{f\max}, \beta_{f\max}, \alpha_{f\min}, \beta_{f\min}\right]$	$[5, 5, -5, -5]\left[\text{deg}\right]$
Constraints on rates of changes of the yaw and pitch angles of the follower	$\left[\dot{\alpha}_{f\max}, \dot{\beta}_{f\max}, \dot{\alpha}_{f\min}, \dot{\beta}_{f\max}\right]$	$[5, 5, -5, -5]\left[\text{deg/h}\right]$
Initial relative position	$[\delta x_0, \delta y_0, \delta z_0]$	[0, 0, 0] [km]
Initial relative velocity	$[\delta \dot{x}_0, \delta \dot{y}_0, \delta \dot{z}_0]$	[0, 0, 0] [km/s]
Initial relative yaw and pitch angles	$[\delta\alpha_0, \delta\beta_0]$	$[0, 0]\left[\text{deg}\right]$
Initial relative rates of changes of the yaw and pitch angles	$\left[\delta\dot{\alpha}_0, \delta\dot{\beta}_0\right]$	$[0, 0]\left[\text{deg/h}\right]$

6.5.1 MPC-Based Coupled Orbit-Attitude Controller for Formation-Keeping

The solar-sail resonant halo orbit with $u_l = 0.2$ in Fig. 6.3 is taken as the nominal orbit for the leader. The dimensional time-invariant and time-varying reference relative trajectories in frame $\mathcal{T}_R$ are assumed to have the following algebraic forms

$$\left[\delta \boldsymbol{r}_{ref}\right]_{\mathcal{T}_R} = [10, 10, 10]^{\mathrm{T}} \text{ km} \tag{6.64}$$

$$\left[\delta \boldsymbol{r}_{ref}\right]_{\mathcal{T}_R} = [10\cos(\omega_r t), 10\sin(\omega_r t), 10\cos(\omega_r t)]^{\mathrm{T}} \text{ km} \tag{6.65}$$

where the frequency of the time-varying reference relative trajectory is set to that of the nominal orbit $\omega_r = 2\pi / T_S$.

The position error between the relative trajectory and the reference one is defined as

$$\boldsymbol{e} = \delta \boldsymbol{r} - \delta \boldsymbol{r}_{ref} \tag{6.66}$$

The tracking results for the time-invariant and time-varying relative trajectories are give in Fig. 6.5 where precise formation-keeping can be achieved in both cases with the proposed MPC-based controller. In particular, the position error along each direction are kept within 0.01 m for the time-invariant case and 20 m for the time-varying case after one orbital period. Besides, the convergent time is within 0.1 times the orbital period in both cases. Figure 6.6 illustrates the time histories of the yaw and pitch angles and their rates of change for the follower in the process of coupled orbit-attitude control. In both cases, the attitude angles and their rates of change are kept within the prescribed ranges which shows the proposed MPC strategy can handle the output constraints in an effective way. The time histories of the control variables are given in Fig. 6.7. Despite of the complex constraints on control variables due to saturation and coupling, the optimal feasible input can be obtained within the framework of the MPC technique. Moreover, the time of fraction that all the RCDs are switched off is 0.13% and on is 0.18% for the follower in case of tracking the time-invariant reference trajectory $[10, 10, 10]^{\mathrm{T}}$ km. As for the time-varying one $[10\cos(\omega_r t), 10\sin(\omega_r t), 10\cos(\omega_r t)]^{\mathrm{T}}$ km, the time of fraction is 0.06% and 0.15%, respectively, where all the RCDs are switched off is and on. Hence, the proposed MPC strategy has strong capability in handling the input constraints. Through the numerical simulations, it can be shown that certain geometric configurations can be maintained for solar sail formation flying in the Earth-Moon system under the proposed MPC-based coupled orbit-attitude controller using the RCDs as actuators.

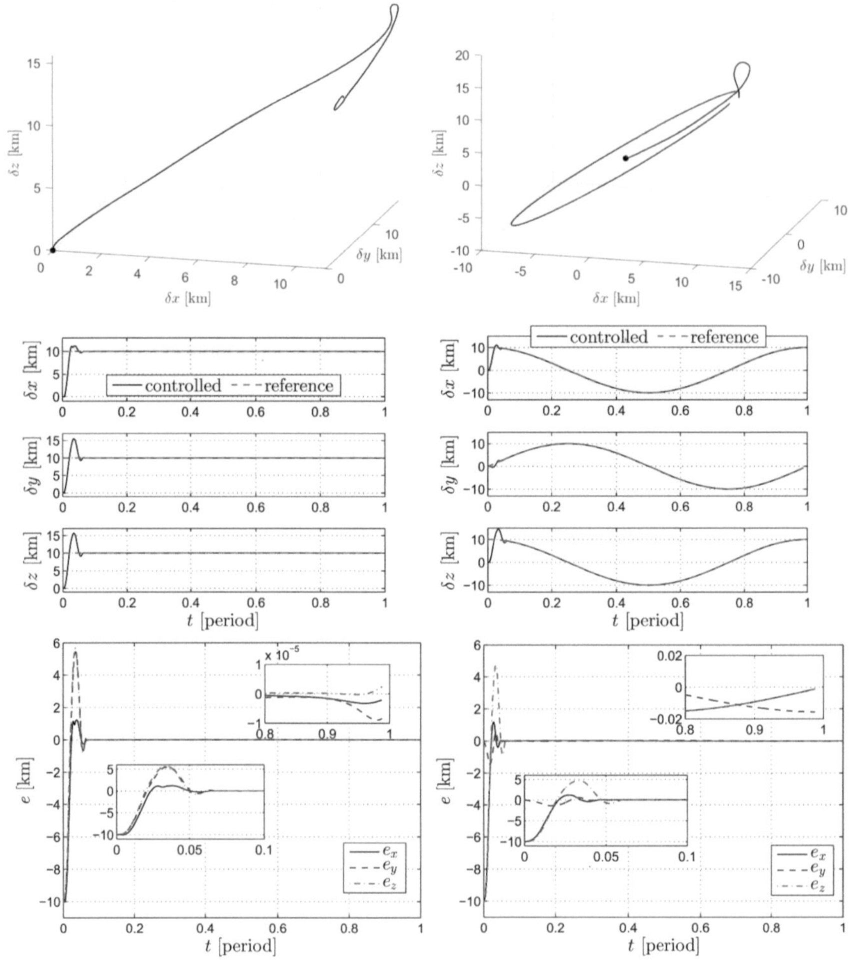

Fig. 6.5 Formation-keeping results for reference relative trajectories. Left: $\left[\delta \boldsymbol{r}_{ref}\right]_{\mathcal{T}_R} = [10, 10, 10]^{\mathrm{T}}$ km; Right: $\left[\delta \boldsymbol{r}_{ref}\right]_{\mathcal{T}_R} = [10\cos(\omega_r t), 10\sin(\omega_r t), 10\cos(\omega_r t)]^{\mathrm{T}}$ km

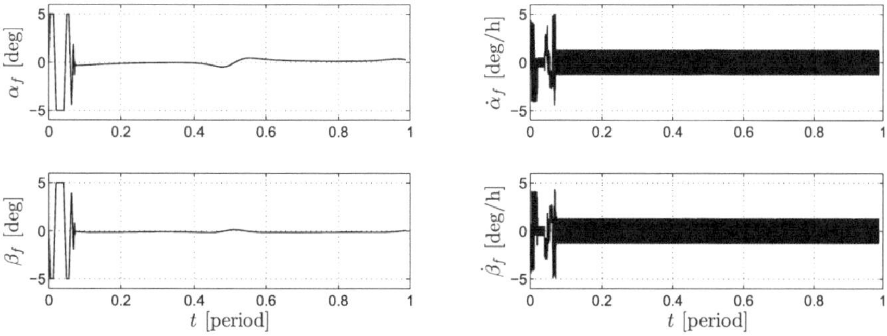

(a) $\left[\delta \boldsymbol{r}_{ref}\right]_{\mathcal{T}_R} = [10, 10, 10]^{\mathrm{T}}$ km

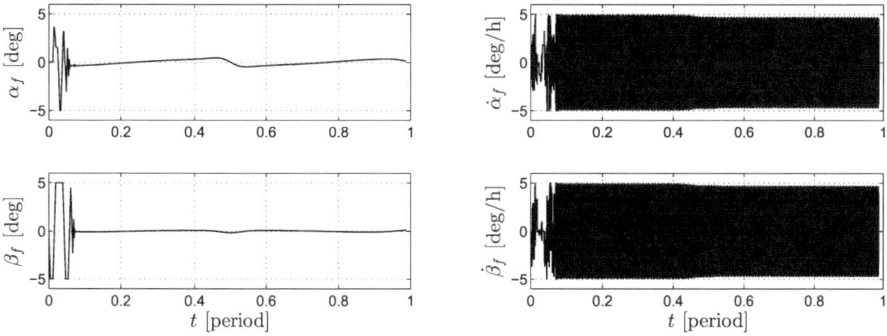

(b) $\left[\delta \boldsymbol{r}_{ref}\right]_{\mathcal{T}_R} = [10\cos(\omega_r t), 10\sin(\omega_r t), 10\cos(\omega_r t)]^{\mathrm{T}}$ km

Fig. 6.6 Attitude angles and their rates of change of the follower corresponding to the formation-keeping cases in Fig. 6.5

6.5.2 Controllability Analysis of the MPC-Based Controller

To demonstrate the effects of nominal orbit and displacement of formation configuration on the controllability of the proposed MPC-based controller, the three solar-sail resonant halo orbits in Fig. 6.3 are taken as the nominal orbits and several different time-invariant reference relative trajectories are tested for each nominal orbit. The tracking results for the three nominal orbits are given in Figs. 6.8, 6.9 and 6.10, respectively. For same nominal orbits, the results indicate that the convergent time increases with the displacement of formation configuration increasing. Once the displacement reaches a threshold, the formation-keeping is not controllable anymore with the control parameters in Table 6.3. In addition, the threshold for the nominal orbit with $u_l = 0.1$ is maximum whereas the one for the nominal orbit with $u_l = 0.3$ is minimum among the three orbits. Therefore, it can be concluded that the controllability of the controller decreases with u of the nominal orbit increasing.

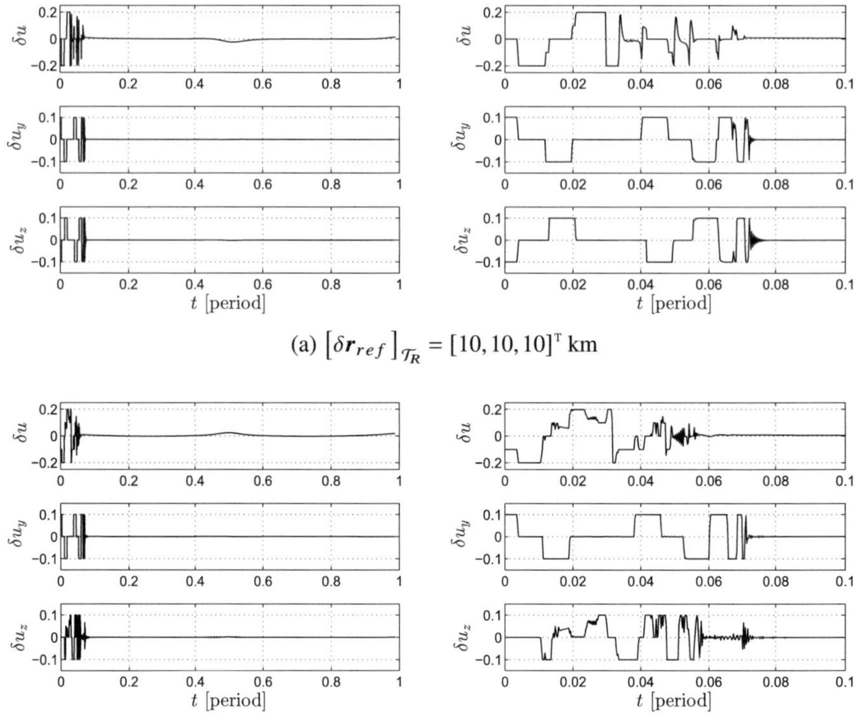

(a) $\left[\delta \boldsymbol{r}_{ref}\right]_{\mathcal{T}_R} = [10, 10, 10]^{\mathrm{T}}$ km

(b) $\left[\delta \boldsymbol{r}_{ref}\right]_{\mathcal{T}_R} = [10\cos(\omega_r t), 10\sin(\omega_r t), 10\cos(\omega_r t)]^{\mathrm{T}}$ km

Fig. 6.7 Control variables corresponding to the formation-keeping cases in Fig. 6.5

This conclusion is coincident with that of Fig. 6.4 because the orbital instability can decrease the controllability of the controller.

When the reference relative trajectory is time-varying, Fig. 6.11 shows the effects of frequency on the controllability of the controller where the displacement is fixed and the solar-sail resonant halo orbit with $u_l = 0.2$ in Fig. 6.3 is taken as the nominal orbit. The results indicate that the proposed MPC-based controller does not work for the time-varying relative trajectory whose frequency exceeds a threshold. Moreover, for the time-varying relative trajectory with low frequency, the tracking precision increases with the frequency decreasing.

In summary, three conclusions can be obtained on the controllability of the proposed MPC-based controller for solar sail formation flying in the Earth-Moon system using the RCDs as actuators: (a) the controllability decreases with the displacement of formation configuration increasing and the formation configuration is not controllable anymore once its displacement exceeds a threshold; (b) the controllability decreases with the instability of the nominal orbit increasing; (c) for time-varying reference relative trajectory, the controllability decreases with frequency increasing.

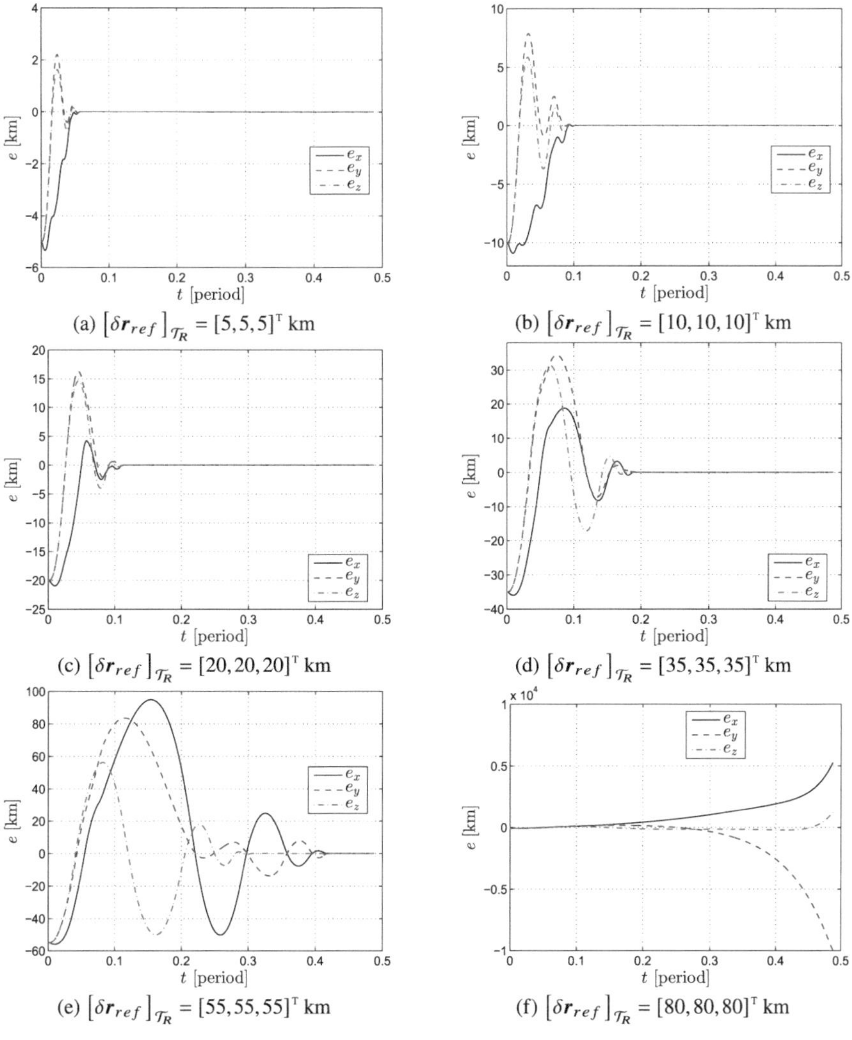

(a) $\left[\delta \boldsymbol{r}_{ref}\right]_{\mathcal{T}_R} = [5, 5, 5]^{\mathrm{T}}$ km
(b) $\left[\delta \boldsymbol{r}_{ref}\right]_{\mathcal{T}_R} = [10, 10, 10]^{\mathrm{T}}$ km
(c) $\left[\delta \boldsymbol{r}_{ref}\right]_{\mathcal{T}_R} = [20, 20, 20]^{\mathrm{T}}$ km
(d) $\left[\delta \boldsymbol{r}_{ref}\right]_{\mathcal{T}_R} = [35, 35, 35]^{\mathrm{T}}$ km
(e) $\left[\delta \boldsymbol{r}_{ref}\right]_{\mathcal{T}_R} = [55, 55, 55]^{\mathrm{T}}$ km
(f) $\left[\delta \boldsymbol{r}_{ref}\right]_{\mathcal{T}_R} = [80, 80, 80]^{\mathrm{T}}$ km

Fig. 6.8 Tracking errors for several different time-invariant $\left[\delta \boldsymbol{r}_{ref}\right]_{\mathcal{T}_R}$ with the solar-sail resonant halo orbit of $u_l = 0.1$ in Fig. 6.3 as the nominal orbit

6.6 Conclusions

The problem of solar sail formation flying around the solar-sail resonant halo orbits in the Earth-Moon system has been investigated in this chapter. The coupled orbit-attitude dynamics have been considered for active joined orbit and attitude control of a spinning solar sail. In particular, the RCDs are used as the actuators for both orbit control and attitude control because of the advantages of a small mass and no fuel consumption. Considering the saturation and coupling of the control variables,

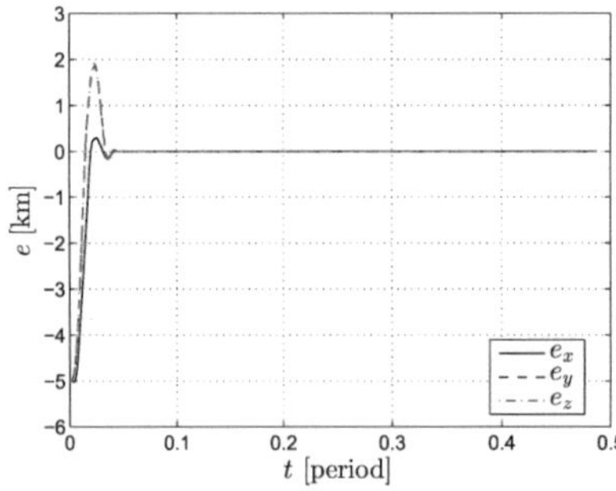

(a) $\left[\delta \boldsymbol{r}_{ref}\right]_{\mathcal{T}_R} = [5,5,5]^{\mathrm{T}}$ km

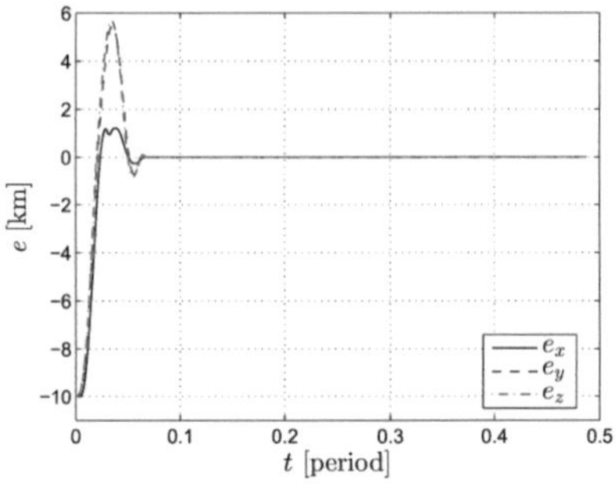

(b) $\left[\delta \boldsymbol{r}_{ref}\right]_{\mathcal{T}_R} = [10,10,10]^{\mathrm{T}}$ km

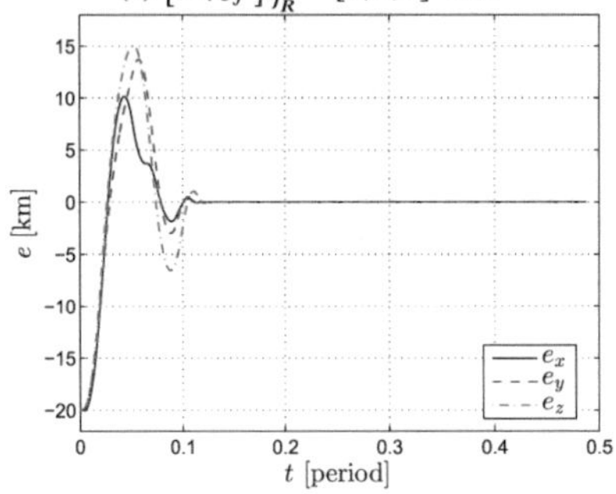

(c) $\left[\delta \boldsymbol{r}_{ref}\right]_{\mathcal{T}_R} = [20,20,20]^{\mathrm{T}}$ km

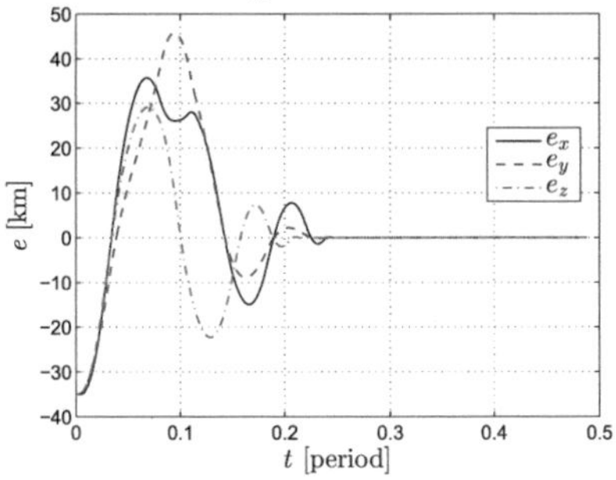

(d) $\left[\delta \boldsymbol{r}_{ref}\right]_{\mathcal{T}_R} = [35,35,35]^{\mathrm{T}}$ km

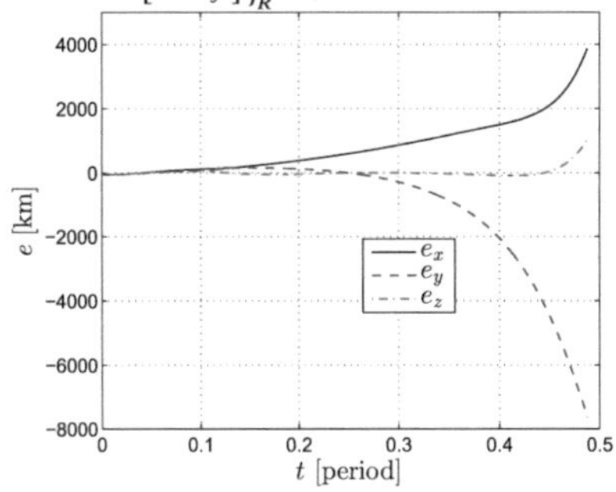

(e) $\left[\delta \boldsymbol{r}_{ref}\right]_{\mathcal{T}_R} = [55,55,55]^{\mathrm{T}}$ km

Fig. 6.9 Tracking errors for several different time-invariant $\left[\delta \boldsymbol{r}_{ref}\right]_{\mathcal{T}_R}$ with the solar-sail resonant halo orbit of $u_l = 0.2$ in Fig. 6.3 as the nominal orbit

constraints on the attitude angles and their rates of change, and the optimal switch regulation of the RCDs, a coupled orbit-attitude controller has been developed within the framework of the model predictive control (MPC) technique. According to the simulation results, the two main conclusions are given in the following:

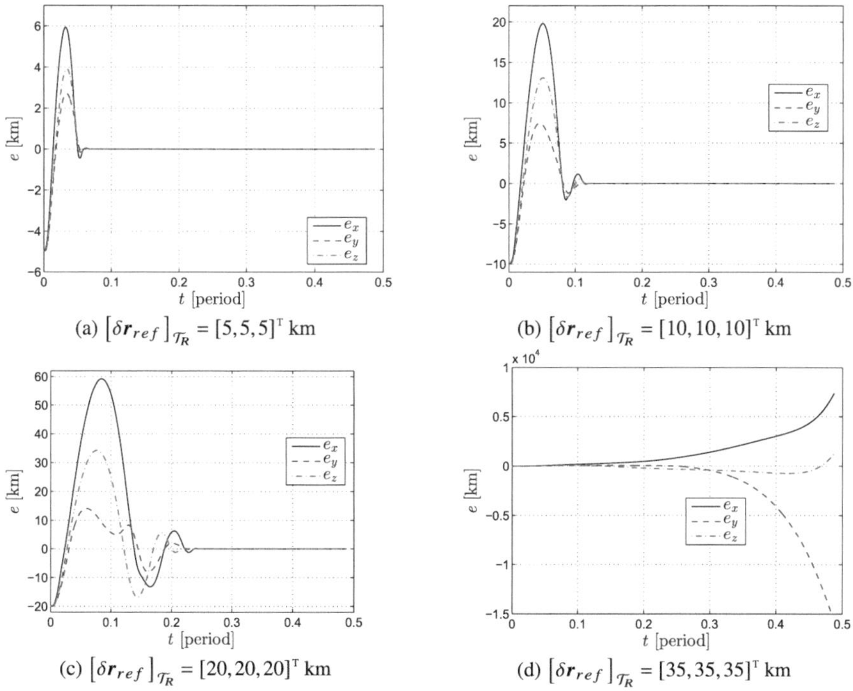

(a) $\left[\delta \boldsymbol{r}_{ref}\right]_{\mathcal{T}_R} = [5,5,5]^{\mathrm{T}}$ km

(b) $\left[\delta \boldsymbol{r}_{ref}\right]_{\mathcal{T}_R} = [10,10,10]^{\mathrm{T}}$ km

(c) $\left[\delta \boldsymbol{r}_{ref}\right]_{\mathcal{T}_R} = [20,20,20]^{\mathrm{T}}$ km

(d) $\left[\delta \boldsymbol{r}_{ref}\right]_{\mathcal{T}_R} = [35,35,35]^{\mathrm{T}}$ km

Fig. 6.10 Tracking errors for several different time-invariant $\left[\delta \boldsymbol{r}_{ref}\right]_{\mathcal{T}_R}$ with the solar-sail resonant halo orbit of $u_l = 0.3$ in Fig. 6.3 as the nominal orbit

- The proposed MPC-based controller has the capacity to achieve high-precision formation-keeping of certain time-invariant and time-varying configurations for solar sail formation flying in the Earth-Moon system. Besides, the MPC-based controller can handle input and output constraints in an effective way and obtain the optimal feasible input.
- The proposed MPC-based controller is only valid for the formation configurations whose displacement and frequency are under certain thresholds. In addition, the controllability decreases with the instability of the nominal orbit increasing.

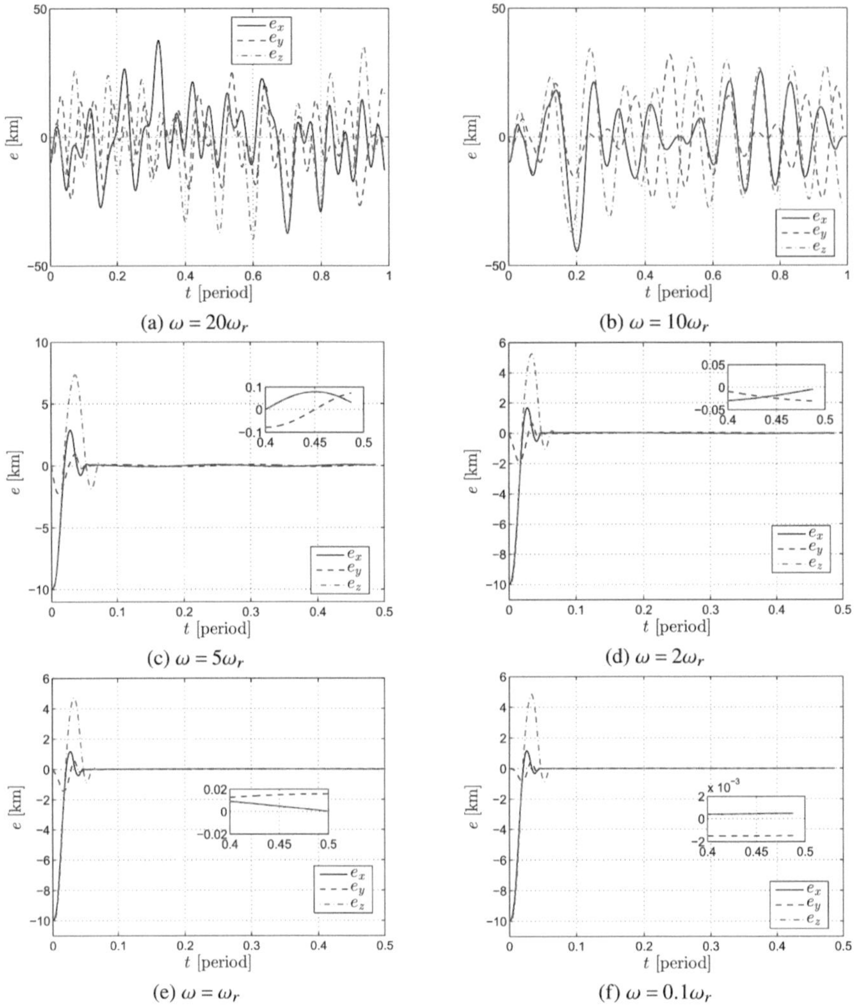

(a) $\omega = 20\omega_r$ (b) $\omega = 10\omega_r$ (c) $\omega = 5\omega_r$ (d) $\omega = 2\omega_r$ (e) $\omega = \omega_r$ (f) $\omega = 0.1\omega_r$

Fig. 6.11 Tracking errors for $\left[\delta \boldsymbol{r}_{ref}\right]_{\mathcal{T}_R} = [10\cos(\omega t), 10\sin(\omega t), 10\cos(\omega t)]^{\mathrm{T}}$ km for different values of frequency ω and with the solar-sail resonant halo orbit of $u_l = 0.2$ in Fig. 6.3 as the nominal orbit

References

1. Simo J, McInnes C R. Asymptotic analysis of displaced lunar orbits[J]. Journal of Guidance, Control, and Dynamics, 2009, 32(5): 1666-1671.
2. Gong S, Li J, Simo J. Orbital motions of a solar sail around the L_2 Earth–Moon libration point[J]. Journal of Guidance, Control, and Dynamics, 2014, 37(4): 1349-1356.
3. Gao C, Wang W, Yuan J, et al. Hybrid sail displaced orbits around L_2 point in the elliptic Earth-Moon system[J]. Journal of Guidance, Control, and Dynamics, 2019, 42(2): 416-424.

4. Heiligers J, Macdonald M, Parker J S. Extension of Earth-Moon libration point orbits with solar sail propulsion[J]. Astrophysics and Space Science, 2016, 361: 1-20.
5. Yuan J, Gao C, Zhang J. Periodic orbits of solar sail equipped with reflectance control device in Earth–Moon system[J]. Astrophysics and Space Science, 2018, 363(2): 1-23.
6. Heiligers J, Parker J S, Macdonald M. Novel solar-sail mission concepts for high-latitude Earth and lunar observation[J]. Journal of Guidance, Control, and Dynamics,2016, 41(1): 212-230.
7. Gao C, Yuan J, Zhang J, et al. Propellant-efficient station-keeping using a hybrid sail in the Earth–Moon system[J]. Nonlinear Dynamics, 2019, 95: 1323-1346.
8. Gao C, Masdemont J J, Gómez G, et al. The web of resonant periodic orbits in the Earth–Moon Quasi-Bicircular Problem including solar radiation pressure[J]. Communications in Nonlinear Science and Numerical Simulation, 2022, 111: 106480.
9. Gao C, Masdemont J J, Gómez, G, et al. Station-keeping for Earth–Moon solar-sail resonant libration point orbits[J]. Communications in Nonlinear Science and Numerical Simulation, 2023, 123: 107274.
10. Ren W. Formation keeping and attitude alignment for multiple spacecraft through local interactions[J]. Journal of Guidance, Control, and Dynamics, 2007, 30(2): 633-638.
11. Gao C, Yuan J, Zhao Y. ADRC for spacecraft attitude and position synchronization in libration point orbits[J]. Acta Astronautica, 2018, 145: 238-249.
12. Mu J, Gong S, Li J. Reflectivity-controlled solar sail formation flying for magnetosphere mission[J]. Aerospace Science and Technology, 2013, 30: 339–348.
13. Mu J, Gong S, Ma P, et al. Dynamics and control of flexible spinning solar sails under reflectivity modulation[J]. Advances in Space Research, 2015, 56(8): 1737-1751.
14. Tsuda Y, Mori O, Funase R, et al. Flight status of IKAROS deep space solar sail demonstrator[J]. Acta Astronautica, 2011, 69: 833-840.
15. Lucking C, Colombo C, McInnes C R. Electrochromic orbit control for smart-dust devices[J]. Journal of Guidance, Control, and Dynamics, 2012, 35(5): 1548–1558.
16. Mu J, Gong S, Li J. Coupled control of reflectivity modulated solar sail for GeoSail formation flying[J]. Journal of Guidance, Control, and Dynamics, 2014, 38(4): 740–751.
17. Hu T, Gong S, Mu J, et al. Switch programming of reflectivity control devices for the coupled dynamics of a solar sail[J]. Advances in Space Research, 2016, 57(5): 1147-1158.
18. Li Q, Yuan J, Zhang B, et al. Model predictive control for autonomous rendezvous and docking with a tumbling target[J]. Aerospace Science and Technology, 2017, 69: 700-711.
19. Li P, Zhu Z, Meguid S A. State dependent model predictive control for orbital rendezvous using pulse-width pulse-frequency modulated thrusters[J]. Advances in Space Research, 2016, 58(1): 64-73.

Chapter 7
Solar Sail Cooperative Formation Flying Around Sun-[Earth+Moon] L_2-Type Artificial Equilibrium Points

Now, we turn our focus from the Earth-Moon system to the Sun-[Earth+Moon] system. For solar sail formation flying around an L_2-type artificial equilibrium point in the Sun-[Earth+Moon] circular restricted three-body problem (CRTBP), a distributed control architecture is proposed in this chapter. Two typical cases, depending on whether the formation structure is leaderless or includes a virtual leader, are investigated. In particular, the virtual leader case is further analyzed according to whether the state information of the virtual leader is available to all of the sails or to a part of the formation structure only. The protocols of the consensus-based algorithms are formulated on a general directed (unidirectional) communication topology, by exploring each available local neighbor-to-neighbor information interaction in a cooperative manner. In that case, a synchronized formation tracking may be achieved while increasing the reliability of the formation system. Illustrative examples show the effectiveness of the proposed approach in a typical mission scenario.

7.1 Introduction

Halo orbits around collinear Lagrange points enable a variety of advanced scientific missions, including China's recent lunar probe Chang'e-4. In particular, continuous relay communications between Earth and Chang'e-4, which reached the south pole region on the far side of the Moon on 3 January 2019, has been obtained via the satellite Queqiao, which entered its Earth-Moon L_2 halo orbit on 15 June 2018. Periodic (or quasi-periodic) orbits around collinear Lagrange points have attracted much attention of the scientific community due to their unique merits. For example, in the Earth-Sun system, the non-Keplerian orbits sunward of the L_1 point are useful for space weather observations, while orbits near the L_2 point in the Earth-Moon system (with a sufficient out-of-plane displacement from it) facilitate constant telecommunication between the Earth and the far side of the Moon [1–3], as in the

© The Author(s), under exclusive license to Springer Nature Singapore Pte Ltd. 2026

C. Gao and W. Wang, *Solar Sailing for Deep-Space Explorations*,
https://doi.org/10.1007/978-981-95-0109-0_7

case of Chang'e-4 mission [4]. From a heliocentric point of view, halo orbits in the vicinity of the Sun-[Earth+Moon] L_2 point naturally avoid solar eclipses (for cosmic microwave measurements), while halo orbits around the L_1 point are never shadowed by the Earth or the Moon (for both Earth's sunlit hemisphere observation and heliophysics studies [5]). The halo orbits around the Sun-[Earth+Moon] L_2 point have been proposed and studied for several scientific missions [6]. For example, the (past) ESA Herschel and Plank space observatories [7], the (present) Gaia probe [8], and the PLATO spacecraft [9].

The intrinsic instability of halo orbits around collinear Lagrange points requires effective methods for station keeping [10] which, usually, necessitate a typical ΔV budget of a few m/s/year or a (very) low propulsive acceleration, when a continuous-thrust propulsion system is considered in the mission design. The low level of necessary thrust, together with the typical requirement of a long mission time to collect a meaningful set of scientific data, make the generation of halo orbits around collinear Lagrange points an interesting option for a solar sail-based spacecraft [11–15], which exploits the natural solar radiation pressure to generate a continuous thrust using the interaction of solar photons with a large reflecting surface, without any propellant consumption. After the striking success of the pioneering Japanese mission IKAROS [16, 17], solar sailing is now considered a promising alternative to conventional (either chemical or electric) propulsion systems [18, 19].

In addition, the use of a propellantless propulsion system, such as the solar sail or the electric solar wind sail [20, 21], allows the collinear equilibrium point to be suitably displaced, so as to obtain a family of artificial equilibrium points (AEPs), whose distance from the Earth may be varied within some limits by properly selecting the spacecraft propulsive acceleration [1, 22]. In this context, the concept of solar sail formation flying has been suggested as an effective means for improving the observation capability (in terms of field of view and resolution) and reducing the overall system cost [23–26]. In fact, the payload mass of each spacecraft may be reduced by disaggregating the functional modules among different sails, with each sail conveying the indispensable mass only. Such a mass reduction results in a significant increase of propulsive acceleration, which in turn mitigates the demanding requirement of guaranteeing a very high area-to-mass ratio in some future scientific missions. In addition, a formation enables multiple solar sails to operate within a limited region to fulfill complex spaceborne tasks, which would be otherwise infeasible for a single monolithic sail. By exploiting a task distribution among smaller, less-expensive spacecraft, the sails in formation are able to interact with each other and operate in a cooperative way, so that the overall system flexibility may be significantly enhanced [27].

The study of spacecraft formation flight is usually addressed by means of a leader-follower formation architecture [24, 26, 28], in which the leader moves along a predefined (nominal) trajectory, and the followers track the leader with some given (possibly time-varying) offset. In that case, the feature of the formation system is defined by specifying the behavior of the leader alone, while the information flow goes from the leader to the followers, without any feedback [30–32]. Such a topological structure introduces an inherent vulnerability, because the leader is a single point of failure for the whole system so that, when the follower is subject to strong perturbations, the formation may barely be maintained. In most studies of the maintenance of a spacecraft formation, the leader state is assumed to be available to each follower, even though this is usually unrealistic due to small communication bandwidths and range limitations. In practice, only a part of the formation has usual access to the state of the reference point and those spacecraft may not have a directed path to all of the other formation agents.

The aim of this chapter is to extend the concept of consensus-based control to a solar sail formation structure flying around an L_2-type AEP in the Sun-[Earth+Moon] circular restricted three-body problem. In particular, two different cases are investigated, depending on whether the formation structure is leaderless or includes a virtual leader. In the latter case the information of the virtual leader state may be available to all of the sails or to a part of the formation structure only. The protocols of the consensus-based algorithms are formulated on a general directed (unidirectional) communication topology, by exploring every available local neighbor-to-neighbor information interactions in a cooperative manner. In that case, a synchronized formation tracking can be achieved while the reliability of the formation system is strengthened.

This chapter is organized as follows. Section 7.2 illustrates the procedure for generating an L_2-type AEP in the Sun-[Earth+Moon] CRTBP using a solar sail. Section 7.3 presents an approximate analytical solution to the nominal orbit via Fourier series expansions, while Sects 7.4 and 7.5 develop the formation design, and the solar sail coordinated control strategies by accommodating information interactions among the spacecraft to account both for the leaderless case and the virtual-leader case. The proposed method is then validated in Sect. 7.6 by numerical simulations. Finally, some concluding remarks are given in Sect. 7.7.

7.2 Generation of L_2-Type Artificial Equilibrium Points

Consider a flat, perfectly reflecting [22], solar sail s with a total reflective area A_s, which moves within the Sun-[Earth+Moon] system. The solar sail is equipped with a reflectivity control device comprised of a number of sufficiently small electrochromic material panels (EMPs) of total area A_p, whose optical coefficients may be changed with the application of a suitable electric voltage. In particular, let the area covered with the switched-off EMPs be denoted by A_{off} [28]; see Fig. 7.1.

According to the simplified mathematical model [33], each EMP acts as a perfectly absorbing (or perfectly reflecting) medium in a power-off (or power-on) mode. Therefore, the magnitude of the sail propulsive acceleration can be modulated, within a suitable range, by selecting the working state of the EMPs between the high reflectivity mode (power-on), and the low reflectivity mode (power-off). The solar sail propulsive acceleration vector $\boldsymbol{a}$ can be written as

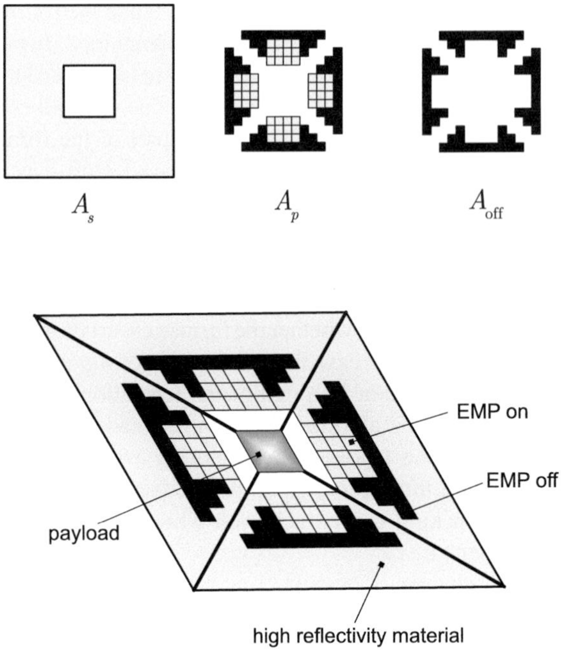

Fig. 7.1 Conceptual scheme of a solar sail with Electrochromic Material Panels

$$\boldsymbol{a} = \frac{\beta\, G\, m_\odot}{2\, r_{\odot s}^2} \left(\hat{\boldsymbol{r}}_{\odot s} \cdot \hat{\boldsymbol{n}}\right) \left[u\, \hat{\boldsymbol{r}}_{\odot s} + 2\, (1-u) \left(\hat{\boldsymbol{r}}_{\odot s} \cdot \hat{\boldsymbol{n}}\right) \hat{\boldsymbol{n}}\right] \tag{7.1}$$

where G is the universal gravitational constant, $m_\odot$ is the Sun's mass, $\boldsymbol{r}_{\odot s}$ is the Sun-sail vector (with $r_{\odot s} \triangleq \|\boldsymbol{r}_{\odot s}\|$ and $\hat{\boldsymbol{r}}_{\odot s} \triangleq \boldsymbol{r}_{\odot s}/r_{\odot s}$), $\hat{\boldsymbol{n}}$ is the unit vector normal to the sail plane in the direction opposite to the Sun, and β is the sail lightness number [22], defined as the ratio of the maximum propulsive acceleration magnitude of a perfectly reflecting sail to the (local) solar gravitational acceleration. In Eq. (7.1), $u \triangleq A_{\text{off}}/A_s$ is the reflectivity modulation ratio which, in this simplified model, is assumed to range with continuity between 0 (when all of EMPs are in power-on mode) and $u_{\max} \triangleq A_p/A_s \leq 1$ (when all of EMPs are in power-off mode). Accordingly, the control variables are the reflectivity modulation ratio u, and the normal unit vector $\hat{\boldsymbol{n}}$. Note that in the special case of $u = 0$, Eq. (7.1) reduces to the formula describing an ideal solar sail [22] (in which all of the incoming photons are specularly reflected), whereas if $u = u_{\max} = 1$, the solar sail acts as a flat black-body of area A_s (all of the incoming photons are absorbed).

Assuming that the [Earth+Moon] system, of mass $m_\oplus$, covers a circular orbit around the Sun with radius $r_\oplus \triangleq 1$ au, the motion of the (massless) solar sail-based spacecraft can be described in a synodic reference frame $\mathcal{T}\left(O;\, \hat{\boldsymbol{x}},\, \hat{\boldsymbol{y}},\, \hat{\boldsymbol{z}}\right)$, with origin at the center-of-mass O of the Sun-[Earth+Moon] system, which rotates synchronously with the revolution of the [Earth+Moon] around the Sun with an angular velocity $\boldsymbol{\omega}$. In this frame, the unit vector $\hat{\boldsymbol{x}}$ points towards the [Earth+Moon] center-of-mass, $\hat{\boldsymbol{z}}$

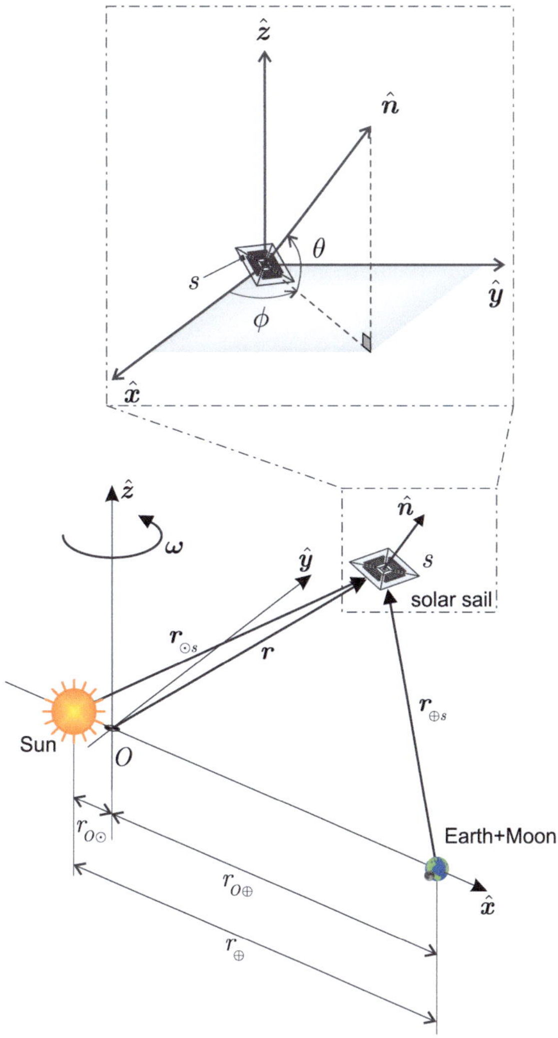

Fig. 7.2 Synodic reference frame and solar sail thrust vector angles

lies in the direction of the angular velocity vector $\boldsymbol{\omega}$, while the plane $(\hat{\boldsymbol{x}}, \hat{\boldsymbol{y}})$ coincides with the ecliptic plane; see Fig. 7.2.

The components of the sail normal unit vector $\hat{\boldsymbol{n}}$ can be expressed, as a function of two attitude angles $\{\theta, \phi\}$, as

$$[\hat{\boldsymbol{n}}]_{\mathcal{T}} = [\cos\theta\ \cos\phi,\ \cos\theta\ \sin\phi,\ \sin\theta]^{\mathrm{T}} \tag{7.2}$$

where $\theta \in [-\pi/2, \pi/2]$ rad is the angle between the direction of $\hat{\boldsymbol{n}}$ and the $(\hat{\boldsymbol{x}}, \hat{\boldsymbol{y}})$ plane, while $\phi \in [-\pi/2, \pi/2]$ rad is the angle (measured counterclockwise) between the direction of $\hat{\boldsymbol{x}}$ and the projection of $\hat{\boldsymbol{n}}$ onto $(\hat{\boldsymbol{x}}, \hat{\boldsymbol{y}})$; see Fig. 7.2.

For convenience, a dimensionless set of units is now introduced, such that the total mass of the primaries ($m_\odot + m_\oplus$), the Sun-[Earth+Moon] reference distance ($r_\oplus$), and the universal gravitation constant (G) are all unitary. Accordingly, the rotational period of the synodic frame is $T = 2\pi/\omega \equiv 2\pi$, the O-Sun distance is $r_{O\odot} = \mu\, r_\oplus \equiv \mu$, the O-[Earth+Moon] distance is $r_{O\oplus} = (1-\mu)\, r_\oplus \equiv (1-\mu)$, where $\mu \triangleq m_\oplus / (m_\oplus + m_\odot) = 3.05425 \times 10^{-6}$ is the dimensionless mass of the [Earth+Moon] system. In particular, since the Sun's mass can be written in a dimensionless form as $m_\odot = (1-\mu)\,(m_\odot + m_\oplus) \equiv (1-\mu)$, the solar sail propulsive acceleration vector $\boldsymbol{a}$, given by Eq. (7.1), becomes

$$\boldsymbol{a} = \frac{\beta\,(1-\mu)}{2\, r_{\odot s}^2} \left(\hat{\boldsymbol{r}}_{\odot s} \cdot \hat{\boldsymbol{n}}\right) \left[u\, \hat{\boldsymbol{r}}_{\odot s} + 2\,(1-u) \left(\hat{\boldsymbol{r}}_{\odot s} \cdot \hat{\boldsymbol{n}}\right) \hat{\boldsymbol{n}}\right] \tag{7.3}$$

Therefore, the (dimensionless) solar sail equation of motion is

$$\ddot{\boldsymbol{r}} + 2\,\boldsymbol{\omega} \times \dot{\boldsymbol{r}} = \frac{\partial\,\Omega}{\partial\,\boldsymbol{r}} + \boldsymbol{a} \tag{7.4}$$

where $\boldsymbol{r}$ is the O-sail position vector, $\boldsymbol{a}$ is given by Eq. (7.3), Ω is the effective potential function, and the angular velocity vector

$$[\boldsymbol{\omega}]_{\mathcal{T}} = [0,\ 0,\ 1]^{\mathrm{T}} \tag{7.5}$$

while the sail position vector

$$[\boldsymbol{r}]_{\mathcal{T}} = [x,\ y,\ z]^{\mathrm{T}} \tag{7.6}$$

With the aid of Fig. 7.2, the potential function Ω can be written as

$$\Omega \triangleq \frac{x^2 + y^2}{2} + \frac{1-\mu}{r_{\odot s}} + \frac{\mu}{r_{\oplus s}} \tag{7.7}$$

where the Sun-sail distance $r_{\odot s}$ and the [Earth+Moon]-sail distance $r_{\oplus s} \triangleq \|\boldsymbol{r}_{\oplus s}\|$ are given by

$$r_{\odot s} = \sqrt{(x+\mu)^2 + y^2 + z^2} \tag{7.8}$$

$$r_{\oplus s} = \sqrt{(x+\mu-1)^2 + y^2 + z^2} \tag{7.9}$$

In the simple case of $\boldsymbol{a} = \boldsymbol{0}$, Eq. (7.4) presents five equilibrium points, that is, the five classical equilibrium points $\{L_1, L_2, L_3, L_4, L_5\}$ of the circular restricted three-body problem, among which L_2 is the (unstable) collinear point along the Sun-[Earth+Moon] line past the [Earth+Moon] system. Assuming $\hat{\boldsymbol{n}} = \hat{\boldsymbol{r}}_{\odot s} = \hat{\boldsymbol{x}}$, a suitable value of the sail lightness number β allows the L_2 equilibrium point to be displaced along the $\hat{\boldsymbol{x}}$-axis of the synodic reference frame $\mathcal{T}$, such as to obtain a sort of L_2-type artificial equilibrium point (AEP) [1]. In other terms, a family of collinear AEPs can be created when a suitable centripetal force is exerted on the solar sail, that is, generated by the gravitational acceleration of the celestial bodies and the propulsive acceleration.

Bearing in mind Eq. (7.3), when $\hat{\boldsymbol{n}} = \hat{\boldsymbol{x}}$ and $\boldsymbol{r}_{\odot s} = (x + \mu)\ \hat{\boldsymbol{x}}$, the dimensionless propulsive acceleration vector becomes

$$\boldsymbol{a} = \frac{\beta\ (1 - \mu)\ (2 - u)}{2\ (x + \mu)^2}\ \hat{\boldsymbol{x}} \tag{7.10}$$

whereas the dimensionless potential function of Eq. (7.7) reduces to

$$\Omega = \frac{x^2}{2} + \frac{1 - \mu}{x + \mu} + \frac{\mu}{|x + \mu - 1|} \tag{7.11}$$

Therefore, the position $x = x_{\text{AEP}}$ of the L_2-type AEP along the $\hat{\boldsymbol{x}}$-axis is obtained by enforcing the conditions

$$\dot{\boldsymbol{r}} = \boldsymbol{0}, \quad \ddot{\boldsymbol{r}} = \boldsymbol{0}, \quad \boldsymbol{r} = x_{\text{AEP}}\ \hat{\boldsymbol{x}} \tag{7.12}$$

with $x_{\text{AEP}} > (1 - \mu)$. Substituting Eq. (7.11) into Eq. (7.4), the result is a single scalar equation

$$x_{\text{AEP}} - \frac{(1 - \mu)\ [1 - \beta\ (1 - u/2)]}{(x_{\text{AEP}} + \mu)^2} - \frac{\mu}{(x_{\text{AEP}} + \mu - 1)^2} = 0 \tag{7.13}$$

whose solution gives the position of the L_2-type AEP along the $\hat{\boldsymbol{x}}$-axis.

For a given value of sail lightness number β and reflectivity modulation ratio u, the location of the AEP requires the determination of the root x_{AEP} of Eq. (7.13), which can be obtained using standard numerical algorithms. For example, when $u = 0.4$ and $\beta = 0.01$, Eq. (7.13) gives $x_{\text{AEP}} \simeq 1.009$. On the other hand, for a given value of $x_{\text{AEP}} > (1 - \mu)$ and u, the required sail lightness number can be obtained from Eq. (7.13) as

$$\beta = \frac{(x_{\text{AEP}} + \mu)^2\ \left[(1 - \mu) / (x_{\text{AEP}} + \mu)^2 + \mu / (\mu - 1 + x_{\text{AEP}})^2 - x_{\text{AEP}}\right]}{(1 - u/2)\ (1 - \mu)} \tag{7.14}$$

The function $\beta = \beta\ (x_{\text{AEP}}, u)$ of Eq. (7.14) is illustrated in Fig. 7.3 for $u = \{0, 0.4, 1\}$. Note that, as expected, the condition $\beta = 0$ gives the position, along the $\hat{\boldsymbol{x}}$-axis, of the classical L_2 point.

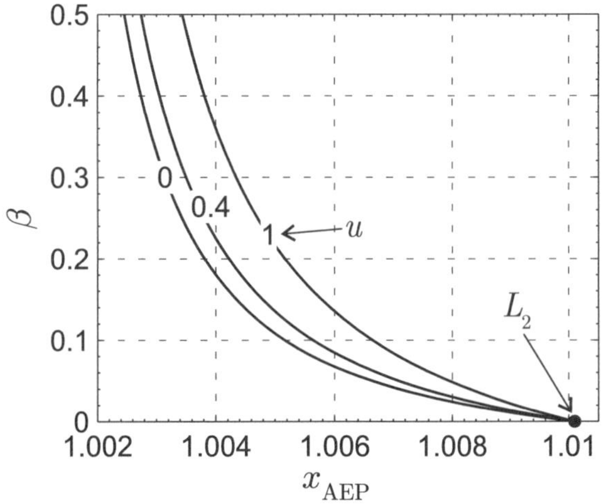

Fig. 7.3 Required sail lightness number as a function of x_{AEP} and u; see Eq. (7.14)

7.3 Nominal Trajectory Selection

The reference halo orbit around the L_2-type AEP is designated here as the nominal trajectory. Nevertheless, other types of periodic or quasi-periodic orbits (e.g. Lissajous orbits) may also be chosen as the nominal trajectory, depending on the specific scientific mission requirements. In general, the properties of the dynamical system modeled by Eqs. (7.3)–(7.9) are highly dependent on the characteristics of the solar sail propulsive acceleration vector $\boldsymbol{a}$, that is, on u and on the components of the normal unit vector $\hat{\boldsymbol{n}}$; see Eq. (7.3).

According to Refs. [11, 29], there exists a large family of halo orbits when the sail attitude is maintained fixed in the rotating reference frame $\mathcal{T}$. In the following analysis, the nominal trajectory will be obtained under the condition $\hat{\boldsymbol{n}} \equiv \hat{\boldsymbol{r}}_{\odot s}$, that is, with the assumption that the solar sail normal vector always lies along the Sun-sail line. This happens in a Sun-facing condition, which can be maintained in a passive way by shaping the sail structure with a slightly conical form (with its apex directed sunward) [34, 35]. A Sun-facing condition can also be obtained using a Sun-pointing smart dust [36–39], that is, a millimeter-scale solar sail [40] whose external surface is covered by an electrochromic reflective film. Accordingly, Eqs. (7.3)–(7.9) give

$$\ddot{x} - 2\,\dot{y} - x = -\frac{(1-\mu)\,[1-\beta\,(1-u/2)]\,(x+\mu)}{\left[(x+\mu)^2+y^2+z^2\right]^{3/2}} - \frac{\mu\,(x+\mu-1)}{\left[(x+\mu-1)^2+y^2+z^2\right]^{3/2}} \tag{7.15}$$

$$\ddot{y} + 2\,\dot{x} - y = -\frac{(1-\mu)\,[1-\beta\,(1-u/2)]\,y}{\left[(x+\mu)^2+y^2+z^2\right]^{3/2}} - \frac{\mu\,y}{\left[(x+\mu-1)^2+y^2+z^2\right]^{3/2}} \tag{7.16}$$

$$\ddot{z} = -\frac{(1-\mu)\,[1-\beta\,(1-u/2)]\,z}{\left[(x+\mu)^2+y^2+z^2\right]^{3/2}} - \frac{\mu\,z}{\left[(x+\mu-1)^2+y^2+z^2\right]^{3/2}} \tag{7.17}$$

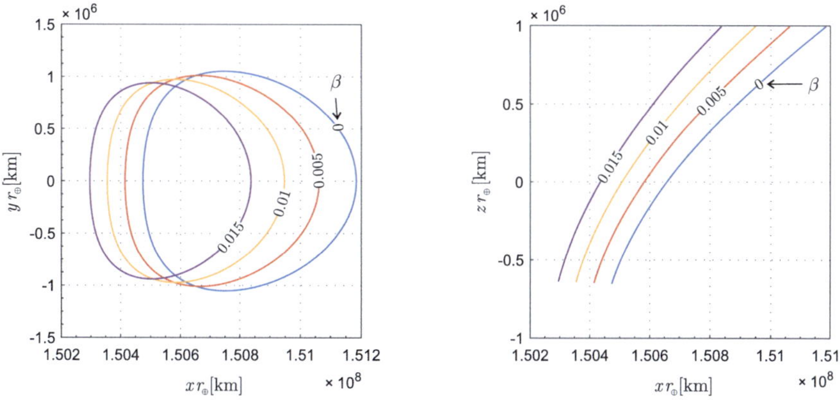

(a) $(\hat{x} - \hat{y})$ projection

(b) $(\hat{x} - \hat{z})$ projection

Fig. 7.4 Families of halo orbits around L_2-type AEPs with $u = 0.4$ and $\beta = \{0, 0.005, 0.01, 0.015\}$.

The linearized form of Eqs. (7.15)–(7.17) [41], or the third-order approximation obtained with the aid of Lindstedt-Poincaré perturbation method [42, 43], is usually used to identify the artificial halo orbits in the neighborhood of the L_2-type AEP point [11]. However, in both cases the resulting solution gives rise to a remarkable deviation from the nominal trajectory when the (dimensionless) propagation time exceeds about one half period ($T/2 = \pi$). For this reason, an alternative approach is advisable. In this context, the differential correction method [14, 44, 45] is first adopted to identify the artificial halo orbit around the designed L_2-type AEP point, whose position is obtained with a given value of the pair $\{\beta, u\}$. For example, the families of halo orbits around L_2-type AEPs with $u = 0.4$ and $\beta = \{0, 0.005, 0.01, 0.015\}$ are shown in Fig. 7.4.

The non-integrable nature of Eqs. (7.15)–(7.17) prevents the artificial halo orbit from any analytical (exact) representation. However, as will be shown later, the solar sail formation control system requires an explicit description of the nominal trajectory. To this end, a Fourier series-based solution with a least-square technique is now proposed to estimate the nominal trajectory with a reduced computational effort [48]. The nominal trajectory is represented with an approximate trigonometric polynomial as

$$\boldsymbol{r}_h = \sum_{k=0}^{n} \left[\boldsymbol{c}_k \cos\left(k\, \omega_h\, t\right) + \boldsymbol{s}_k \sin\left(k\, \omega_h\, t\right)\right] \tag{7.18}$$

where ω_h is the angular frequency of the nominal halo orbit, and $\boldsymbol{r}_h$ is the position vector of a generic point h at time t, with $[\boldsymbol{r}_h]_{\mathcal{T}_R} = [x_h,\ y_h,\ z_h]^{\mathrm{T}}$; see Fig. 7.5. $\boldsymbol{\ell} = \{\boldsymbol{c}_k, \boldsymbol{s}_k\}$ is the set of coefficients to be evaluated. For a given order n of the truncated series (7.18), the value of $\boldsymbol{\ell}$ may be found using an unconstrained least-square procedure, with the minimization index defined as

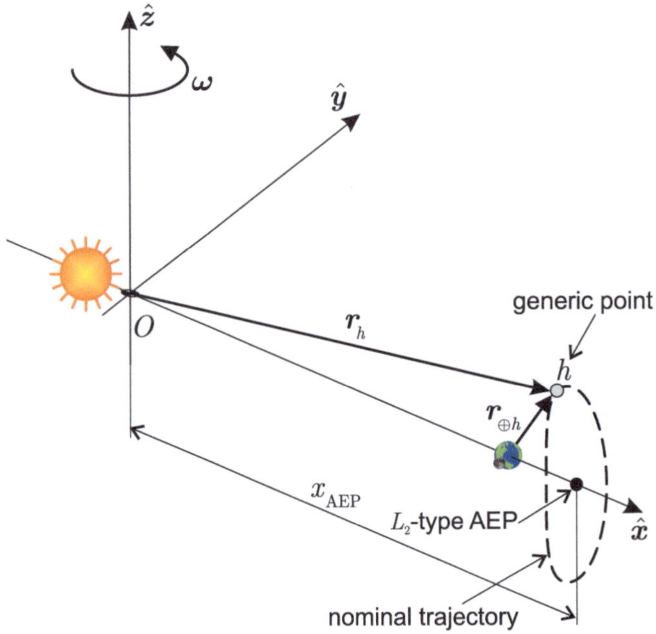

Fig. 7.5 L_2-type AEP and nominal trajectory.

$$\boldsymbol{\ell} = \arg\min \left\{ \int_0^{T_h} \|\boldsymbol{X} - \boldsymbol{X}_h\|^2 \, \mathrm{d}\, t \right\} \tag{7.19}$$

where $\boldsymbol{X} = \left[\boldsymbol{r}^{\mathrm{T}}, \, \dot{\boldsymbol{r}}^{\mathrm{T}}\right]^{\mathrm{T}}$ is the state vector along the trajectory, and $T_h = 2\,\pi/\omega_h$ is the period of the nominal halo orbit.

7.4 Solar Sail Formation Design

Having analyzed the generation (and analytical approximation) of the nominal trajectory, we are now in a position to study the relative motion of a solar sail-based spacecraft formation. Let $\boldsymbol{\rho}_i = \boldsymbol{r}_i - \boldsymbol{r}_h$ denote the relative dimensionless position vector of the ith solar sail s_i with respect to the corresponding point h on the nominal trajectory. In particular, the nominal trajectory is obtained with an assigned value of both the sail lightness number $\beta = \beta_h$, and the reflectivity modulation ratio $u = u_h$. Note that the approximation of the vector $\boldsymbol{r}_h$ is given by Eq. (7.18). Moreover, let $\boldsymbol{r}_{\odot h}$ and $\boldsymbol{r}_{\oplus h}$ denote the position vectors from the Sun and the Earth to a generic point h on the nominal trajectory, respectively; see Fig. 7.6.

The dimensionless relative equation of motion of the spacecraft s_i, in the rotating reference frame $\mathcal{T}_R$, can be written as

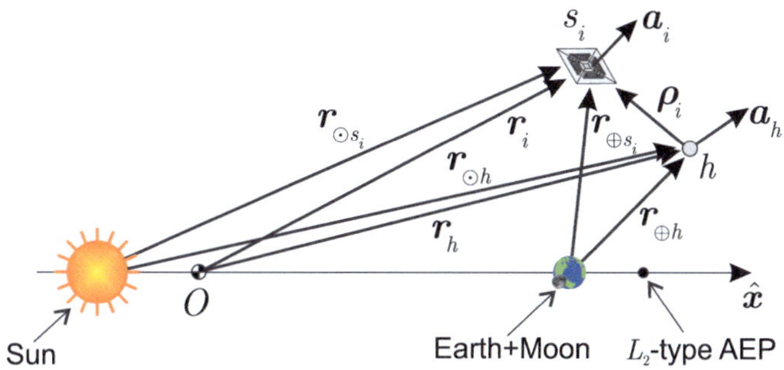

Fig. 7.6 Schematization of the relative position vector between sail s_i and the nominal trajectory

$$\ddot{\boldsymbol{\rho}}_i + 2\,\boldsymbol{\omega} \times \dot{\boldsymbol{\rho}}_i = \frac{\partial\,\Omega_i}{\partial\,\boldsymbol{r}_i} - \frac{\partial\,\Omega_h}{\partial\,\boldsymbol{r}_h} + \boldsymbol{a}_i - \boldsymbol{a}_h \tag{7.20}$$

where $[\boldsymbol{\omega}]_{\mathcal{T}_R} = [0,\ 0,\ 1]^{\mathrm{T}}$ and $\boldsymbol{a}_h$ is the acceleration required by the sail at point h to follow the nominal trajectory with a Sun-facing attitude, that is

$$\boldsymbol{a}_h = \frac{\beta_h\ (1-\mu)\ (1-u_h/2)}{r_{\odot h}^2}\,\hat{\boldsymbol{r}}_{\odot h} \tag{7.21}$$

Because the solar sails operate in the neighborhood of the nominal trajectory, the relative distance $\rho_i = \left\|\boldsymbol{\rho}_i\right\|$ is much smaller than $r_{\odot h} = \|\boldsymbol{r}_{\odot h}\|$. Moreover, the magnitude of the relative unit normal vector $\Delta\,\hat{\boldsymbol{n}}_i = \hat{\boldsymbol{n}}_i - \hat{\boldsymbol{n}}_h$, and the relative reflectivity ratio $\Delta\,u_i = u_i - u_h$ are both negligible to first-order. Accordingly, the relevant terms in Eq. (7.20) can be linearized around those of the nominal trajectory as

$$\boldsymbol{a}_i \simeq \boldsymbol{a}_h + \frac{\partial\,\boldsymbol{a}_h}{\partial\,\boldsymbol{r}_{\odot h}}\,\boldsymbol{\rho}_i + \frac{\partial\,\boldsymbol{a}_h}{\partial\,\hat{\boldsymbol{n}}_h}\,\Delta\,\hat{\boldsymbol{n}}_i + \frac{\partial\,\boldsymbol{a}_h}{\partial\,u_h}\,\Delta\,u_i \tag{7.22}$$

$$\frac{\partial\,\Omega_i}{\partial\,\boldsymbol{r}_i} \simeq \frac{\partial\,\Omega_h}{\partial\,\boldsymbol{r}_h} + \frac{\partial^2\,\Omega_h}{\partial\,\boldsymbol{r}_h^2}\,\boldsymbol{\rho}_i \tag{7.23}$$

where

$$\frac{\partial\,\boldsymbol{a}_h}{\partial\,\boldsymbol{r}_{\odot h}} = \frac{\beta_h\ (1-\mu)}{r_{\odot h}^3}\left\{\frac{u_h}{2}\left[\left(\hat{\boldsymbol{r}}_{\odot h}\,\hat{\boldsymbol{n}}_h^{\mathrm{T}}\right) + \left(\hat{\boldsymbol{n}}_h\cdot\hat{\boldsymbol{r}}_{\odot h}\right)\mathbb{I}_3 - \left(\hat{\boldsymbol{n}}_h\cdot\hat{\boldsymbol{r}}_{\odot h}\right)\left(\hat{\boldsymbol{r}}_{\odot h}\,\hat{\boldsymbol{r}}_{\odot h}^{\mathrm{T}}\right)\right]\right.$$
$$\left. + 2\,(1-u_h)\left[\left(\hat{\boldsymbol{n}}_h\cdot\hat{\boldsymbol{r}}_{\odot h}\right)\left(\hat{\boldsymbol{n}}_h\,\hat{\boldsymbol{n}}_h^{\mathrm{T}}\right) - 2\left(\hat{\boldsymbol{n}}_h\cdot\hat{\boldsymbol{r}}_{\odot h}\right)^2\left(\hat{\boldsymbol{n}}_h\,\hat{\boldsymbol{r}}_{\odot h}^{\mathrm{T}}\right)\right]\right\} \tag{7.24}$$

$$\frac{\partial\,\boldsymbol{a}_h}{\partial\,\hat{\boldsymbol{n}}_h} = \frac{\beta_h\ (1-\mu)}{r_{\odot h}^2}\left\{\frac{u_h}{2}\left(\hat{\boldsymbol{r}}_{\odot h}\,\hat{\boldsymbol{r}}_{\odot h}^{\mathrm{T}}\right) + (1-u_h)\left[2\left(\hat{\boldsymbol{n}}_h\cdot\hat{\boldsymbol{r}}_{\odot h}\right)\left(\hat{\boldsymbol{n}}_h\,\hat{\boldsymbol{r}}_{\odot h}^{\mathrm{T}}\right) + \left(\hat{\boldsymbol{n}}_h\cdot\hat{\boldsymbol{r}}_{\odot h}\right)^2\mathbb{I}_3\right]\right\} \tag{7.25}$$

$$\frac{\partial\,\boldsymbol{a}_h}{\partial\,u_h} = \frac{\beta_h\ (1-\mu)}{r_{\odot h}^2}\left[\frac{1}{2}\left(\hat{\boldsymbol{n}}_h\cdot\hat{\boldsymbol{r}}_{\odot h}\right)\hat{\boldsymbol{r}}_{\odot h} - \left(\hat{\boldsymbol{n}}_h\cdot\hat{\boldsymbol{r}}_{\odot h}\right)^2\hat{\boldsymbol{n}}_h\right] \tag{7.26}$$

$$\frac{\partial^2 \Omega_h}{\partial \boldsymbol{r}_h^2} = \frac{(1-\mu)}{r_{\odot h}^3} \left(3\,\hat{\boldsymbol{r}}_{\odot h}\,\hat{\boldsymbol{r}}_{\odot h}^{\mathrm{T}} - \mathbb{I}_3\right) + \frac{\mu}{r_{\oplus h}^3} \left(3\,\hat{\boldsymbol{r}}_{\oplus h}\,\hat{\boldsymbol{r}}_{\oplus h}^{\mathrm{T}} - \mathbb{I}_3\right) + \mathrm{diag}\,(1,\ 1,\ 0) \tag{7.27}$$

The nominal trajectory around the L_2-type AEP is designed assuming the solar sail normal vector to lie along the Sun-sail line, hence

$$\left[\hat{\boldsymbol{n}}_h\right]_{\mathcal{T}} \equiv \left[\hat{\boldsymbol{r}}_{\odot h}\right]_{\mathcal{T}} = [x_h + \mu,\ y_h,\ z_h]^{\mathrm{T}} \tag{7.28}$$

$$\left[\hat{\boldsymbol{r}}_{\oplus h}\right]_{\mathcal{T}} = [x_h + \mu - 1,\ y_h,\ z_h]^{\mathrm{T}} \tag{7.29}$$

Recalling Eq. (7.2), the attitude angles necessary for the nominal trajectory are written as

$$\theta_h = \arcsin\left(\hat{\boldsymbol{n}}_h \cdot \hat{\boldsymbol{z}}\right), \qquad \phi_h = \arcsin\left[\left(\hat{\boldsymbol{n}}_h \times \hat{\boldsymbol{z}}\right) / \left\|\hat{\boldsymbol{n}}_h \times \hat{\boldsymbol{z}}\right\| \cdot \hat{\boldsymbol{x}}\right] \tag{7.30}$$

where $\hat{\boldsymbol{n}}_h$ is given by Eq. (7.28). Accordingly, the relative normal vector $\Delta\,\hat{\boldsymbol{n}}_i$ in Eq. (7.22) can be further linearized as

$$\Delta\,\hat{\boldsymbol{n}}_i \simeq \frac{\partial\,\hat{\boldsymbol{n}}_h}{\partial\,[\theta_h,\ \phi_h]}\,[\Delta\,\theta_i,\ \Delta\,\phi_i]^{\mathrm{T}} \tag{7.31}$$

where $\Delta\,\theta_i = \theta_i - \theta_h$ and $\Delta\,\phi_i = \phi_i - \phi_h$ are the relative sail attitude angles, while the Jacobian matrix is

$$\frac{\partial\,\hat{\boldsymbol{n}}_h}{\partial\,[\theta_h,\ \phi_h]} = \begin{bmatrix} -\sin\theta_h\,\cos\phi_h & -\cos\theta_h\,\sin\phi_h \\ -\sin\theta_h\,\sin\phi_h & \cos\theta_h\,\cos\phi_h \\ \cos\theta_h & 0 \end{bmatrix} \tag{7.32}$$

Substituting Eqs. (7.22)–(7.32) into Eq. (7.20), the (dimensionless) linear differential equation of the relative motion of the ith sail s_i, with respect to a generic point h on the nominal trajectory, can be written in a compact form as

$$\ddot{\boldsymbol{\rho}}_i + 2\,\mathbb{M}_v\,\dot{\boldsymbol{\rho}}_i + \mathbb{M}_p\,\boldsymbol{\rho}_i = \mathbb{M}_h\,\boldsymbol{u}_i \tag{7.33}$$

where

$$\mathbb{M}_v = \begin{bmatrix} 0 & -1 & 0 \\ 1 & 0 & 0 \\ 0 & 0 & 0 \end{bmatrix} \tag{7.34}$$

$$\mathbb{M}_p = -\frac{\partial^2\,\Omega_h}{\partial\,\boldsymbol{r}_h^2} - \frac{\partial\,\boldsymbol{a}_h}{\partial\,\boldsymbol{r}_{\odot h}} \tag{7.35}$$

$$\mathbb{M}_h = \left[\frac{\partial\,\boldsymbol{a}_h}{\partial\,\hat{\boldsymbol{n}}_h}\,\frac{\partial\,\hat{\boldsymbol{n}}_h}{\partial\,[\theta_h,\ \phi_h]},\ \frac{\partial\,\boldsymbol{a}_h}{\partial\,u_h}\right] \tag{7.36}$$

while $\boldsymbol{u}_i$ is the control input of a generic solar sail s_i, defined as

$$\boldsymbol{u}_i = [\Delta\,\theta_i,\ \ \Delta\,\phi_i,\ \ \Delta\,u_i]^{\mathrm{T}} \tag{7.37}$$

In the analysis to follow, the cooperative strategies for the formation system described by Eq. (7.33) will be designed such that all of the solar sails in formation reach consensus as $t \to +\infty$.

7.5 Cooperative Formation Control

The distributed coordinated control algorithms are now considered for a formation structure with $N \geq 3$ solar sail-based spacecraft, relying on local neighbor-to-neighbor interactions. In the case of data transmission through local sensing [46, 47], the solar sails in a formation are likely to be equipped with sensors with a limited field of view only, which induce unidirectional communication topologies. To model such a behavior, the information exchange among the spacecraft in the formation is assumed to be characterized by a generic directed graph.

To proceed, a few pertinent concepts of algebraic graph theory are first briefly summarized for the sake of completeness [49]. A directed graph $\mathcal{G}$ consists of a finite nonempty vertex set $\mathcal{V} = \{\upsilon_1, \ldots, \upsilon_N\}$, an edge set $\mathcal{E} = \{(\upsilon_1, \upsilon_2), \ldots, (\upsilon_{N-1}, \upsilon_N)\} \subseteq \mathcal{V} \times \mathcal{V}$, and a weighted adjacency matrix $\mathbb{W} = \left[w_{ij}\right] \in \mathbb{R}^{N\times N}$. An edge $\left(\upsilon_i, \upsilon_j\right) \in \mathcal{E}$ in a directed graph implies that the information flows from vertex υ_i to υ_j, but not necessarily vice versa. When there is an edge from vertex υ_i to vertex υ_j, then υ_i is defined as the parent vertex, while υ_j is defined as the child vertex. The generic entry of a weighted adjacency matrix $\mathbb{W}$ is $w_{ij} > 0$, $\forall \left(\upsilon_j, \upsilon_i\right) \in \mathcal{E}$ with $i \neq j$, and $w_{ii} = 0$. The Laplacian matrix $\mathbb{L} = \left[l_{ij}\right] \in \mathbb{R}^{N\times N}$ associated with $\mathbb{W}$ is defined such that $l_{ij} = -w_{ij}$ and $l_{ii} = \sum_{j=1}^{N} w_{ij}$, $\forall i \neq j$. A rooted directed spanning tree is constituted by edges that connect all the vertices; see Fig. 7.7. Note that a directed graph has a rooted spanning tree if and only if there exists at least one vertex having a directed path to all of the other vertices. Finally, the topology is assumed to be fixed throughout the chapter and the Laplacian matrix $\mathbb{L}$ is constant.

In modeling the topology of the solar sail relative motion, each sail is represented by a vertex, while the interactions between two generic sails are characterized by a weighted directed edge. In the next sections, distributed control strategies will be developed in both the leaderless case and the virtual-leader case, respectively, such that overall consensus will be reached.

7.5.1 *Leaderless Case*

Consider first the case in which a centralized leadership is absent and the solar sail-based spacecraft in formation are required to maintain a desired relative configura-

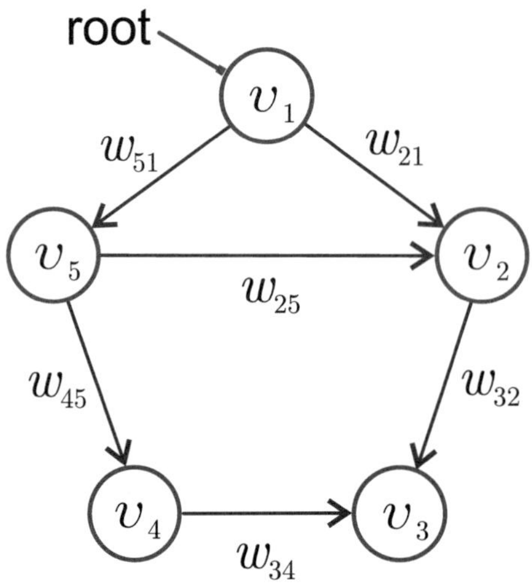

Fig. 7.7 Schematization of a directed topology graph including a rooted spanning tree

tion. Such a leaderless formation structure is said to follow a behavior approach [50], which is typical in cluster flight or swarm missions [51–53], where the control law is designed for each vehicle such that the solar sails have a common formation center c. In this case, more flexibility, reliability, and robustness can be guaranteed than those in a centralized leader-follower configuration [30].

Let $\{\boldsymbol{\rho}_{ci}, \dot{\boldsymbol{\rho}}_{ci}\} \in \mathbb{R}^3 \times \mathbb{R}^3$, with $i = 1, \dots, N$, be the desired relative state of the solar sail s_i with respect to the formation center c; see Fig. 7.8. If subscript $j \neq i$ denotes the sail s_j, a consensus (the control objective) is said to be reached when

$$(\boldsymbol{\rho}_i - \boldsymbol{\rho}_{ci}) \to (\boldsymbol{\rho}_j - \boldsymbol{\rho}_{cj}) \to \boldsymbol{\rho}_c \quad \cap \quad (\dot{\boldsymbol{\rho}}_i - \dot{\boldsymbol{\rho}}_{ci}) \to (\dot{\boldsymbol{\rho}}_j - \dot{\boldsymbol{\rho}}_{cj}) \to \mathbf{0}_3 \tag{7.38}$$

where $\boldsymbol{\rho}_c$ is the relative position vector of the formation center with respect to the point h on the nominal (artificial halo orbit) trajectory; see Fig. 7.8. In this case, the formation center is assumed to be initially unspecified. Indeed, it is determined by all of the solar sails through a negotiation process.

For convenience, introduce the auxiliary vectors $\boldsymbol{q}_i = \boldsymbol{\rho}_i - \boldsymbol{\rho}_{ci}$ and $\dot{\boldsymbol{q}}_i = \dot{\boldsymbol{\rho}}_i - \dot{\boldsymbol{\rho}}_{ci}$, so that Eq. (7.38) can be rewritten as

$$\boldsymbol{q}_i \to \boldsymbol{q}_j \to \boldsymbol{\rho}_c \quad \cap \quad \dot{\boldsymbol{q}}_i \to \dot{\boldsymbol{q}}_j \to \mathbf{0}_3 \tag{7.39}$$

and assume a cooperative control law in Eq. (7.33), defined as

$$\boldsymbol{u}_i = \mathbb{M}_h^{-1} \left\{ \ddot{\boldsymbol{\rho}}_{ci} + 2\,\mathbb{M}_v\, \dot{\boldsymbol{\rho}}_i + \mathbb{M}_p\, \boldsymbol{\rho}_i - \alpha\, \dot{\boldsymbol{q}}_i - \sum_{j=1}^{N} w_{ij} \left[\left(\boldsymbol{q}_i - \boldsymbol{q}_j \right) + \gamma \left(\dot{\boldsymbol{q}}_i - \dot{\boldsymbol{q}}_j \right) \right] \right\} \tag{7.40}$$

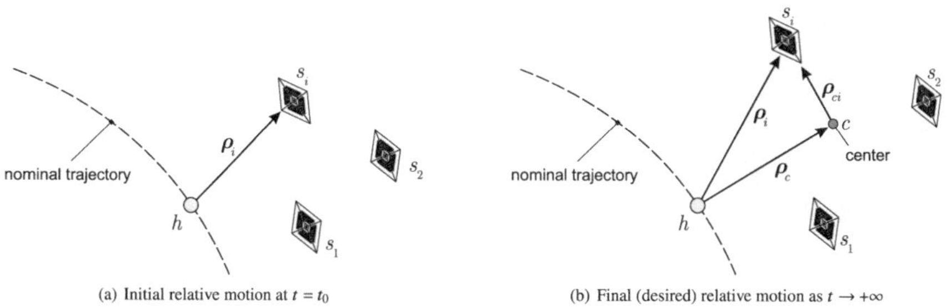

Fig. 7.8 Schematic illustration of the leaderless solar sail formation structure

where $\{\alpha, \gamma\} \in \mathbb{R}^+$, while the matrices $\mathbb{M}_p$, $\mathbb{M}_v$, and $\mathbb{M}_h$ are given by Eqs. (7.34)–(7.36).

Theorem 1 *Using the control law* (7.40) *with the solar sail formation structure represented by Eq.* (7.33)*, consensus (that is, conditions* (7.39)*) is asymptotically achieved if the topology graph* $\mathcal{G}$ *includes (at least) a spanning tree, and the coefficient* γ *satisfies*

$$\gamma > \max_{\{i \,|\, \mathrm{Re}(\lambda_i)<0\}} \sqrt{\frac{2}{|\lambda_i| \cos\left[\dfrac{\pi}{2} - \arctan\dfrac{-\mathrm{Re}(\lambda_i)}{\mathrm{Im}(\lambda_i)}\right]}} \tag{7.41}$$

where λ_i *is the* i*th eigenvalue of matrix* $-\mathbb{L}$.

Proof With the control input of Eq. (7.40), the second-order differential equation for $\boldsymbol{q}_i$ can be written as

$$\ddot{\boldsymbol{q}}_i + \alpha\,\dot{\boldsymbol{q}}_i + \sum_{j=1}^{N} w_{ij}\,\left[(\boldsymbol{q}_i - \boldsymbol{q}_j) + \gamma\,(\dot{\boldsymbol{q}}_i - \dot{\boldsymbol{q}}_j)\right] = \boldsymbol{0}_3 \tag{7.42}$$

which can be collected into a compact state-space form by introducing the auxiliary vector $\boldsymbol{q} = \left[\boldsymbol{q}_1^{\mathrm{T}}, \ldots, \boldsymbol{q}_N^{\mathrm{T}}\right]^{\mathrm{T}}$, viz.,

$$\left[\dot{\boldsymbol{q}}^{\mathrm{T}}, \ddot{\boldsymbol{q}}^{\mathrm{T}}\right]^{\mathrm{T}} = (\mathbb{Q} \otimes \mathbb{I}_3)\,\left[\boldsymbol{q}^{\mathrm{T}}, \dot{\boldsymbol{q}}^{\mathrm{T}}\right]^{\mathrm{T}} \tag{7.43}$$

where $\otimes$ denotes the Kronecker product, and the matrix $\mathbb{Q} \in \mathbb{R}^{2N\times 2N}$ is defined as

$$\mathbb{Q} = \begin{bmatrix} \mathbb{O}_N & \mathbb{I}_N \\ -\mathbb{L} & -\alpha\,\mathbb{I}_N - \gamma\,\mathbb{L} \end{bmatrix} \tag{7.44}$$

The eigenvalues of $\mathbb{Q}$ are found by enforcing the condition $\det(\eta\,\mathbb{I}_{2N} - \mathbb{Q}) = 0$, which leads to

$$\det\left[\left(\eta^2 + \alpha\,\eta\right)\mathbb{I}_N + (1 + \gamma\,\eta)\,\mathbb{L}\right] = 0 \tag{7.45}$$

The roots of Eq. (7.45) are calculated by solving $\eta^2 + \alpha\,\eta = \lambda_i\,(1 + \gamma\,\eta)$, and the eigenvalues of $\mathbb{Q}$ are given by

$$\eta_{i\pm} = \frac{\gamma\,\lambda_i - \alpha \pm \sqrt{(\gamma\,\lambda_i - \alpha)^2 + 4\,\lambda_i}}{2} \tag{7.46}$$

If there exists a spanning tree in the communication topology, then $-\mathbb{L}$ has a simple zero eigenvalue, while all other eigenvalues have negative real parts [47]. In this case, it can be deduced from Eq. (7.46) that all of the eigenvalues of $\mathbb{Q}$ have negative real parts, i.e. $\mathrm{Re}\,(\eta_i) < 0$, and the system is asymptotically stable if the inequality in Eq. (7.41) holds true.

Having discussed the conditions under which a consensus may be reached, another important problem is to find the equilibrium state for the solar sails. To that end, consider the linear system represented by Eq. (7.43), whose state-transition matrix is

$$\mathrm{e}^{(\mathbb{Q}\otimes\mathbb{I}_3)\,t} = \mathrm{e}^{(\mathbb{Q}\,t)\otimes\mathbb{I}_3} = \mathrm{e}^{\mathbb{Q}\,t}\otimes\mathbb{I}_3 \tag{7.47}$$

As is shown in Ref. [47], as $t \to +\infty$ the following relationship holds

$$\lim_{t\to+\infty} \mathrm{e}^{\mathbb{Q}\,t} = \begin{bmatrix} \mathbf{1}_N\,\boldsymbol{p}^{\mathrm{T}} & \frac{1}{\alpha}\,\mathbf{1}_N\,\boldsymbol{p}^{\mathrm{T}} \\ \mathbb{O}_N & \mathbb{O}_N \end{bmatrix} \tag{7.48}$$

where $\mathbf{1}_N \in \mathbb{R}^N$ denotes the column vector of all ones, and $\boldsymbol{p} \in \mathbb{R}^N$ is a left eigenvector of the matrix $-\mathbb{L}$ associated with eigenvalue 0, such that $\boldsymbol{p}\cdot\mathbf{1}_N = 1$. According to Eqs. (7.43), (7.47) and (7.48), it follows that

$$\boldsymbol{q} \to \left[\left(\mathbf{1}_N\,\boldsymbol{p}^{\mathrm{T}}\right)\otimes\mathbb{I}_3\right]\boldsymbol{q}_c + \frac{1}{\alpha}\left[\left(\mathbf{1}_N\,\boldsymbol{p}^{\mathrm{T}}\right)\otimes\mathbb{I}_3\right]\dot{\boldsymbol{q}}_c, \qquad \dot{\boldsymbol{q}} \to \mathbf{0}_{3\,N} \tag{7.49}$$

where $\boldsymbol{q}_c = \left[\boldsymbol{q}_{c1}^{\mathrm{T}}, \ldots, \boldsymbol{q}_{cN}^{\mathrm{T}}\right]^{\mathrm{T}}$ is the initial value of $\boldsymbol{q}$.

Let p_i be the ith component of the left eigenvector $\boldsymbol{p}$, it can be verified from Eq. (7.49) that the final position vector $\boldsymbol{\rho}_c$ of the formation center c relative to a generic point h on the nominal trajectory is

$$\boldsymbol{\rho}_c = \lim_{t\to+\infty} \boldsymbol{q} = \sum_{i=1}^{N} p_i\left(\boldsymbol{q}_{ci} + \frac{1}{\alpha}\,\dot{\boldsymbol{q}}_{ci}\right) \tag{7.50}$$

which amounts to stating that consensus is guaranteed in the sense of the conditions given by Eq. (7.39). □

Note that with the control law given by Eq. (7.40), the final location of the formation center is defined by a weighted average of the initial states of all solar sails. **Theorem** 1 also indicates that $\left(\boldsymbol{\rho}_i - \boldsymbol{\rho}_j\right) \to \left(\boldsymbol{\rho}_{ci} - \boldsymbol{\rho}_{cj}\right)$. Therefore, $\boldsymbol{\rho}_{ci}$ and $\boldsymbol{\rho}_{cj}$ can be selected so as to acquire a desired separation between the solar sails s_i and s_j.

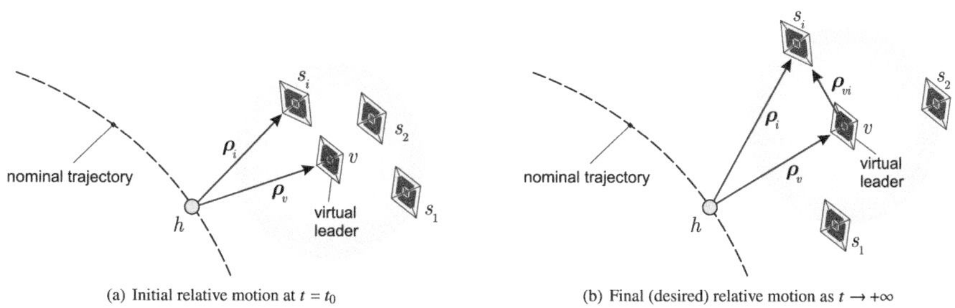

Fig. 7.9 Schematic illustration of the virtual-leader solar sail formation flying

7.5.2 *Virtual-Leader Case*

In some scientific missions, such as those involving synthesizing interferometers [27, 53], it is necessary to prescribe the state relative to some reference point for each vehicle, and treat the entire formation as a single entity [31, 54]. In those cases, a reference point plays the role of virtual leader (v) of the solar sail formation structure, around which a desired relative configuration can be formed by properly designing the cooperative control strategy. To that end, a distributed control algorithm gives good performance especially when the state of the virtual leader cannot be known precisely due to the limitations on the communication bandwidth and the sensing range. Such a control system will now be designed in two different subcases, depending on whether i) all of the sails or ii) only a subgroup of the sails can share information of the virtual leader state.

Let $\boldsymbol{\rho}_v$ denote the position vector of the virtual leader v relative to a generic point h on the nominal trajectory, and $\{\boldsymbol{\rho}_{vi}, \dot{\boldsymbol{\rho}}_{vi}\} \in \mathbb{R}^3 \times \mathbb{R}^3$ be the desired relative state of the ith solar sail s_i with respect to the virtual leader; see Fig. 7.9. In contrast to the leaderless case, the position vector $\boldsymbol{\rho}_v$ is here specified in advance and is possibly time-varying. The goal of the cooperative control law is to drive each solar sail in formation toward the desired relative trajectory, while guaranteeing a group synchronization during the maneuver, viz.

$$\boldsymbol{\rho}_i - \boldsymbol{\rho}_{vi} \to \boldsymbol{\rho}_j - \boldsymbol{\rho}_{vj} \to \boldsymbol{\rho}_v \quad \cap \quad \dot{\boldsymbol{\rho}}_i - \dot{\boldsymbol{\rho}}_{vi} \to \dot{\boldsymbol{\rho}}_j - \dot{\boldsymbol{\rho}}_{vj} \to \boldsymbol{\rho}_v \tag{7.51}$$

7.5.2.1 Full Access to the Virtual Leader State

When the information of the virtual leader state is available to all of the solar sails, the consensus protocol is designed as

$$\boldsymbol{u}_i = \mathbb{M}_h^{-1} \left\{ \ddot{\boldsymbol{\rho}}_v + \ddot{\boldsymbol{\rho}}_{vi} + 2\,\mathbb{M}_v\,\dot{\boldsymbol{\rho}}_i + \mathbb{M}_p\,\boldsymbol{\rho}_i - \sigma\,\left(\boldsymbol{e}_i + \zeta\,\dot{\boldsymbol{e}}_i\right) - \sum_{j=1}^{N} w_{ij}\,\left[\left(\boldsymbol{e}_i - \boldsymbol{e}_j\right) + \zeta\,\left(\dot{\boldsymbol{e}}_i - \dot{\boldsymbol{e}}_j\right)\right] \right\} \tag{7.52}$$

where the parameters $\{\sigma, \zeta\} \in \mathbb{R}^+$, while the position error and the velocity error of the sail s_i are $\boldsymbol{e}_i = \boldsymbol{\rho}_i - \boldsymbol{\rho}_v - \boldsymbol{\rho}_{vi}$ and $\dot{\boldsymbol{e}}_i = \dot{\boldsymbol{\rho}}_i - \dot{\boldsymbol{\rho}}_v - \dot{\boldsymbol{\rho}}_{vi}$, respectively.

Theorem 2 *Using the control law given by Eq. (7.52) for the solar sail-based formation structure described by Eq. (7.33), consensus (that is,* $\boldsymbol{e}_i \to \boldsymbol{e}_j \to \boldsymbol{0}_3$ *and* $\dot{\boldsymbol{e}}_i \to \dot{\boldsymbol{e}}_j \to \boldsymbol{0}_3$*) is asymptotically achieved if the topology graph* $\mathcal{G}$ *includes (at least) a spanning tree and the coefficient* ζ *satisfies*

$$\zeta > \max_{i=1,\ldots,N} \sqrt{\frac{2}{|\nu_i|\,\cos\left[\dfrac{\pi}{2} - \arctan\dfrac{-\operatorname{Re}(\nu_i)}{\operatorname{Im}(\nu_i)}\right]}} \tag{7.53}$$

where $\nu_i = -\sigma + \lambda_i$ *is the* i*th eigenvalue of* $-\left(\sigma\,\mathbb{I}_N + \mathbb{L}\right)$.

Proof Using the control law (7.52) and bearing in mind the relative motion dynamics given by Eq. (7.33), the differential equation for the position error $\boldsymbol{e}_i$ can be written as

$$\ddot{\boldsymbol{e}}_i + \sigma\,\left(\boldsymbol{e}_i + \zeta\,\dot{\boldsymbol{e}}_i\right) + \sum_{j=1}^{N} w_{ij}\,\left[\left(\boldsymbol{e}_i - \boldsymbol{e}_j\right) + \zeta\,\left(\dot{\boldsymbol{e}}_i - \dot{\boldsymbol{e}}_j\right)\right] = \boldsymbol{0} \tag{7.54}$$

By introducing the auxiliary vector $\boldsymbol{e} = \left[\boldsymbol{e}_1^{\mathrm{T}}, \ldots, \boldsymbol{e}_N^{\mathrm{T}}\right]^{\mathrm{T}}$, a compact form of Eq. (7.54) is given by

$$\left[\dot{\boldsymbol{e}}^{\mathrm{T}}, \ddot{\boldsymbol{e}}^{\mathrm{T}}\right]^{\mathrm{T}} = \left(\mathbb{E} \otimes \mathbb{I}_3\right)\,\left[\boldsymbol{e}^{\mathrm{T}}, \dot{\boldsymbol{e}}^{\mathrm{T}}\right]^{\mathrm{T}} \tag{7.55}$$

where the matrix $\mathbb{E} \in \mathbb{R}^{2N \times 2N}$ is defined as

$$\mathbb{E} = \begin{bmatrix} \mathbb{O}_N & \mathbb{I}_N \\ -\left(\sigma\,\mathbb{I}_N + \mathbb{L}\right) & -\zeta\,\left(\sigma\,\mathbb{I}_N + \mathbb{L}\right) \end{bmatrix} \tag{7.56}$$

The eigenvalues of $\mathbb{E}$ are

$$\xi_{i\pm} = \frac{\zeta\,\nu_i \pm \sqrt{\zeta^2\,\nu_i^2 + 4\,\nu_i}}{2} \tag{7.57}$$

Paralleling the approach in the leaderless case, it is straightforward to see that each eigenvalue ξ_i (with $i = 1, \ldots, N$) has a negative real part if Eq. (7.53) is satisfied, which implies $\boldsymbol{e}_i \to \boldsymbol{0}_3$ and $\dot{\boldsymbol{e}}_i \to \boldsymbol{0}_3$. In particular, a synchronized transition, i.e. $\boldsymbol{e}_i \to \boldsymbol{e}_j \to \boldsymbol{0}_3$ and $\dot{\boldsymbol{e}}_i \to \dot{\boldsymbol{e}}_j \to \boldsymbol{0}_3$, can be guaranteed when the information exchange topology has a spanning tree due to the state feedback couplings among the sails [47].

7.5.2.2 Partial Access to the Virtual Leader State

The cooperative control algorithm given by Eq. (7.52) assumes the virtual leader state to be available to each solar sail, which is usually unrealistic in practice. Considering the fact that the solar sail-based spacecraft usually conveys a rather limited payload and communication instrument mass, it is possible that the information of the virtual leader state can be measured only by the onboard sensors of a part of the sails in formation, and these sails may not have a directed path to all the others. Such a circumstance necessitates an alternative algorithm accounting for the general case where only a part of the formation has access to the virtual leader state. To this end, the consensus strategy is proposed in the form

$$
\begin{aligned}
\boldsymbol{u}_i = \mathbb{M}_h^{-1} \Bigg\{ & \ddot{\boldsymbol{\rho}}_{vi} + 2\,\mathbb{M}_v\, \dot{\boldsymbol{\rho}}_i + \mathbb{M}_p\, \boldsymbol{\rho}_i + \frac{1}{\sum_{j=0}^{N} w_{ij}} \sum_{j=1}^{N} w_{ij} \left[\left(\ddot{\boldsymbol{\rho}}_j - \ddot{\boldsymbol{\rho}}_{vj} \right) - \delta \left(\boldsymbol{e}_i - \boldsymbol{e}_j \right) - \kappa \left(\dot{\boldsymbol{e}}_i - \dot{\boldsymbol{e}}_j \right) \right] \\
& + \frac{1}{\sum_{j=0}^{N} w_{ij}} w_{vi} \left(\ddot{\boldsymbol{\rho}}_v - \delta\, \boldsymbol{e}_i - \kappa\, \dot{\boldsymbol{e}}_i \right) \Bigg\}
\end{aligned}
\tag{7.58}
$$

where $\{\delta, \kappa\} \in \mathbb{R}^+$, $w_{i0} > 0$, $\forall i = 0, \ldots, N$, if the sail s_i has access to the state of the virtual leader and $w_{i0} = 0$ otherwise. Note that the virtual leader is treated as a virtual solar sail with index 0.

Theorem 3 *Using the control law of Eq. (7.58) for the solar sail-based formation system described by Eq. (7.33), consensus (that is,* $\boldsymbol{e}_i \to \boldsymbol{e}_j \to \boldsymbol{0}_3$ *and* $\dot{\boldsymbol{e}}_i \to \dot{\boldsymbol{e}}_j \to \boldsymbol{0}_3$*) is asymptotically achieved if the virtual leader* v *is the root that has a path to each solar sail in formation.*

Proof With the control input of Eq. (7.58), it can be derived from Eq. (7.33) that

$$
\begin{aligned}
\sum_{j=0}^{N} w_{ij} \left(\ddot{\boldsymbol{\rho}}_i - \ddot{\boldsymbol{\rho}}_{vi} \right) &= \sum_{j=1}^{N} w_{ij} \left[\left(\ddot{\boldsymbol{\rho}}_j - \ddot{\boldsymbol{\rho}}_{vj} \right) - \delta \left(\boldsymbol{e}_i - \boldsymbol{e}_j \right) - \kappa \left(\dot{\boldsymbol{e}}_i - \dot{\boldsymbol{e}}_j \right) \right] + w_{i0} \left(\ddot{\boldsymbol{\rho}}_v - \delta\, \boldsymbol{e}_i - \kappa\, \dot{\boldsymbol{e}}_i \right) \\
&= \sum_{j=0}^{N} w_{ij} \left[\left(\ddot{\boldsymbol{\rho}}_j - \ddot{\boldsymbol{\rho}}_{vj} \right) - \delta \left(\boldsymbol{e}_i - \boldsymbol{e}_j \right) - \kappa \left(\dot{\boldsymbol{e}}_i - \dot{\boldsymbol{e}}_j \right) \right]
\end{aligned}
\tag{7.59}
$$

Let $\boldsymbol{\epsilon}_i = \sum_{j=0}^{N} w_{ij} \left(\boldsymbol{e}_i - \boldsymbol{e}_j \right)$, and recall the identity $\boldsymbol{\rho}_i - \boldsymbol{\rho}_{vi} - \boldsymbol{\rho}_j + \boldsymbol{\rho}_{vj} = \boldsymbol{e}_i - \boldsymbol{e}_j$. Then Eq. (7.59) becomes

$$
\ddot{\boldsymbol{\epsilon}}_i + \delta\, \dot{\boldsymbol{\epsilon}}_i + \kappa\, \boldsymbol{\epsilon}_i = \boldsymbol{0}
\tag{7.60}
$$

Since $\delta > 0$ and $\kappa > 0$, it follows from Eq. (7.60) that $\boldsymbol{\epsilon}_i \to \boldsymbol{0}_3$ and $\dot{\boldsymbol{\epsilon}}_i \to \boldsymbol{0}_3$ as $t \to +\infty$. For convenience, introduce the vector $\boldsymbol{l} = [l_{10}, \ldots, l_{N0}]^{\mathrm{T}} \in \mathbb{R}^N$, with

$l_{i0} = -w_{i0}$, and let $\widetilde{\mathbb{L}} = \left[\mathbf{0}_N^{\mathrm{T}},\ 0;\ \mathbb{L},\ \boldsymbol{l}\right] \in \mathbb{R}^{(N+1)\times(N+1)}$ be the extended Laplacian matrix involving the virtual leader. It is straightforward to verify that $\left(\widetilde{\mathbb{L}} \otimes \mathbb{I}_3\right) \tilde{\boldsymbol{e}} \to \mathbf{0}_{3N}$ and $\left(\widetilde{\mathbb{L}} \otimes \mathbb{I}_3\right) \dot{\tilde{\boldsymbol{e}}} \to \mathbf{0}_{3N}$, where $\tilde{\boldsymbol{e}} = \left[\boldsymbol{e}_0^{\mathrm{T}},\ \boldsymbol{e}^{\mathrm{T}}\right]^{\mathrm{T}}$. Note that for the virtual leader, the state errors $\boldsymbol{e}_0 \equiv \mathbf{0}_3$ and $\dot{\boldsymbol{e}}_0 \equiv \mathbf{0}_3$. If the virtual leader has a path to each solar sail, then the information exchange topology involving the virtual leader includes a rooted spanning tree, which further results in $\boldsymbol{e}_i \to \boldsymbol{e}_j \to \boldsymbol{e}_0 = \mathbf{0}_3$ and $\dot{\boldsymbol{e}}_i \to \dot{\boldsymbol{e}}_j \to \boldsymbol{e}_0 = \mathbf{0}_3$.

In contrast to the case in which the virtual leader state is accessible to all of the solar sails, Eq. (7.58) requires the acceleration of the neighbor sail, which may be calculated by numerical differentiation when onboard measurement is unavailable. For example, in the simplest case, $\ddot{\boldsymbol{\rho}}_j$ on the right side of Eq. (7.58) can be approximated by $\left(\dot{\boldsymbol{\rho}}_j\ [k+1] - \dot{\boldsymbol{\rho}}_j\ [k]\right)/\tau$, where k is the discrete-time index, and τ is the sampling interval. Other sampling algorithms can also be adopted such as the Runge-Kutta method.

7.6 Mission Application

To illustrate the performance of the proposed cooperative control laws, a formation mission scenario consisting of three solar sails ($N = 3$) around an artificial halo orbit centered at the Sun-[Earth+Moon] L_2-type AEP is now discussed. The nominal trajectory is obtained with $u_h = 0.4$ and $\beta_h = 0.01$. In this example, the solar sail relative trajectories with respect to the formation center track a projected circular orbit (PCO) with a radius $R = 10\,\text{km}$. Accordingly, for the leaderless case (see Fig. 7.8), the PCO parametric representation of the sail s_i (with $i \in \{1, 2, 3\}$) is assumed to have the following algebraic form

$$\boldsymbol{\rho}_{ci} = (R/r_{\oplus})\ [0,\ \cos(\omega_h t + \varphi_i),\ \sin(\omega_h t + \varphi_i)]^{\mathrm{T}} \tag{7.61}$$

whereas for the virtual-leader case (see Fig. 7.9), the PCO parametric representation is

$$\boldsymbol{\rho}_{vi} = (R/r_{\oplus})\ [0,\ \cos(\omega_h t + \varphi_i),\ \sin(\omega_h t + \varphi_i)]^{\mathrm{T}} \tag{7.62}$$

where $\varphi_i = 2\,(i-1)\,\pi/3$ is the phase angle of the PCO.

In this case, the approximation of the nominal trajectory by a 6th-order Fourier series ($n = 6$) is illustrated in Fig. 7.10, while the dimensionless coefficients $\{\boldsymbol{c}_k, \boldsymbol{s}_k\}$ of Eq. (7.18), evaluated in the least-square sense, are listed in Table 7.1.

The accuracy of the 6th-order Fourier series approximation is confirmed by Fig. 7.11, which shows the time variation of the dimensionless (distance) error $\Delta\tilde{r} = (\|\boldsymbol{r}\| - \|\boldsymbol{r}_h\|)\,/\,\|\boldsymbol{r}\|$. According to Fig. 7.11, the maximum value of $|\Delta\tilde{r}|$ is less than 5×10^{-7}, indicating a good accuracy of the analytical approximation.

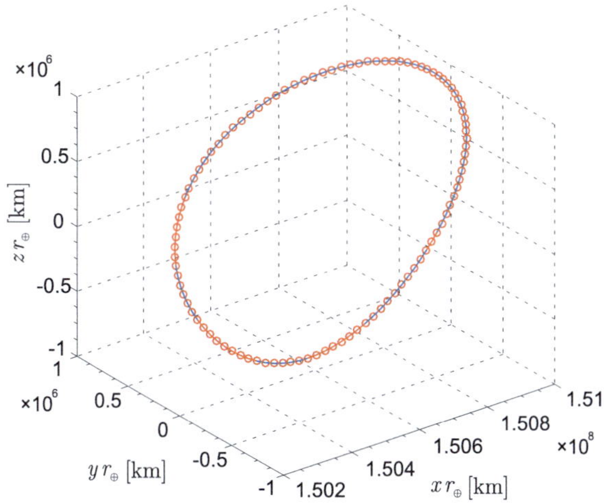

Fig. 7.10 Nominal trajectory: approximate form (circles) vs. the numerical solution (solid line)

Table 7.1 Coefficients of the 6th-order Fourier series approximation of the nominal trajectory, see Eq. (7.18)

k	$\boldsymbol{c}_k$	$\boldsymbol{s}_k$
0	$\left[1.0070,\ 3.0854 \times 10^{-8},\ 1.7412 \times 10^{-3}\right]^{\mathrm{T}}$	$[0,\ 0,\ 0]^{\mathrm{T}}$
1	$\left[2.0039 \times 10^{-3},\ 3.3922 \times 10^{-7},\ 5.3430 \times 10^{-3}\right]^{\mathrm{T}}$	$\left[-2.2928 \times 10^{-9},\ -6.5287 \times 10^{-3},\ 1.2619 \times 10^{-7}\right]^{\mathrm{T}}$
2	$\left[6.5432 \times 10^{-5},\ -5.3794 \times 10^{-8},\ -5.3612 \times 10^{-4}\right]^{\mathrm{T}}$	$\left[1.9804 \times 10^{-9},\ 5.2543 \times 10^{-4},\ -3.2178 \times 10^{-8}\right]^{\mathrm{T}}$
3	$\left[-2.2999 \times 10^{-5},\ 3.5717 \times 10^{-8},\ 1.0509 \times 10^{-4}\right]^{\mathrm{T}}$	$\left[-1.8103 \times 10^{-9},\ -1.0514 \times 10^{-4},\ 1.4718 \times 10^{-8}\right]^{\mathrm{T}}$
4	$\left[6.8071 \times 10^{-6},\ -6.7082 \times 10^{-9},\ -2.4949 \times 10^{-5}\right]^{\mathrm{T}}$	$\left[-1.8024 \times 10^{-9},\ 2.5228 \times 10^{-5},\ -9.8589 \times 10^{-9}\right]^{\mathrm{T}}$
5	$\left[-2.0513 \times 10^{-6},\ 2.1746 \times 10^{-8},\ 6.5864 \times 10^{-6}\right]^{\mathrm{T}}$	$\left[-1.7928 \times 10^{-9},\ -6.7798 \times 10^{-6},\ -8.8916 \times 10^{-9}\right]^{\mathrm{T}}$
6	$\left[5.6290 \times 10^{-7},\ -2.2537 \times 10^{-9},\ -1.8496 \times 10^{-6}\right]^{\mathrm{T}}$	$\left[1.7893 \times 10^{-9},\ 1.8775 \times 10^{-6},\ -8.9881 \times 10^{-9}\right]^{\mathrm{T}}$

Fig. 7.11 Dimensionless distance error of the 6th-order Fourier series-based approximation with respect to the numerical solution

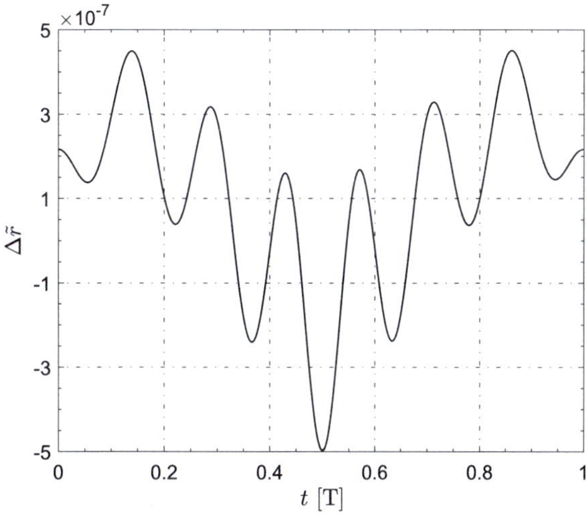

Table 7.2 Initial errors of three solar sails in the leaderless case

	$e_{ij0} \cdot \hat{x}$ [km]	$e_{ij0} \cdot \hat{y}$ [km]	$e_{ij0} \cdot \hat{z}$ [km]	$\dot{e}_{ij0} \cdot \hat{x}$ [m/s]	$\dot{e}_{ij0} \cdot \hat{y}$ [m/s]	$\dot{e}_{ij0} \cdot \hat{z}$ [m/s]
$i = 1, j = 2$	1	2.5	−2	1×10^{-3}	-2×10^{-3}	2×10^{-3}
$i = 2, j = 3$	−2.5	−1.5	2.5	-1×10^{-3}	2×10^{-3}	-2×10^{-3}
$i = 3, j = 1$	1.5	−1	−0.5	-1×10^{-3}	1×10^{-3}	2×10^{-3}

7.6.1 Leaderless Case

Consider first the leaderless case in which a centralized leadership is absent and the formation center is not a priori specified. To evaluate the controller performance of Eq. (7.40), introduce the error indices $\boldsymbol{e}_{ij} = \boldsymbol{q}_i - \boldsymbol{q}_j$ and $\dot{\boldsymbol{e}}_{ij} = \dot{\boldsymbol{q}}_i - \dot{\boldsymbol{q}}_j$, with initial values reported in Table 7.2.

The information exchange topology characterizing the mutual interaction of the three solar sails is assumed to follow a cyclic pursuit structure, as is illustrated in Fig. 7.12. In this case, the weighted adjacency matrix $\mathbb{W}$ is given by

$$\mathbb{W} = \begin{bmatrix} 0 & 8 \times 10^4 & 0 \\ 0 & 0 & 8 \times 10^4 \\ 8 \times 10^4 & 0 & 0 \end{bmatrix} \tag{7.63}$$

For a favorable control performance (in terms of convergence time and tracking accuracy), the dimensionless parameters in Eq. (7.40) are selected as $\alpha = 2.5 \times 10^2$ and $\gamma = 5 \times 10^{-3}$. Note that γ satisfies the constraint of Eq. (7.41), i.e. $\gamma > 4.1 \times 10^{-3}$.

The position and velocity errors of the three solar sails are shown in Figs. 7.13 and 7.14 respectively, while the time variations of the (relative) control variables $\Delta\,\theta_i$, $\Delta\,\phi_i$ and $\Delta\,u_i$ are shown in Fig. 7.15.

Figure 7.16 illustrates the relative transfer trajectories of the three sails, whose initial positions are denoted with the star. It may be seen from the figures that with the control law (7.40), all errors converge to zero within 1.5 days and a good synchronization is maintained along the whole maneuver. Moreover, the final consensus guarantees a similar transition time to all of the solar sails, which is a crucial point for practical applications.

7.6.2 Virtual-Leader Case

Consider now the case where a virtual leader v is included in the solar sail-based formation structure. The initial conditions of the three solar sails are slightly different from the desired trajectories, with the initial errors listed in Table 7.3. The location of

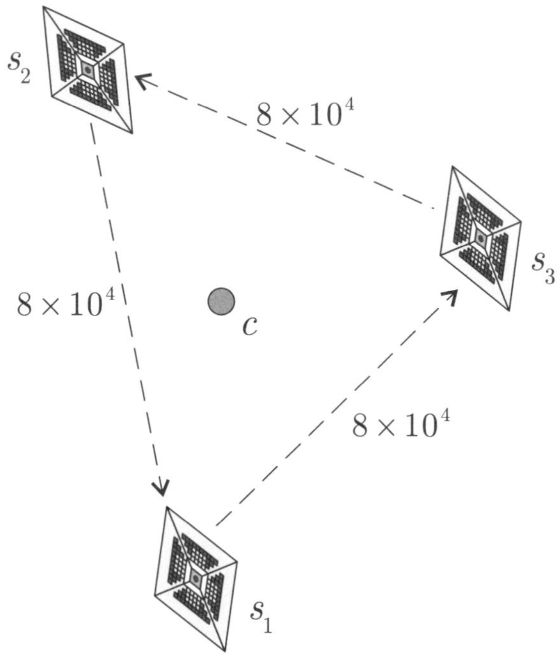

Fig. 7.12 Information exchange topology of the three solar sails in the leaderless case

the virtual leader has a small (constant) bias relative to the nominal trajectory. In the following examples, it is assumed that $\left[\boldsymbol{\rho}_v\right]_{\mathcal{T}} = [1,\ 0,\ 0]^{\mathrm{T}}$ km and $\left[\dot{\boldsymbol{\rho}}_v\right]_{\mathcal{T}} = \mathbf{0}_3$. In particular, illustrative examples will be discussed to verify the proposed control laws that account for the two previous subcases, where the virtual leader state is either accessible to all of the solar sails or to a subset of sails (that is, a part of the formation structure).

7.6.2.1 Full Access to the Virtual Leader State

When all of the solar sails in formation have access to the data from the virtual leader v, the communication topology is as illustrated in Fig. 7.17, where the information flows among the three spacecraft follow a cyclic pursuit pattern. In this case, the weighted adjacency matrix $\mathbb{W}$ is given by

$$\mathbb{W} = \begin{bmatrix} 0 & 1 & 0 \\ 0 & 0 & 1 \\ 1 & 0 & 0 \end{bmatrix} \tag{7.64}$$

The performance of the controller described by Eq. (7.52) is governed by the dimensionless parameters $\{\sigma,\ \zeta\}$, whose values are chosen as $\sigma = 10^5$ and $\zeta = 6.4 \times 10^{-3}$. Note that ζ satisfies the constraint of Eq. (7.53), i.e. $\zeta > 4.47 \times 10^{-3}$.

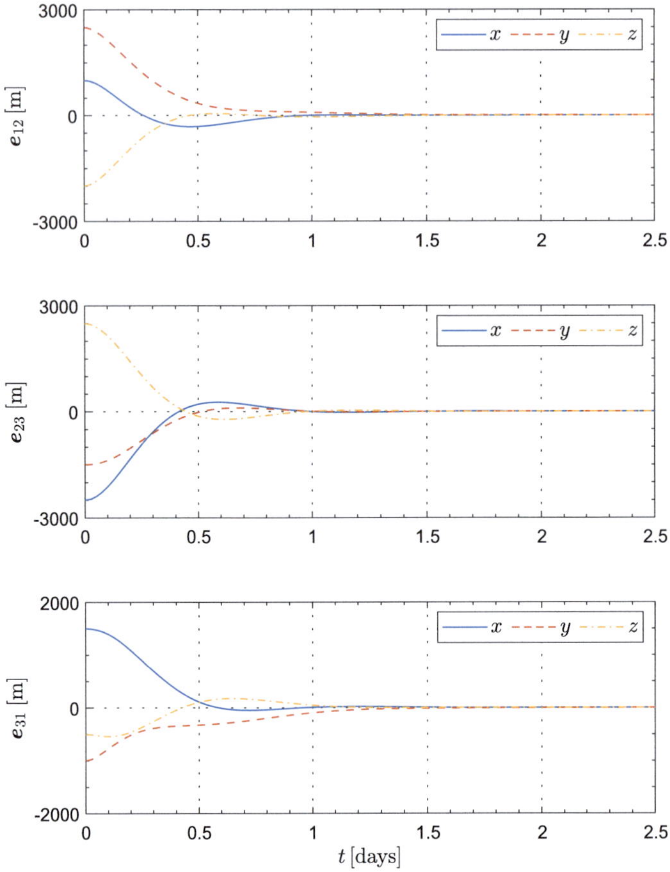

Fig. 7.13 Time history of the position errors $\boldsymbol{e}_{ij}$ in a leaderless case

To illustrate the performance of the control law given by Eq. (7.52), the time variations of the relative position and velocity errors are plotted in Figs. 7.18 and 7.19, for a time interval of 2.5 days. The corresponding time histories of the required control variables $\Delta\,\theta_i$, $\Delta\,\phi_i$ and $\Delta\,u_i$ are shown in Fig. 7.20, while the relative trajectories of the three sails are illustrated in Fig. 7.21. It can be seen that the transient errors of the three solar sails gradually converge to zero after about 1.5 days.

7.6.2.2 Partial Access to the Virtual Leader State

The control law given by Eq. (7.58) is now validated for the case in which the virtual leader state is available to only a part of the solar sail formation structure. Without loss of generality, only the solar sail s_1 is now assumed to have access to the virtual

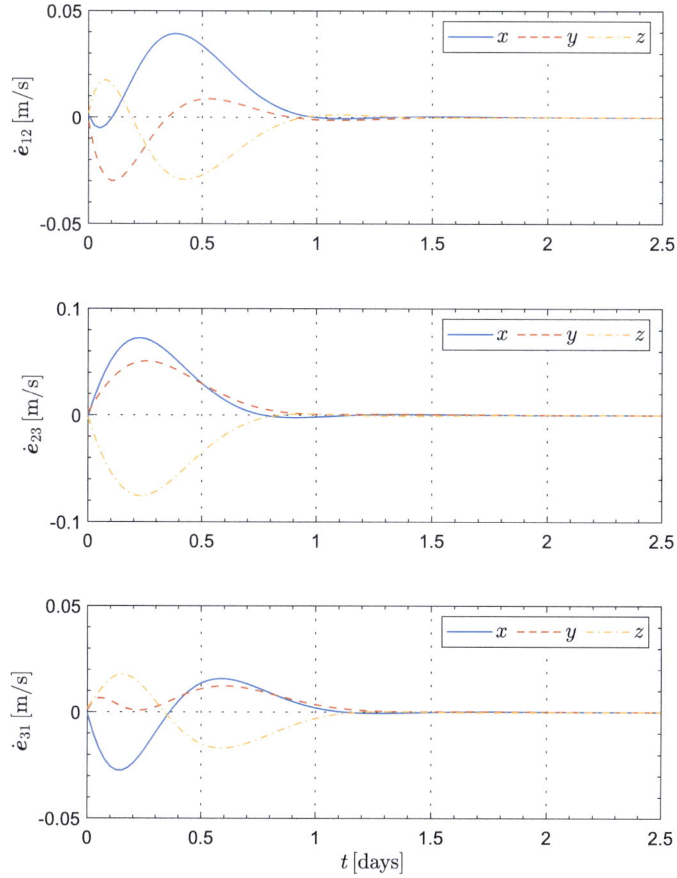

Fig. 7.14 Time history of the velocity errors $\dot{\boldsymbol{e}}_{ij}$ in a leaderless case

leader state, see Fig. 7.22. The initial errors take the same value as those reported in Table 7.3, and the weighted adjacent matrix is given by Eq. (7.64).

The parameters in Eq. (7.58) are selected as $\delta = 10^5$ and $\kappa = 7 \times 10^2$, and the weight of the directed edge between the virtual leader and the sail s_1 is $w_{10} = 1$. A first-order Eulerian differentiation method is adopted in the simulation, and the sampling interval is chosen to be $\tau = 5 \times 10^{-4}$ day.

The position and velocity errors of the three sails are illustrated in Figs. 7.23 and 7.24, respectively, while the control input $\boldsymbol{u}_i$ is shown in Fig. 7.25. The relative trajectories of the three sails are illustrated in Fig. 7.26. Note that, even in this case of partial access to the virtual leader state, the steady state errors of the three sails all converge to zero, while consensus is guaranteed during the whole transition phase.

The illustrative examples show that, to maintain a formation with a 10-km size, a propulsive acceleration with a magnitude of 0.05 mm/s^2 is required for each solar

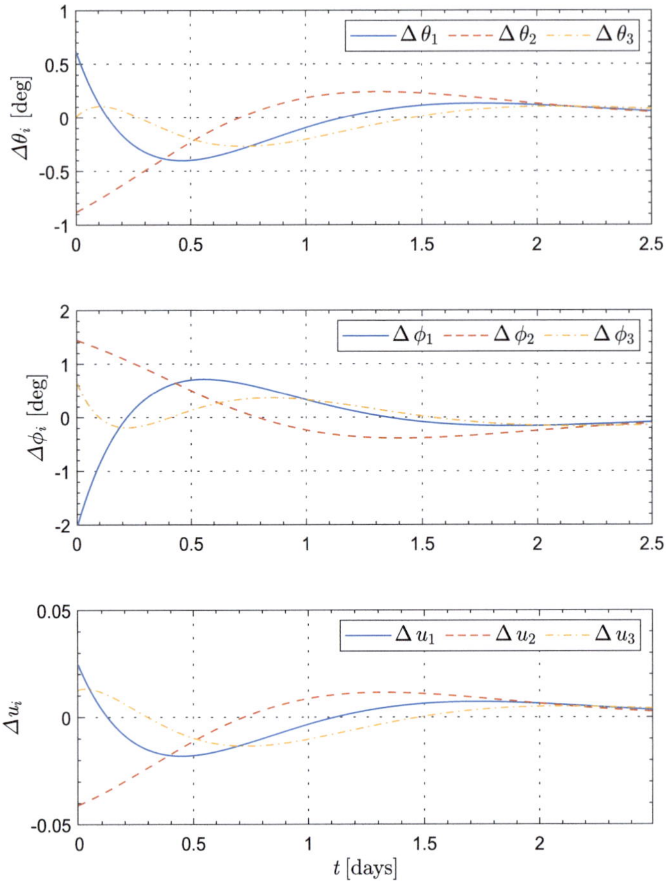

Fig. 7.15 Time history of the control input $\Delta\,\boldsymbol{u}_i$ of the three solar sails, see Eq. (7.40)

sail. Such a performance level is achievable by current technology. The previous theorems on cooperative formation control do not take into account any constraint on the initial errors, which however significantly affect the magnitude of the control input $\Delta\,\boldsymbol{u}_i$. In particular, for a fixed set of control parameters, an actuator saturation may occur when the magnitude of the initial errors exceeds some maximum value (to be found by simulation). For this reason, the choice of the control parameters is usually the output of a trial and error procedure. Nevertheless, some general remarks may simplify the design. For example, in the leaderless case it is preferable to choose a small value of γ in Eq. (7.40) to mitigate the saturation problem. In the virtual leader case, instead, the numerical simulations suggest decreasing ζ in Eq. (7.52) and decreasing κ in Eq. (7.58), when a saturation of control input occurs.

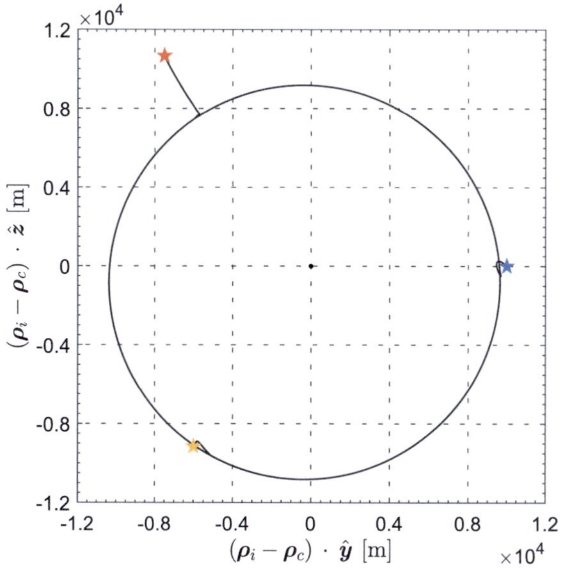

Fig. 7.16 Relative trajectories of the three solar sails in a leaderless case (star denotes initial position)

Table 7.3 Initial errors of three solar sails in the virtual-leader case

	$\boldsymbol{e}_{x0}$ [km]	$\boldsymbol{e}_{y0}$ [km]	$\boldsymbol{e}_{z0}$ [km]	$\dot{\boldsymbol{e}}_{x0}$ [m/s]	$\dot{\boldsymbol{e}}_{y0}$ [m/s]	$\dot{\boldsymbol{e}}_{z0}$ [m/s]
s_1	1	−2.5	2	1×10^{-3}	-2×10^{-3}	2×10^{-3}
s_2	2.5	−2	−1	-1×10^{-3}	2×10^{-3}	-2×10^{-3}
s_3	−2.5	2	1	-1×10^{-3}	1×10^{-3}	2×10^{-3}

Fig. 7.17 Information exchange topology where all the sails have access to the virtual leader

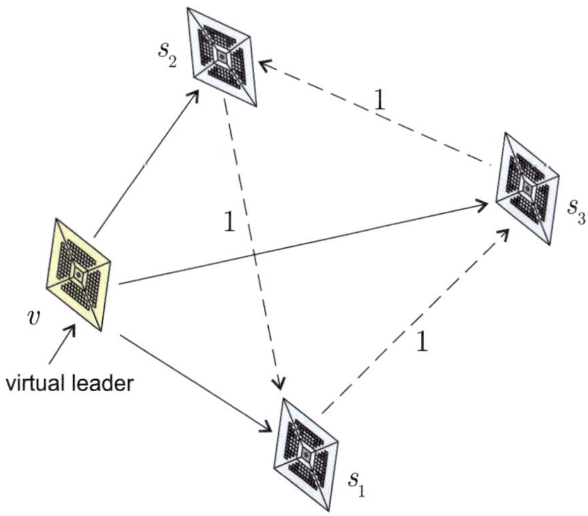

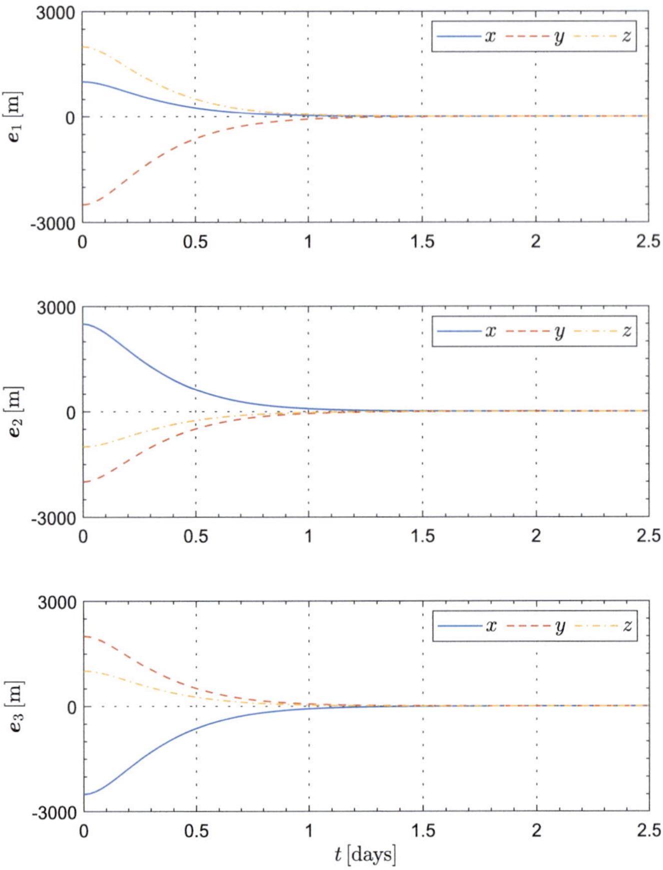

Fig. 7.18 Time history of the position errors $\boldsymbol{e}_i$ when all of the sails have access to the virtual leader

7.7 Conclusions

The problem of solar sail cooperative formation flying around L_2-type artificial equilibrium point orbits in the Sun-[Earth+Moon] system has been investigated. A truncated Fourier series has been used to parameterize the artificial halo orbit in an explicit (albeit approximate) form, which is used as a nominal trajectory for the solar sail formation. Distributed control strategies have been proposed to account for two typical cases, according to whether the formation system is leaderless or accommodates a virtual leader. The latter case has been further discussed when the state information of the virtual leader is available to all of the solar sails or to a part of the formation structure only.

The constitution and maintenance of the solar sail formation relies on the consensus protocols that are formulated on a general directed communication topology

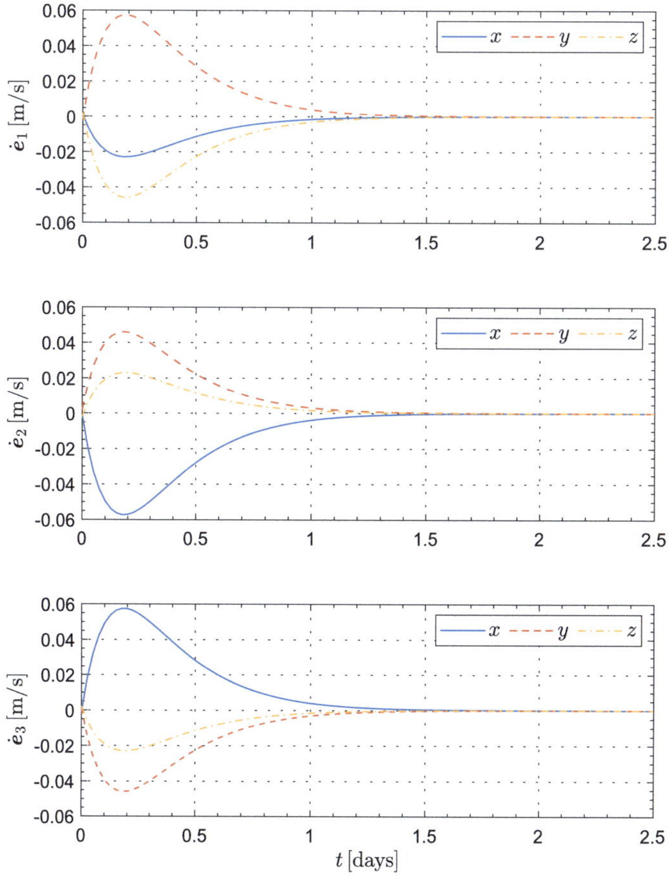

Fig. 7.19 Time history of the velocity errors $\dot{\boldsymbol{e}}_i$ when all of the sails have access to the virtual leader

with local neighbor-to-neighbor information networks. Such a cooperative formation structure drives each solar sail toward a desired trajectory in a coordinated way, so that the synchronized formation tracking can be achieved while improving the formation reliability.

Illustrative examples have shown that the proposed consensus-based algorithms also guarantee a time-balanced maneuver. Compared with a leader-follower topological structure, more flexility and robustness among the formation are incorporated both for the leaderless and the virtual-leader case, in the sense that onboard sensors with a limited field of view can be implemented, and data package loss for some interaction links can be accounted for.

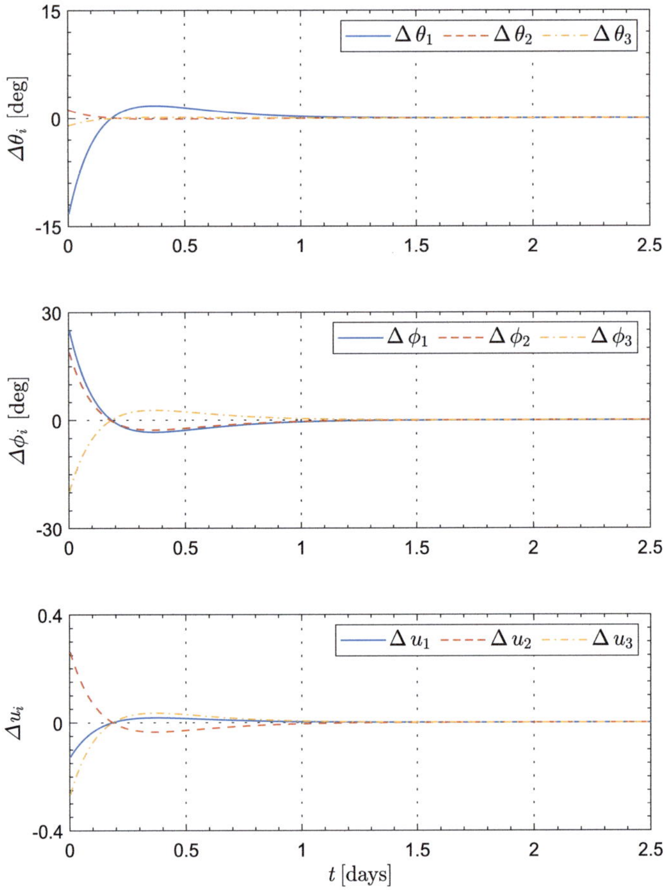

Fig. 7.20 Time history of the control input $\Delta\,\boldsymbol{u}_i$ of the three sails, see Eq. (7.52)

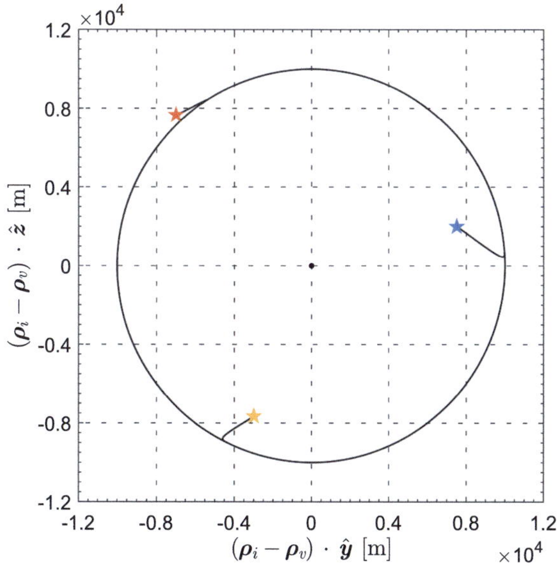

Fig. 7.21 Relative trajectories of the three sails when all of the sails have access to the virtual leader

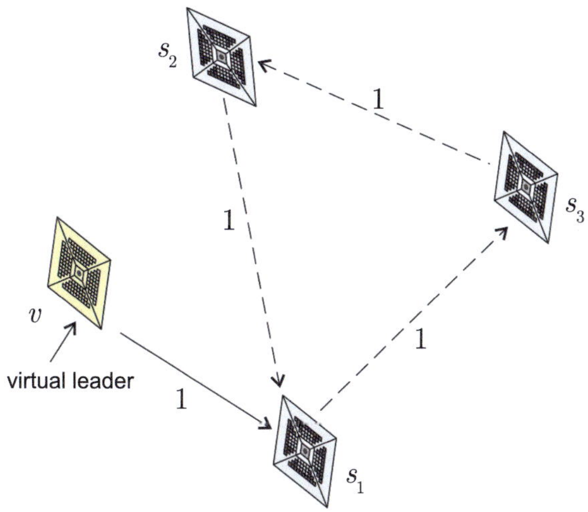

Fig. 7.22 Information exchange topology where only a part of the formation have access to the virtual leader

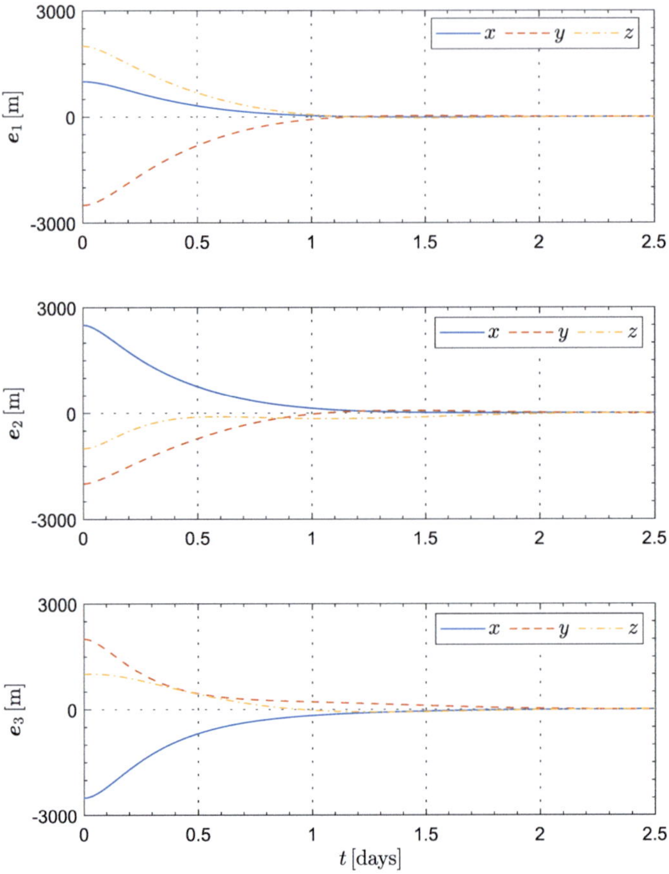

Fig. 7.23 Time history of the position errors $\boldsymbol{e}_i$ when the sail s_1 only has access to the virtual leader

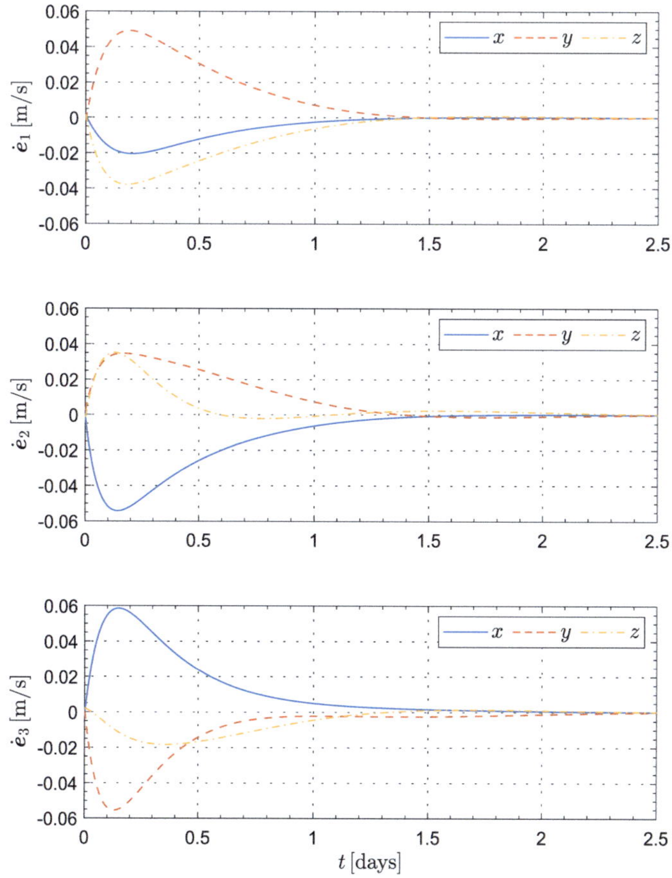

Fig. 7.24 Time history of the velocity errors $\dot{\boldsymbol{e}}_i$ when the sail s_1 only has access to the virtual leader

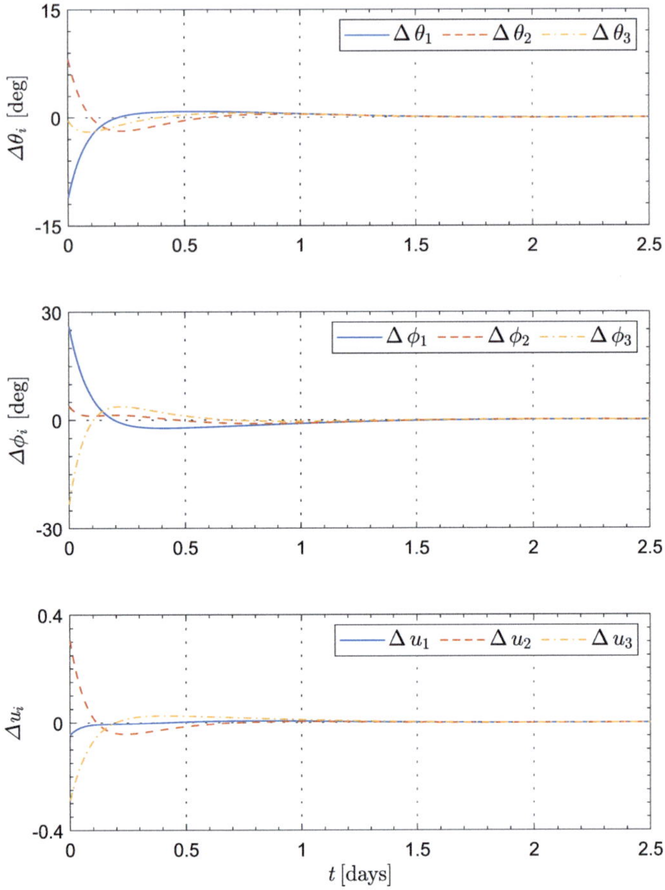

Fig. 7.25 Time history of the control input $\Delta \boldsymbol{u}_i$ of the three sails when only s_1 has access to the virtual leader

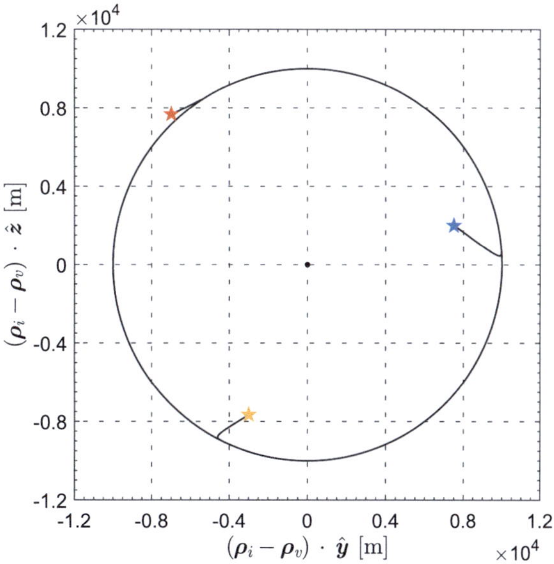

Fig. 7.26 Relative trajectories of the three sails when the sail s_1 only has access to the virtual leader

References

1. McInnes C R. Solar sail trajectories at the lunar L_2 Lagrange point [J]. Journal of Spacecraft and Rockets, 1993, 30 (6): 782–784.
2. Heiligers J, Parker J S, Macdonald, M. Novel solar-sail mission concepts for high-latitude Earth and lunar observation [J]. Journal of Guidance, Control, and Dynamics, 2018, 41 (1): 212–230.
3. Ozimek M T, Grebow D J, Howell, K C. Design of solar sail trajectories with applications to lunar south pole coverage [J]. Journal of Guidance, Control, and Dynamics, 2009, 32 (6): 1884–1897.
4. Jia Y, Zou Y, Ping J, et al. The scientific objectives and payloads of Chang'E-4 mission [J]. Planetary and Space Science, 2018, 162: 207–215.
5. Domingo V, Fleck B, Poland A I, et al. The SOHO mission: an overview [J]. Solar Physics, 1995, 162 (1–2): 1–37.
6. Hechler M, Mora M B, Nogales, M S, et al. Orbit concepts at L_2 for Soyuz launches from Kourou [J]. Acta Astronautica, 2008, 62 (2–3): 140–150.
7. Landgraf M, Renk F, De Vogeleer B. Mission design and analysis of European astrophysics missions orbiting libration points [J]. Acta Astronautica, 2013, 84: 49–55.
8. Milligan D, Rudolph A, Whitehead G, et al. ESA's billion star surveyor-Flight operations experience from Gaia's first 1.5 years [J]. Acta Astronautica, 2016, 127: 394–403.
9. Ragazzoni R, Rauer H, Catala C, et al. A one meter class eye for the planetary transit and oscillation spacecraft [J]. Acta Astronautica, 2015, 115: 18–23.
10. Shirobokov M, Trofimov S, Ovchinnikov M. Survey of station-keeping techniques for libration point orbits [J]. Journal of Guidance, Control, and Dynamics, 2017, 40 (5): 1085–1105.
11. Baoyin H, McInnes C R. Solar sail halo orbits at the Sun-Earth artificial L_1 point [J]. Celestial Mechanics and Dynamical Astronomy, 2006, 94 (2): 155–171.
12. Heiligers J, Hiddink S, Noomen R, et al. Solar sail Lyapunov and halo orbits in the Earth-Moon three-body problem [J]. Acta Astronautica, 2015, 116: 25–35.
13. Farrés A, Jorba Á. Dynamics of a solar sail near a Halo orbit [J]. Acta Astronautica, 2010, 67 (7–8): 979–990.

14. Heiligers J, Macdonald M, Parker J S. Extension of Earth-Moon libration point orbits with solar sail propulsion [J]. Astrophysics and Space Science, 2016, 361: 1–20.
15. Verrier P, Waters T, Sieber J. Evolution of the L_1 halo family in the radial solar sail circular restricted three-body problem [J]. Celestial Mechanics and Dynamical Astronomy, 2014, 120 (4): 373–400.
16. Tsuda Y, Mori O, Funase R, et al. Flight status of IKAROS deep space solar sail demonstrator [J]. Acta Astronautica, 2011, 69 (9–10): 833–840.
17. Tsuda Y, Mori O, Funase R, et al. Achievement of IKAROS–Japanese deep space solar sail demonstration mission [J]. Acta Astronautica, 2013, 82 (2): 183–188.
18. Johnson L, Whorton M, Heaton A, et al. NanoSail-D: A solar sail demonstration mission [J]. Acta Astronautica, 2011, 68 (5–6): 571–575.
19. Johnson L, Young R, Montgomery E, et al. Status of solar sail technology within NASA [J]. Advances in Space Research, 2011, 48 (11): 1687–1694.
20. Janhunen P. Electric sail for spacecraft propulsion [J]. Journal of Propulsion and Power, 2004, 20 (4): 763–764.
21. Janhunen P, Quarta A A, Mengali G. Electric solar wind sail mass budget model [J]. Geoscientific Instrumentation, Methods and Data Systems, 2013, 2 (1): 85–95.
22. McInnes C R. Solar Sailing: Technology, Dynamics and Mission Applications [M]. Springer-Praxis Series in Space Science and Technology, Springer-Verlag, London, Berlin, 1999.
23. Wang W, Yuan J, Mengali G, et al. Invariant manifold and bounds of relative motion between heliocentric displaced orbits [J]. Journal of Guidance, Control, and Dynamics, 2016, 39 (8): 1764–1776.
24. Gong S, Baoyin H, Li J. Solar sail formation flying around displaced solar orbits [J]. Journal of Guidance, Control, and Dynamics, 2007, 30 (4): 1147–1151.
25. Biggs J D, McInnes C R. Solar sail formation flying for deep-space remote sensing [J]. Journal of Spacecraft and Rockets, 2009, 46 (3): 670–678.
26. Parsay K, Schaub H, Designing solar sail formations in sun-synchronous orbits for geomagnetic tail exploration [J]. Acta Astronautica, 2015, 107 (1-2): 218–233.
27. Bandyopadhyay S, Foust R, Subramanian G P, et al. Review of formation flying and constellation missions using nanosatellites [J]. Journal of Spacecraft and Rockets, 2016, 53 (3): 567–578.
28. Wang W, Mengali G, Quarta A A, Yuan J. Multiple solar sail formation flying around heliocentric displaced orbit via consensus [J]. Acta Astronautica, 2019, 154: 256–267.
29. Salazar F J T, McInnes C R, Winter O C. Periodic orbits for space-based reflectors in the circular restricted three-body problem [J]. Celestial Mechanics and Dynamical Astronomy, 2017, 128 (1): 95–113.
30. Beard R W, Lawton J, Hadaegh F Y. A coordination architecture for spacecraft formation control [J]. IEEE Transactions on Control Systems Technology, 2001, 9 (6): 777–790.
31. Ren W, Beard R W. Formation feedback control for multiple spacecraft via virtual structures [J]. IEE Proceedings-Control Theory and Applications, 2004, 151 (3): 357–368.
32. Ren W. Formation keeping and attitude alignment for multiple spacecraft through local interactions [J]. Journal of Guidance, Control, and Dynamics, 2007, 30 (2): 633–638.
33. Mu J, Gong S, Li J. Reflectivity-controlled solar sail formation flying for magnetosphere mission [J]. Aerospace Science and Technology, 2013, 30 (1): 339–348.
34. McInnes C R. Orbits in a generalized two-body problem [J]. Journal of Guidance, Control, and Dynamics, 2003, 26 (5): 743–749.
35. Mengali G, Quarta A A. Optimal control laws for axially symmetric solar sails [J]. Journal of Spacecraft and Rockets, 2005, 42 (6): 1130–1133.
36. Colombo C, McInnes C R. Orbit design for future SpaceChip swarm missions in a planetary atmosphere [J]. Acta Astronautica, 2012, 75: 25–41.
37. Colombo C, Lücking C, McInnes C R. Orbit evolution, maintenance and disposal of SpaceChip swarms through electrochromic control [J]. Acta Astronautica, 2013, 82 (1): 25–37.
38. Quarta A A, Mengali G, Denti E. Optimal in-orbit repositioning of Sun-pointing smart dust [J]. Acta Astronautica, 2019, 154: 278–285.

39. Niccolai L, Bassetto M, Quarta A A, et al. A review of smart dust architecture, dynamics, and mission applications [J]. Progress of Aerospace Science, 2019, 106: 1–14.
40. Atchison J A, Peck M A, A passive, sun-pointing, millimeter-scale solar sail [J]. Acta Astronautica, 2010, 67 (1-2): 108–121.
41. Shahid K, Kumar K D. Formation control at the Sun-Earth L_2 libration point using solar radiation pressure [J]. Journal of Spacecraft and Rockets, 2010, 47 (4): 614–626.
42. Richardson D L. Analytic construction of periodic orbits about the collinear points [J]. Celestial Mechanics, 1980, 22 (3): 241–253.
43. Richardson D L. Halo orbit formulation for the ISEE-3 Mission [J]. Journal of Guidance, Control, and Dynamics, 1980, 3 (6): 543–548.
44. Howell K C. Three-dimensional, periodic, 'halo' orbits [J]. Celestial Mechanics, 1984, 32 (1): 53–71.
45. Waters T J, McInnes C R. Periodic orbits above the ecliptic in the solar-sail restricted three-body problem [J]. Journal of Guidance, Control, and Dynamics, 2007, 30 (3): 687–693.
46. Ren W. Multi-vehicle consensus with a time-varying reference state [J]. System Control Letters, 2007, 56 (7–8): 474–483.
47. Ren W, Atkins E. Distributed multi-vehicle coordinated control via local information exchange [J]. International Journal of Robust and Nonlinear Control, 2007, 17 (11): 1002–1033.
48. Wang W, Mengali G, Quarta A A, et al. Distributed adaptive synchronization for multiple spacecraft formation flying around Lagrange point orbits [J]. Aerospace Science and Technology, 2018, 74: 93–103.
49. Chung F R K. Spectral Graph Theory [M]. Regional Conference Series in Mathematics, American Mathematical Society, Providence, RI, 1997.
50. Balch T, Arkin R C. Behavior-based formation control for multirobot teams [J]. IEEE Transaction on Robotic Automation, 1998, 14 (6): 926–939.
51. Zhang H, Gurfil P. Distributed control for satellite cluster flight under different communication topologies [J]. Journal of Guidance, Control, and Dynamics, 2016, 39 (3), 617–627.
52. Thakur D, Hernandez S, Akella M R. Spacecraft swarm finite-thrust cooperative control for common orbit convergence [J]. Journal of Guidance, Control, and Dynamics, 2015, 38 (3): 478–488.
53. Liu G, Zhang S. A survey on formation control of small satellites [J]. Proceedings of the IEEE, 2018, 106 (3), 440–457.
54. Ren W, Beard R W, Decentralized scheme for spacecraft formation flying via the virtual structure approach [J]. Journal of Guidance, Control, and Dynamics, 2014, 27 (1): 73–82.

Chapter 8
Solar Sail Formation Flying Around Heliocentric Displaced Orbit via Consensus

This chapter investigates the problem of multiple solar sail-based spacecraft formation flying in which the chief follows a heliocentric displaced orbit, whereas each deputy adjusts the sail propulsive acceleration so as to track a desired (relative) trajectory with respect to the chief. In particular, coordinated control strategies are presented for both the full state feedback case and the relative velocity unavailability case, respectively. The developed consensus-based algorithms rely on the protocols formulated on an undirected communication topology with information link couplings, utilizing every available neighbor-to-neighbor information data such that the reliability of the formation system can be enhanced. Illustrative examples show the validity the proposed method.

8.1 Introduction

Some scientific tasks, such as the GeoSail magnetosphere mission [1, 2], require a continuous multi-aspect observation and a high-imaging resolution, thus necessitating the deployment of multiple spacecraft in a formation, so as to constitute a (virtual) synthetic aperture radar. In fact, by distributing the payload among multiple spacecraft while each one carries the indispensable functional modules only, a mass decrease (and size reduction) may be achieved for a single spacecraft, thereby significantly improving the system performance in terms of propulsive acceleration. In this context, the concept of solar sail formation flying has been proposed in some mission scenarios [3, 4]. Most of the available literature involving spacecraft formation flying is focused on a classical chief-deputy architecture, where the information data flows from the chief to each deputy, without any mutual negotiation among deputies. One inherent weakness of such a chief-deputy topology is that the whole system is not able to effectively respond to unexpected situations or risks, since the chief is a single point of failure for the whole system [5]. In case of chief malfunction or

© The Author(s), under exclusive license to Springer Nature Singapore Pte Ltd. 2026

C. Gao and W. Wang, *Solar Sailing for Deep-Space Explorations*,
https://doi.org/10.1007/978-981-95-0109-0_8

when a fault takes place in the chief-deputy communication links, the whole mission is at risk of failure. A possible solution to this problem consists in incorporating the deputy-to-deputy information exchange into the feedback control system, so as to increase the group robustness [7]. More precisely, the formation flying problem presented in Ref. [7] involves electric solar wind sail-based spacecraft [8, 9], i.e., a propulsion system that is substantially different from a solar sail and, more importantly, the relative velocity of each neighboring spacecraft is assumed to be exactly known.

The aim of this chapter is to present a distributed solar sail formation architecture around a heliocentric displaced orbit via a consensus concept. The coordinated cooperative control algorithms are used for solar sail formation maintenance, where each deputy adjusts its sail attitude and reflectivity modulation ratio so as to follow a prescribed relative geometry with respect to the chief. By exploiting the available neighbor-to-neighbor information data, the underlying consensus concept facilitates each deputy spacecraft to update its state in a cooperative way, (i.e., using information taken from its local neighbors) in such a way that the final tracking error of each spacecraft converges to zero. The chapter also discusses the case in which the relative velocity information from the neighbor is not available. In this sense, the approach discussed in this chapter improves the model described in Ref. [7].

The chapter is organized as follows. Section 9.1 investigates the conditions of generating a heliocentric, elliptic, displaced orbit for a solar sail-based spacecraft equipped with a reflectivity control device, and discusses the feasible regions for a given set of orbital parameters. Section 9.2 analyzes the mathematical model of the relative dynamics for solar sail formation flying around a heliocentric displaced orbit, which is then used as a basis for designing consensus-based distributed control algorithms in Sect. 9.3. The effectiveness of the control system is illustrated in Sect. 9.4 by means of some numerical simulations. Finally, some concluding remarks are given in Sect. 9.5.

8.2 Mathematical Model

Consider a mission scenario in which multiple solar sail-based spacecraft, flying in formation, provide a continuous observation of the polar region of a celestial body as, for example, a planet (subscript P). Using the heliocentric (Keplerian) trajectory of the planet as the reference orbit of eccentricity e_P, introduce a perifocal reference frame $\mathcal{T}_P\left(O; \hat{\boldsymbol{x}}_P, \hat{\boldsymbol{y}}_P, \hat{\boldsymbol{z}}_P\right)$, in which the origin O is at the Sun's center-of-mass, the plane $\left(\hat{\boldsymbol{x}}_P, \hat{\boldsymbol{y}}_P\right)$ coincides with the planet orbital plane $\mathcal{P}$, axis $\hat{\boldsymbol{x}}_P$ points to the perihelion, and $\hat{\boldsymbol{z}}_P$ is positive in the direction of the planet angular momentum vector $\boldsymbol{\omega}_P$, see Fig. 8.1.

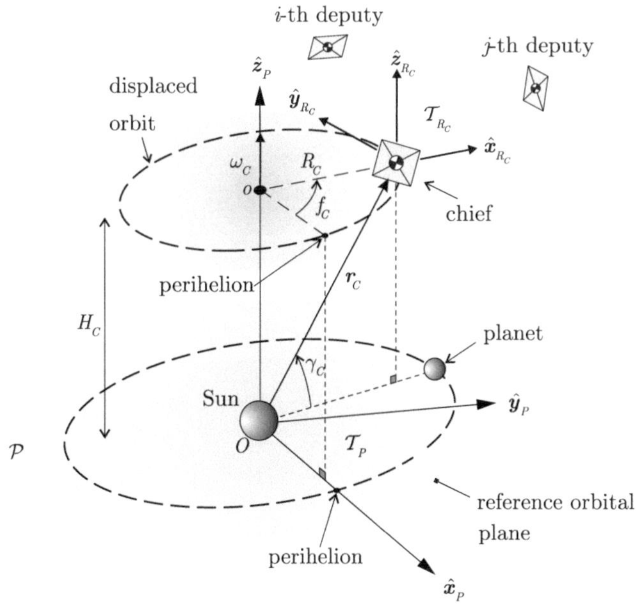

Fig. 8.1 Reference frames

8.2.1 Conditions for Maintaining an Elliptic Displaced Orbit

In order to observe the polar region of the planet, the chief spacecraft (subscript C) covers an elliptic displaced orbit of semimajor axis a_C and eccentricity e_C, whose orbital plane is parallel to $\mathcal{P}$ and is displaced at a constant distance H_C (with respect to $\mathcal{P}$), see Fig. 8.1. The displaced orbit is maintained by exploiting a solar sail covers with the reflectivity control device (RCD) [10, 11], whose mathematical model is consistent with that proposed in Ref. [12]. More precisely, the sail surface can be thought of as being constituted by two parts. The first one is covered with high reflectivity material for which a specular reflection model without degradation [13, 14] is assumed (i.e., an ideal force model), whereas the second part is covered by a number of electrochromic material panels (EMPs) that are used to modulate the propulsive acceleration modulus within a suitable (small) range, see Fig. 8.2.

Each EMP is able to vary its reflectivity as a function of the applied voltage and, in a simplified analysis, when the EMP is in its power-on mode the solar photons are specularly reflected, whereas in power-off mode the incoming photons are totally absorbed. Note that this behaviour is consistent with the simplified mathematical model discussed in Ref. [12]. In order to avoid the introduction of additional torque on the spacecraft, it is assumed that a symmetry exists in the distribution of switched on/off panels with respect to the vehicle center of mass. Considering the generic solar sail-based spacecraft of the formation architecture, let the total sail area be denoted by A_s, while A_p is the sail area covered with EMPs. It is useful to introduce the

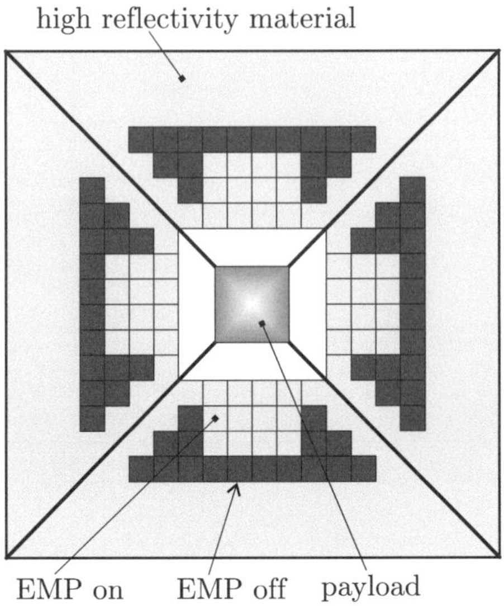

Fig. 8.2 Conceptual scheme of a solar sail with electrochromic material panels

reflectivity modulation ratio $u \triangleq A_{\text{off}}/A_s$, where A_{off} is the area of the switched-off panels calculated at a given time instant [12]. Accordingly, the range of the reflectivity modulation ratio is $u \in [0,\ u_{\max}]$, where $u_{\max} = A_p/A_s$ corresponds to the case where all of EMPs are in power-off mode. In particular, the ratio u is assumed to be varied with continuity, within its admissible range, by modulating the amount of on/off time of EMP devices.

For a continuous polar region observation, the chief orbital angular velocity ($\boldsymbol{\omega}_C$) and eccentricity (e_C) should match those of the planet, that is, $\boldsymbol{\omega}_C = \boldsymbol{\omega}_P$ and $e_C = e_P$. As a result, the chief lies, at any given time, in the same vertical plane passing through $\hat{\boldsymbol{z}}_P$ and containing the planet. A heliocentric displaced orbit with such an arrangement is referred to as Planet Following Displaced Orbit (PFDO) [6]. Figure 8.1 also shows a second reference frame $\mathcal{T}_{R_C}(o;\hat{\boldsymbol{x}}_{R_C},\hat{\boldsymbol{y}}_{R_C},\hat{\boldsymbol{z}}_{R_C})$, i.e., a chief-fixed rotating frame in which $\hat{\boldsymbol{x}}_{R_C}$ is directed from the focus o of the displaced orbit to the chief spacecraft, while $\hat{\boldsymbol{z}}_{R_C}$ is along the direction of $\boldsymbol{\omega}_C$, that is, the direction of $\hat{\boldsymbol{z}}_{R_C}$ is orthogonal to the (displaced) orbital plane.

The equation of motion of the chief spacecraft in the rotating reference frame $\mathcal{T}_{R_C}$ can be written as [7]

$$\ddot{\boldsymbol{r}}_C + 2\,\boldsymbol{\omega}_C \times \dot{\boldsymbol{r}}_C + \dot{\boldsymbol{\omega}}_C \times \boldsymbol{r}_C + \boldsymbol{\omega}_C \times (\boldsymbol{\omega}_C \times \boldsymbol{r}_C) = -\frac{\mu_\odot}{r_C^3}\,\boldsymbol{r}_C + \boldsymbol{a}_C \tag{8.1}$$

where $\boldsymbol{a}_C$ is the SRP acceleration, $\mu_\odot$ is the Sun's gravitational parameter, and $\boldsymbol{r}_C$ is the Sun-chief vector (with $r_C = \|\boldsymbol{r}_C\|$ being the Sun-chief distance) whose components in the rotating reference frame $\mathcal{T}_{R_C}$ are (see Fig. 8.1)

$$[\boldsymbol{r}_C]_{\mathcal{T}_{R_C}} = [R_C,\ 0,\ H_C]^{\mathrm{T}} \tag{8.2}$$

and R_C is the o-chief distance given by

$$R_C = \frac{a_C\,(1 - e_C^2)}{1 + e_C\,\cos f_C} \tag{8.3}$$

where a_C is the displaced orbit semimajor axis, and f_C is the chief true anomaly measured counterclockwise from the direction of the displaced orbit periapsis. Note that, since the chief is assumed to be synchronous with the planet, the components of the chief angular velocity $\boldsymbol{\omega}_C$ in frame $\mathcal{T}_{R_C}$ are

$$[\boldsymbol{\omega}_C]_{\mathcal{T}_{R_C}} = \left[0,\ 0,\ \dot{f}_C\right]^{\mathrm{T}} \tag{8.4}$$

with

$$\dot{f}_C = \frac{n_P\,(1 + e_C\,\cos f_C)^2}{\sqrt{(1 - e_C^2)^3}} \tag{8.5}$$

where $n_P = \sqrt{\mu_\odot / a_P^3}$ is the planet mean motion.

Taking into account the effect of RCD and using the approach of Ref. [12], the expression of the (chief) SRP acceleration $\boldsymbol{a}_C$ can be written as

$$\boldsymbol{a}_C = \frac{\beta_C\,\mu_\odot}{2\,r_C^2}\,(\hat{\boldsymbol{r}}_C \cdot \hat{\boldsymbol{n}}_C)\left[u_C\,\hat{\boldsymbol{r}}_C + 2\,(1 - u_C)\,(\hat{\boldsymbol{r}}_C \cdot \hat{\boldsymbol{n}}_C)\,\hat{\boldsymbol{n}}_C\right] \tag{8.6}$$

where $\hat{\boldsymbol{r}}_C = \boldsymbol{r}_C / r_C$ is the Sun-chief unit vector, u_C is the chief reflectivity modulation ratio, β_C is the chief lightness number, defined as the ratio of the maximum propulsive acceleration modulus to the (local) solar gravitational acceleration ($\mu_\odot / r_C^2$), and $\hat{\boldsymbol{n}}_C$ is the sail unit normal vector in the direction opposite to the Sun. Note that the components of $\hat{\boldsymbol{n}}_C$ in frame $\mathcal{T}_{R_C}$ are

$$\left[\hat{\boldsymbol{n}}_C\right]_{\mathcal{T}_{R_C}} = \left[\cos(\alpha_C + \gamma_C),\ 0,\ \sin(\alpha_C + \gamma_C)\right]^{\mathrm{T}} \tag{8.7}$$

where $\gamma_C = \arctan(H_C / R_C) \subset [0,\ \pi/2]$ rad is the elevation angle, that is, the angle between the Sun-chief line and plane $\mathcal{P}$, and $\alpha_C = \arccos\left(\hat{\boldsymbol{r}}_C \cdot \hat{\boldsymbol{n}}_C\right) \in [0,\ \pi/2]$ rad is the sail cone angle. In particular, bearing in mind Eq. (8.3), the elevation angle can be written as a function of the displaced orbit characteristics $\{a_C, e_C \equiv e_P\}$ and the chief angular position f_C as

$$\gamma_C = \arctan\left[\frac{H_C\,(1 + e_C\,\cos f_C)}{a_C\,(1 - e_C^2)}\right] \tag{8.8}$$

On the other hand, the sail unit normal vector can be written as a function of the sail attitude as

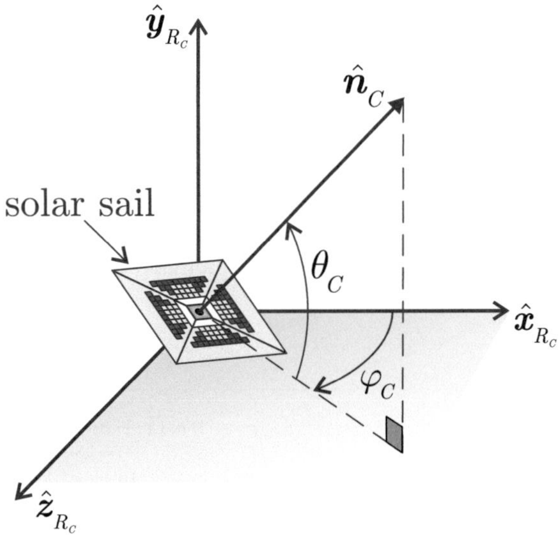

Fig. 8.3 Attitude angles of the solar sail-based chief spacecraft

$$[\hat{\boldsymbol{n}}_C]_{\mathcal{T}_{R_C}} = [\cos\theta_C \, \cos\varphi_C, \; \sin\theta_C, \; \cos\theta_C \, \sin\varphi_C]^{\mathrm{T}} \tag{8.9}$$

where θ_C is the angle between $\hat{\boldsymbol{n}}$ and the $(\hat{\boldsymbol{x}}_{R_C}, \hat{\boldsymbol{z}}_{R_C})$ plane, and φ_C is the angle measured from $\hat{\boldsymbol{x}}_{R_C}$ to the projection of $\hat{\boldsymbol{n}}_C$ onto the $(\hat{\boldsymbol{x}}_{R_C}, \hat{\boldsymbol{z}}_{R_C})$ plane, see Fig. 8.3. In particular, from Eqs. (8.7) and (8.9), the chief attitude angles are $\theta_C = 0$ and $\varphi_C = \alpha_C + \gamma_C$.

According to Eq. (8.1), and taking into account Eqs. (8.2)–(8.7), the normal (i.e. along the $\hat{\boldsymbol{z}}_{R_C}$ direction) and radial (i.e. along the $\hat{\boldsymbol{x}}_{R_C}$ direction) components necessary to maintain a PFDO are

$$\frac{\mu_\odot}{r_C^2} \sin\gamma_C = \frac{\beta_C \, \mu_\odot}{2\, r_C^2} \cos\alpha_C \left[u_C \, \sin\gamma_C + 2\,(1 - u_C)\, \cos\alpha_C \, \sin(\alpha_C + \gamma_C)\right] \tag{8.10}$$

$$\ddot{R}_C + \frac{\mu_\odot}{r_C^2} \cos\gamma_C - R_C \, \dot{f}_C^2 = \frac{\beta_C \, \mu_\odot}{2\, r_C^2} \cos\alpha_C \left[u_C \, \cos\gamma_C + 2\,(1 - u_C)\, \cos\alpha_C \, \cos(\alpha_C + \gamma_C)\right] \tag{8.11}$$

After some algebraic manipulations (here omitted for the sake of brevity) on Eqs. (8.10) and (8.11), the value of the sail cone angle α_C required for maintaining a PFDO can be found by solving the following sextic equation in the variable $\tan\alpha_C$ as

$$c_6 \, \tan^6\alpha_C + c_5 \, \tan^5\alpha_C + c_4 \, \tan^4\alpha_C + c_3 \, \tan^3\alpha_C + c_2 \, \tan^2\alpha_C + c_1 \, \tan\alpha_C + c_0 = 0 \tag{8.12}$$

where the coefficients $\{c_0, c_1, c_2, c_3, c_4, c_5, c_6\}$ are given by

$$c_6 = \frac{a_C^4\, H_C^2\, \left(1 - e_C^2\, \cos^2\, f_C\right)^2}{a_P^6\, \left(1 - e_C^2\right)^2} \tag{8.13}$$

$$c_5 = \frac{4\, a_C^3}{a_P^3} \left[\frac{H_C\, \left(1 - e_C^2\, \cos^2\, f_C\right)}{\sqrt{a_C^2\, \left(1 - e_C^2\right)^2 + H_C^2\, (1 + e_C\, \cos f_C)^2}} - \sqrt{c_6}\, (1 - e_C\, \cos f_C) \right] \tag{8.14}$$

$$c_4 = 4 \left[\frac{a_C\, \left(1 - e_C^2\right)}{\sqrt{a_C^2\, \left(1 - e_C^2\right)^2 + H_C^2\, (1 + e_C\, \cos f_C)^2}} - \frac{a_C^3\, (1 - e_C\, \cos f_C)}{a_P^3} \right]^2 - c_6 \tag{8.15}$$

$$c_3 = 0 \tag{8.16}$$

$$c_2 = c_4 - \frac{\beta_C^2\, a_C^2\, \left(1 - e_C^2\right)^2}{a_C^2\, \left(1 - e_C^2\right)^2 + H_C^2\, (1 + e_C\, \cos f_C)^2} \tag{8.17}$$

$$c_1 = -c_5 \tag{8.18}$$

$$c_0 = c_6 \tag{8.19}$$

Those coefficients are function of the sail lightness number β_C, the chief angular position f_C, the planet orbit characteristics $\{a_P, e_P \equiv e_C\}$, and the displaced orbit characteristics $\{a_C, H_C\}$. The corresponding value of the reflectivity modulation ratio u_C is given by

$$u_C = \frac{2\, \left[\sin \gamma_C - \beta_C\, \cos^2 \alpha_C\, \sin\left(\alpha_C + \gamma_C\right)\right]}{\beta_C\, \cos \alpha_C\, \left[\sin \gamma_C - 2\, \cos \alpha_C\, \sin\left(\alpha_C + \gamma_C\right)\right]} \tag{8.20}$$

which is a function of β_C, the sail cone angle α_C, and the elevation angle γ_C.

For illustrative purposes, consider now a chief spacecraft covering an Earth-synchronized elliptic PFDO with an eccentricity $e_C \equiv e_P = 0.0167$. In this scenario, Figs. 8.4 and 8.5 show the feasible regions with different constraints on the maximum value of the reflectivity modulation ratio $u_{C_{\max}}$, when the sail lightness number is $\beta_C = \{0.3, 0.6\}$. Note that the area of the feasible region enlarges as the values of $u_{C_{\max}}$ and β_C increase.

Consider, for example, an elliptic displaced orbit with a semimajor axis $a_C = 0.95$ au and a displacement $H_C = 0.05$ au. Figure 8.6 shows that the cone angle α_C and the reflectivity modulation ratio u_C reach their maximum values when the chief is at the aphelion, whereas the minimum values are obtained at the perihelion of the displaced orbit.

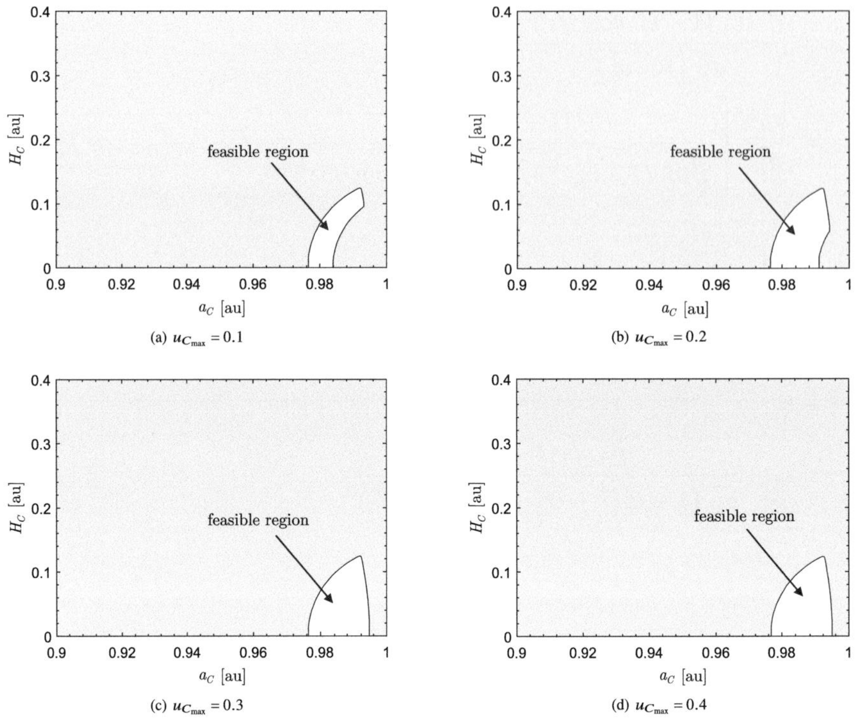

Fig. 8.4 Feasible regions as a function of $\{a_C, H_C\}$ with $\beta_C = 0.3$ and $u_{C_{\max}} = \{0.1, 0.2, 0.3, 0.4\}$

8.2.2 Chief-Deputy Relative Motion

Having analyzed the generation of an elliptic displaced orbit for the chief, we are now in a position to study the relative motion between chief and deputies. Taking into account Eq. (8.1), the equation of relative motion can be written as

$$\ddot{\boldsymbol{\rho}}_i + 2\,\boldsymbol{\omega}_C \times \dot{\boldsymbol{\rho}}_i + \dot{\boldsymbol{\omega}}_C \times \boldsymbol{\rho}_i + \boldsymbol{\omega}_C \times \left(\boldsymbol{\omega}_C \times \boldsymbol{\rho}_i\right) = -\frac{\mu_\odot}{r_i^3}\,\boldsymbol{r}_i + \frac{\mu_\odot}{r_C^3}\,\boldsymbol{r}_C + \boldsymbol{a}_i - \boldsymbol{a}_C \tag{8.21}$$

where $\boldsymbol{\rho}_i = \boldsymbol{r}_i - \boldsymbol{r}_C$ is the relative position vector between the i-th deputy and the chief spacecraft, see Fig. 8.7.

Since the deputy sails operate in close vicinity of the chief, the relative distances among the solar sails are very small compared with the Sun-sail distance. Accordingly, paralleling the approach adopted in Ref. [3] and bearing in mind Eq. (8.6), the deputy propulsive and gravitational accelerations in Eq. (8.21) can be linearized around those of the chief as

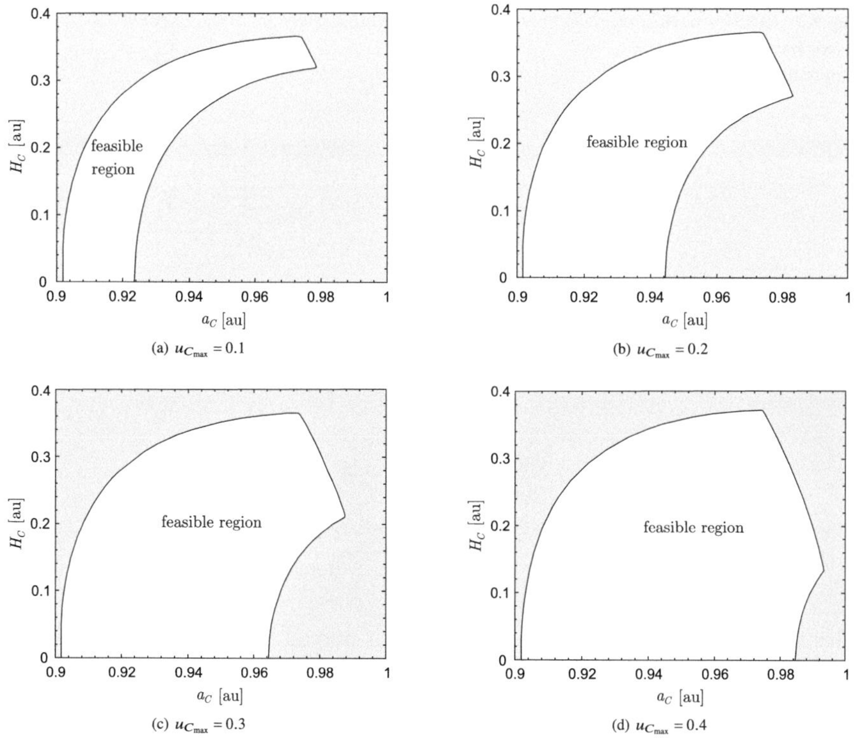

Fig. 8.5 Feasible regions as a function of $\{a_C, H_C\}$ with $\beta_C = 0.6$ and $u_{C_{\max}} = \{0.1, 0.2, 0.3, 0.4\}$

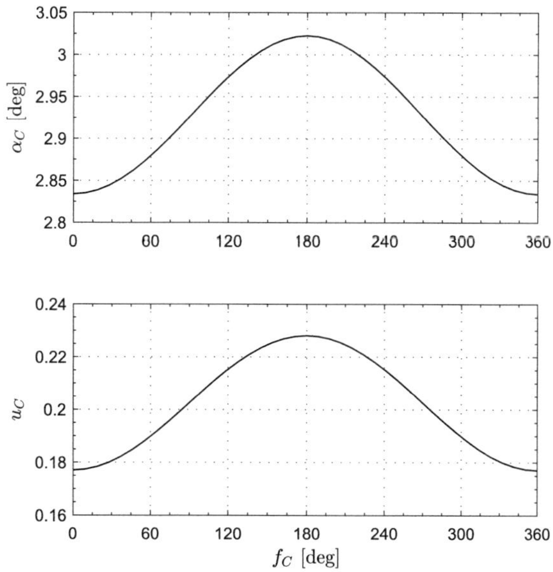

Fig. 8.6 Cone angle and reflectivity modulation ratio of the chief spacecraft (with $\beta_C = 0.6$) during a single revolution

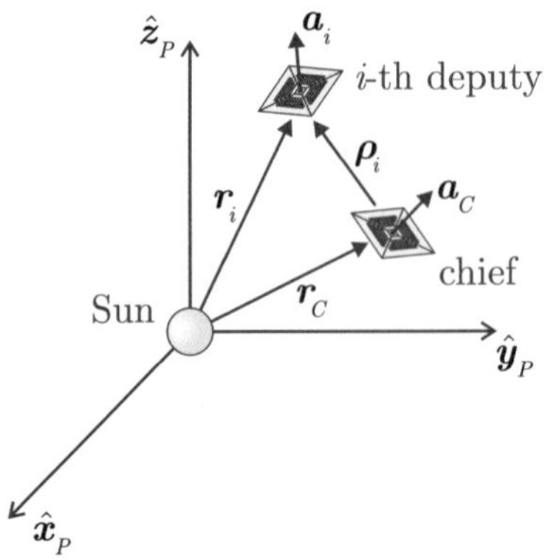

Fig. 8.7 Relative position vector between the i-th deputy and the chief

$$a_i \simeq a_C + \frac{\partial\, a_C}{\partial\, r_C}\, \rho_i + \frac{\partial\, a_C}{\partial\, \hat{n}_C}\, \Delta\, \hat{n}_i + \frac{\partial\, a_C}{\partial\, u_C}\, \Delta\, u_i \tag{8.22}$$

$$-\frac{\mu_\odot}{r_i^3}\, r_i \simeq -\frac{\mu_\odot}{r_C^3}\, r_C + \frac{\partial\,(-\mu_\odot\, r_C/r_C^3)}{\partial\, r_C}\, \rho_i \tag{8.23}$$

where $\Delta\, \hat{n}_i \triangleq \hat{n}_i - \hat{n}_C$, and $\hat{n}_i$ is the sail unit normal vector of the i-th deputy, $\Delta\, u_i \triangleq u_i - u_C$, and u_i is the reflectivity modulation ratio of the i-th deputy. Also

$$\frac{\partial\, a_C}{\partial\, r_C} = \frac{\beta_C\, \mu_\odot}{r_C^4} \left\{ \frac{u_C}{2} \left[\left(r_C\, \hat{n}_C^{\mathrm{T}} \right) + \left(\hat{n}_C \cdot r_C \right) \mathbb{I}_3 - \frac{\left(\hat{n}_C \cdot r_C \right) \left(r_C\, r_C^{\mathrm{T}} \right)}{r_C^2} \right] \right.$$
$$\left. +2\,(1-u_C) \left[\left(\hat{n}_C \cdot r_C \right) \left(\hat{n}_C\, \hat{n}_C^{\mathrm{T}} \right) - \frac{2\, \left(\hat{n}_C \cdot r_C \right)^2 \left(\hat{n}_C\, r_C^{\mathrm{T}} \right)}{r_C^2} \right] \right\} \tag{8.24}$$

$$\frac{\partial\, a_C}{\partial\, \hat{n}_C} = \frac{\beta_C\, \mu_\odot}{r_C^4} \left\{ \frac{u_C}{2}\, r_C\, r_C^{\mathrm{T}} + (1-u_C) \left[2\, \left(\hat{n}_C \cdot r_C \right) \left(\hat{n}_C\, r_C^{\mathrm{T}} \right) + \left(\hat{n}_C \cdot r_C \right)^2 \mathbb{I}_3 \right] \right\} \tag{8.25}$$

$$\frac{\partial\, a_C}{\partial\, u_C} = \frac{\beta_C\, \mu_\odot}{r_C^4} \left[\frac{1}{2} \left(\hat{n}_C \cdot r_C \right) r_C - \left(\hat{n}_C \cdot r_C \right)^2 \hat{n}_C \right] \tag{8.26}$$

$$\frac{\partial\, \left(-\mu_\odot\, r_C/r_C^3 \right)}{\partial\, r_C} = \frac{3\, \mu_\odot}{r_C^5}\, r_C\, r_C^{\mathrm{T}} - \frac{\mu_\odot}{r_C^3}\, \mathbb{I}_3 \tag{8.27}$$

where $\mathbb{I}_3$ is the 3×3 identity matrix.

Bearing in mind that $\hat{n}_i$ can be written in terms of φ_i and θ_i, see Eq. (8.9) and Fig. 8.3, it is possible to further linearize $\Delta\, \hat{n}_i$ as

$$\Delta\, \hat{n}_i \simeq \frac{\partial\, \hat{n}_C}{\partial\, [\varphi_C,\ \theta_C]}\, [\Delta\, \varphi_i,\ \Delta\, \theta_i]^{\mathrm{T}} \tag{8.28}$$

where $\Delta\varphi_i \triangleq \varphi_i - \varphi_C$ and $\Delta\theta_i \triangleq \theta_i - \theta_C$ are the relative sail attitude angles, and the Jacobian matrix is given by

$$\frac{\partial\, \hat{\boldsymbol{n}}_C}{\partial\, [\varphi_C,\ \theta_C]} = \begin{bmatrix} -\sin\varphi_C\ \cos\theta_C & -\cos\varphi_C\ \sin\theta_C \\ 0 & \cos\theta_C \\ \cos\varphi_C\ \cos\theta_C & -\sin\varphi_C\ \sin\theta_C \end{bmatrix} \tag{8.29}$$

Substituting Eqs. (8.22)–(8.29) into Eq. (8.21), the linear differential equation of the relative motion between the i-th deputy and the chief is

$$\ddot{\boldsymbol{\rho}}_i + 2\,\mathbb{M}_v\,\dot{\boldsymbol{\rho}}_i + \mathbb{M}_p\,\boldsymbol{\rho}_i = \mathbb{M}_c\,\boldsymbol{u}_i \tag{8.30}$$

where $\boldsymbol{u}_i$ is the control input of the i-th deputy, given by

$$\boldsymbol{u}_i = [\Delta\varphi_i,\ \Delta\theta_i,\ \Delta u_i]^{\mathrm{T}} \tag{8.31}$$

whereas $\mathbb{M}_v$, $\mathbb{M}_p$, and $\mathbb{M}_c$ are the coefficient matrices, defined as

$$\mathbb{M}_v = \begin{bmatrix} 0 & -\omega_C & 0 \\ \omega_C & 0 & 0 \\ 0 & 0 & 0 \end{bmatrix} \tag{8.32}$$

$$\mathbb{M}_p = \dot{\mathbb{M}}_v - \mathrm{diag}\left(\omega_C^2,\ \omega_C^2,\ 0\right) - \frac{\partial\, \boldsymbol{a}_C}{\partial\, \boldsymbol{r}_C} - \frac{\partial\left(-\mu_\odot\, \boldsymbol{r}_C / r_C^3\right)}{\partial\, \boldsymbol{r}_C} \tag{8.33}$$

$$\mathbb{M}_c = \left[\frac{\partial\, \boldsymbol{a}_C}{\partial\, \hat{\boldsymbol{n}}_C}\, \frac{\partial\, \hat{\boldsymbol{n}}_C}{\partial\, [\varphi_C,\ \theta_C]},\ \frac{\partial\, \boldsymbol{a}_C}{\partial u_C}\right] \tag{8.34}$$

Note that, from Eqs. (8.25) and (8.26), it can be verified that the matrix $\mathbb{M}_c$ in Eq. (8.34) is nonsingular provided $\beta_C \neq 0$.

In the following analysis, the relative trajectory tracking problem involving the formation system described by Eq. (8.30) will be addressed. In particular, distributed cooperative control strategies are presented for two important cases: (1) with full state feedback; (2) without relative velocity measurement. In both cases the formation maintenance error eventually approaches zero while consensus is preserved during transition.

8.3 Solar Sail Formation Flying via Consensus Control

The consensus control algorithms are now developed for a spacecraft formation architecture with a single chief and $N \geq 2$ deputies, based on a local neighbor-to-neighbor information exchange. In particular, the communication topology of the relative motion is described by an undirected graph, under the assumption that the information exchange is bidirectional.

A few pertinent concepts of graph theory are first reviewed, according to Ref. [5]. A weighted undirected graph $\mathcal{G}$ consists of a finite non-empty vertex set $\mathcal{V} = \{\upsilon_1, \ldots, \upsilon_N\}$, an edge set $\mathcal{E} = \{(\upsilon_1, \upsilon_2), \ldots, (\upsilon_{N-1}, \upsilon_N)\} \subseteq \mathcal{V} \times \mathcal{V}$, and a weighted adjacency matrix $\mathbb{W} = [w_{ij}] \in \mathbb{R}^{N\times N}$. The matrix $\mathbb{W}$ is defined such that its generic entry $w_{ij} = w_{ij} \in \mathbb{R}^+$ if $(\upsilon_i, \upsilon_j) \in \mathcal{E}$ with $i \neq j$, whereas $w_{ii} = 0$. Note that the matric $\mathbb{W}$ is symmetric, and $w_{ij} = 0$ if $(\upsilon_i, \upsilon_j) \notin \mathcal{E}$. The Laplacian matrix $\mathbb{L} = [l_{ij}] \in \mathbb{R}^{N\times N}$ associated with $\mathbb{W}$ is defined as $l_{ij} = -w_{ij}$ and $l_{ii} = \sum_{j=1}^{N} w_{ij}, \forall i \neq j$. In addition, the matrix $\mathbb{L}$ is symmetric positive semi-definite, and $\sum_{j=1}^{N} l_{ij} = 0, \forall i \neq j$. Finally, an undirected graph is called connected if there exists a path from every vertex to every other vertex.

In the following discussion, emphasis is focused on the cases where the information exchange topology is fixed and the weights w_{ij} are constant. In modeling the topology of the solar sail-based formation architecture during the relative motion, each deputy spacecraft is represented by a vertex, while the data flow between any two deputies is characterized by a weighted edge. In particular, assume that each deputy has access to the state information of the chief, that is, the topology graph $\mathcal{G}$ is connected.

8.3.1 Consensus Control with Full State Feedback

Consider first the case where the consensus protocol is formulated via full state feedback, i.e., both the relative position and relative velocity information are available when there is a path between two deputies (i.e., two vertices). Let $(\boldsymbol{\rho}_i^\star, \dot{\boldsymbol{\rho}}_i^\star) \in \mathbb{R}^3 \times \mathbb{R}^3$ denote the desired relative trajectory of the i-th deputy with respect to the chief, and introduce the position error $\boldsymbol{e}_i = \boldsymbol{\rho}_i - \boldsymbol{\rho}_i^\star$ and the velocity error $\dot{\boldsymbol{e}}_i = \dot{\boldsymbol{\rho}}_i - \dot{\boldsymbol{\rho}}_i^\star$. The aim of the developed controller is to drive each deputy towards the desired relative trajectory, viz. $\boldsymbol{e}_i \to \boldsymbol{0}$ and $\dot{\boldsymbol{e}}_i \to \boldsymbol{0}$, while consensus (i.e. $\boldsymbol{e}_i \to \boldsymbol{e}_j$ and $\dot{\boldsymbol{e}}_i \to \dot{\boldsymbol{e}}_j$) is guaranteed during the transition phase by exploiting the available information data transmitted from the local neighbors. To this end, the cooperative control law for the system represented by Eq. (8.30) is designed as

$$\boldsymbol{u}_i = \mathbb{M}_c^{-1} \left\{ \ddot{\boldsymbol{\rho}}_i^\star + 2\,\mathbb{M}_\upsilon\, \dot{\boldsymbol{\rho}}_i^\star + \mathbb{M}_p\, \boldsymbol{\rho}_i - \lambda_i^p\, \boldsymbol{e}_i - \lambda_i^\upsilon\, \dot{\boldsymbol{e}}_i - \sum_{j=1}^{N} w_{ij}^p \left(\boldsymbol{e}_i - \boldsymbol{e}_j\right) - \sum_{j=1}^{N} w_{ij}^\upsilon \left(\dot{\boldsymbol{e}}_i - \dot{\boldsymbol{e}}_j\right) \right\} \tag{8.35}$$

where $\{\lambda_i^p, \lambda_i^\upsilon\} \in \mathbb{R}^+$, w_{ij}^p and w_{ij}^υ are the (i, j) entries of the weighted adjacency matrices $\mathbb{W}_p$ and $\mathbb{W}_\upsilon$, respectively. Note that $\mathbb{W}_p$ and $\mathbb{W}_\upsilon$ are allowed to be different and, in particular, the last two terms in Eq. (8.35) are designed to maintain the formation during the transition phase, while the other terms are selected in order to drive each deputy towards the desired relative trajectory.

Theorem 1 *Using the control law given by Eq.* (8.35) *for the system represented by Eq.* (8.30), *consensus tracking is asymptotically achieved, that is,* $\boldsymbol{e}_i \to \boldsymbol{e}_j \to \mathbf{0}$ *and* $\dot{\boldsymbol{e}}_i \to \dot{\boldsymbol{e}}_j \to \mathbf{0}$ *as* $t \to \infty$.

Proof Consider the candidate Lyapunov function V defined as

$$V = \frac{1}{2} \sum_{i=1}^{N} \lambda_i^p \, \boldsymbol{e}_i^{\mathrm{T}} \boldsymbol{e}_i + \frac{1}{2} \sum_{i=1}^{N} \dot{\boldsymbol{e}}_i^{\mathrm{T}} \dot{\boldsymbol{e}}_i + \frac{1}{4} \sum_{i=1}^{N} \sum_{j=1}^{N} w_{ij}^p \left(\boldsymbol{e}_i - \boldsymbol{e}_j\right)^{\mathrm{T}} \left(\boldsymbol{e}_i - \boldsymbol{e}_j\right) \tag{8.36}$$

Note that V is positive definite, and its time derivative is given by

$$\begin{aligned} \dot{V} = & \sum_{i=1}^{N} \lambda_i^p \, \boldsymbol{e}_i^{\mathrm{T}} \dot{\boldsymbol{e}}_i + \sum_{i=1}^{N} \dot{\boldsymbol{e}}_i^{\mathrm{T}} \left[-2\,\mathbb{M}_v \, \dot{\boldsymbol{e}}_i - \lambda_i^p \, \boldsymbol{e}_i - \lambda_i^v \, \dot{\boldsymbol{e}}_i - \sum_{j=1}^{N} w_{ij}^p \left(\boldsymbol{e}_i - \boldsymbol{e}_j\right) - \sum_{j=1}^{N} w_{ij}^v \left(\dot{\boldsymbol{e}}_i - \dot{\boldsymbol{e}}_j\right) \right] \\ & + \frac{1}{2} \sum_{i=1}^{N} \sum_{j=1}^{N} w_{ij}^p \left(\boldsymbol{e}_i - \boldsymbol{e}_j\right)^{\mathrm{T}} \left(\dot{\boldsymbol{e}}_i - \dot{\boldsymbol{e}}_j\right) \end{aligned} \tag{8.37}$$

Bearing in mind Eq. (8.32), the matrix $\mathbb{M}_v$ is skew symmetric and, therefore, $\dot{\boldsymbol{e}}_i^{\mathrm{T}} \, \mathbb{M}_v \, \dot{\boldsymbol{e}}_i = 0$. Recalling that $w_{ij}^p = w_{ji}^p$, the following relationship holds

$$\begin{aligned} \sum_{i=1}^{N} \sum_{j=1}^{N} \dot{\boldsymbol{e}}_i^{\mathrm{T}} \, w_{ij}^p \left(\boldsymbol{e}_i - \boldsymbol{e}_j\right) &= \frac{1}{2} \sum_{i=1}^{N} \sum_{j=1}^{N} \dot{\boldsymbol{e}}_i^{\mathrm{T}} \, w_{ij}^p \left(\boldsymbol{e}_i - \boldsymbol{e}_j\right) + \frac{1}{2} \sum_{j=1}^{N} \sum_{i=1}^{N} \dot{\boldsymbol{e}}_j^{\mathrm{T}} \, w_{ji}^p \left(\boldsymbol{e}_j - \boldsymbol{e}_i\right) \\ &= \frac{1}{2} \sum_{i=1}^{N} \sum_{j=1}^{N} w_{ij}^p \left(\dot{\boldsymbol{e}}_i - \dot{\boldsymbol{e}}_j\right)^{\mathrm{T}} \left(\boldsymbol{e}_i - \boldsymbol{e}_j\right) \end{aligned} \tag{8.38}$$

so that

$$\sum_{i=1}^{N} \sum_{j=1}^{N} \dot{\boldsymbol{e}}_i^{\mathrm{T}} \, w_{ij}^v \left(\dot{\boldsymbol{e}}_i - \dot{\boldsymbol{e}}_j\right) = \frac{1}{2} \sum_{i=1}^{N} \sum_{j=1}^{N} w_{ij}^v \left(\dot{\boldsymbol{e}}_i - \dot{\boldsymbol{e}}_j\right)^{\mathrm{T}} \left(\dot{\boldsymbol{e}}_i - \dot{\boldsymbol{e}}_j\right) \tag{8.39}$$

Substituting Eqs. (8.38)–(8.39) into Eq. (8.37), the time derivative of the Lyapunov function becomes

$$\dot{V} = -\sum_{i=1}^{N} \lambda_i^v \, \dot{\boldsymbol{e}}_i^{\mathrm{T}} \dot{\boldsymbol{e}}_i - \frac{1}{2} \sum_{i=1}^{N} \sum_{j=1}^{N} w_{ij}^v \left(\dot{\boldsymbol{e}}_i - \dot{\boldsymbol{e}}_j\right)^{\mathrm{T}} \left(\dot{\boldsymbol{e}}_i - \dot{\boldsymbol{e}}_j\right) \leq 0 \tag{8.40}$$

Note that the formation system described by Eq. (8.30) is non-autonomous, since its coefficients are time dependent. Therefore, Matrosov's theorem [15] is here adopted for convergence analysis. To that end, let

$$F\left(\boldsymbol{e}_i, \dot{\boldsymbol{e}}_i\right) = \sum_{i=1}^{N} \boldsymbol{e}_i^{\mathrm{T}} \dot{\boldsymbol{e}}_i \tag{8.41}$$

Since $\dot{V}$ is semi-negative definite, see Eq. (8.40), it follows that $\{\boldsymbol{e}_i, \dot{\boldsymbol{e}}_i\}$ is uniformly bounded and so is the function F. Also, $\dot{V} = 0$ implies $\dot{\boldsymbol{e}}_i = \boldsymbol{0}$, therefore the time derivative of F on domain $\mathcal{D} = \left\{(\boldsymbol{e}_i, \dot{\boldsymbol{e}}_i) \mid \dot{V} = 0\right\} = \{(\boldsymbol{e}_i, \dot{\boldsymbol{e}}_i) \mid \dot{\boldsymbol{e}}_i = \boldsymbol{0}\}$ can be written as

$$\dot{F} = -\sum_{i=1}^{N} \lambda_i^p \boldsymbol{e}_i^{\mathrm{T}} \boldsymbol{e}_i - \sum_{i=1}^{N} \sum_{j=1}^{N} w_{ij}^p \boldsymbol{e}_i^{\mathrm{T}} \left(\boldsymbol{e}_i - \boldsymbol{e}_j\right) = -\sum_{i=1}^{N} \lambda_i^p \boldsymbol{e}_i^{\mathrm{T}} \boldsymbol{e}_i - \frac{1}{2} \sum_{i=1}^{N} \sum_{j=1}^{N} w_{ij}^p \left(\boldsymbol{e}_i - \boldsymbol{e}_j\right)^{\mathrm{T}} \left(\boldsymbol{e}_i - \boldsymbol{e}_j\right) \leq 0 \tag{8.42}$$

Note that $\left|\dot{F}\right|$ is positive definite for any $\boldsymbol{e}_i \neq \boldsymbol{0}$. As a result, there exists a class $\mathcal{K}$ function $\mathfrak{R}$, such that $\left|\dot{F}\right| \geq \mathfrak{R}\left(\|\boldsymbol{e}_i\|\right)$. According to Matrosov's theorem, the equilibrium of the system under the control law Eq. (8.35) is uniformly asymptotically stable, that is, $\boldsymbol{e}_i \to \boldsymbol{e}_j \to \boldsymbol{0}$ and $\dot{\boldsymbol{e}}_i \to \dot{\boldsymbol{e}}_j \to \boldsymbol{0}$ as $t \to \infty$. This complete the proof. □

8.3.2 Consensus Without Relative Velocity Measurement

The distributed control law given by Eq. (8.35) requires each deputy to know the relative velocities of its local neighbors. However, in practical operations such an information is not available, especially for solar sail-based spacecraft with a small payload and instrument mass. Moreover, even when the spacecraft is equipped with a (relative) velocity sensor, the precision still remains questionable for large baseline formation missions. These two issues require finding an alternative algorithm accounting for the unavailability of relative velocity measurement. To this end, the consensus strategy is proposed as [16]

$$\dot{\boldsymbol{\xi}}_i = \boldsymbol{\Phi}\, \boldsymbol{\xi}_i + \lambda_i^v \boldsymbol{e}_i + \sum_{j=1}^{N} w_{ij}^v \left(\boldsymbol{e}_i - \boldsymbol{e}_j\right) \tag{8.43}$$

$$\boldsymbol{\zeta}_i = \boldsymbol{\Gamma} \left[\boldsymbol{\Phi}\, \boldsymbol{\xi}_i + \lambda_i^v \boldsymbol{e}_i + \sum_{j=1}^{N} w_{ij}^v \left(\boldsymbol{e}_i - \boldsymbol{e}_j\right)\right] \tag{8.44}$$

$$\boldsymbol{u}_i = \mathbb{M}_c^{-1} \left\{\ddot{\boldsymbol{\rho}}_i^\star + 2\,\mathbb{M}_v\, \dot{\boldsymbol{\rho}}_i^\star + \mathbb{M}_p\, \boldsymbol{\rho}_i - \lambda_i^p \boldsymbol{e}_i - \sum_{j=1}^{N} w_{ij}^p \left(\boldsymbol{e}_i - \boldsymbol{e}_j\right) - \boldsymbol{\zeta}_i\right\} \tag{8.45}$$

where $\left\{\boldsymbol{\xi}_i, \boldsymbol{\zeta}_i\right\} \in \mathbb{R}^3$, $\left\{\lambda_i^p, \lambda_i^v, w_{ij}^p, w_{ij}^v\right\}$ are defined as in Eq. (8.35), the matrix $\boldsymbol{\Phi} \in \mathbb{R}^{3\times3}$ is Hurwitz, and $\boldsymbol{\Gamma} = \boldsymbol{\Gamma}^{\mathrm{T}} \in \mathbb{R}^{3\times3}$ is the positive-definite solution to the Lyapunov equation $\boldsymbol{\Phi}^{\mathrm{T}}\, \boldsymbol{\Gamma} + \boldsymbol{\Gamma}\, \boldsymbol{\Phi} = -\boldsymbol{\Theta}$ where $\boldsymbol{\Theta} = \boldsymbol{\Theta}^{\mathrm{T}} \in \mathbb{R}^{3\times3}$ is positive-definite. In particular, since no relative velocity information is used, w_{ij}^v can be thought of playing a similar role as in Eq. (8.35). Note that the terms involving the relative velocity requirement in Eq. (8.35) are now replaced by the output of a passive filter, as is shown in Eqs. (8.43) and (8.44).

Theorem 2 *Using the control law given by Eq. (8.45) for the system represented by Eq. (8.30), consensus is asymptotically reached, that is,* $\boldsymbol{e}_i \rightarrow \boldsymbol{e}_j \rightarrow \mathbf{0}$ *and* $\dot{\boldsymbol{e}}_i \rightarrow \dot{\boldsymbol{e}}_j \rightarrow \mathbf{0}$ *as* $t \rightarrow \infty$.

Proof Consider the candidate Lyapunov function V defined as

$$V = \frac{1}{2}\sum_{i=1}^{N} \lambda_i^p \boldsymbol{e}_i^{\mathrm{T}} \boldsymbol{e}_i + \frac{1}{2}\sum_{i=1}^{N} \dot{\boldsymbol{e}}_i^{\mathrm{T}} \dot{\boldsymbol{e}}_i + \frac{1}{4}\sum_{i=1}^{N}\sum_{j=1}^{N} w_{ij}^p \left(\boldsymbol{e}_i - \boldsymbol{e}_j\right)^{\mathrm{T}} \left(\boldsymbol{e}_i - \boldsymbol{e}_j\right) + \frac{1}{2}\dot{\boldsymbol{\xi}}^{\mathrm{T}} \left(\bar{\mathbb{L}}_v \otimes \mathbb{I}_3\right)^{-1} \left(\mathbb{I}_N \otimes \boldsymbol{\Gamma}\right) \dot{\boldsymbol{\xi}} \tag{8.46}$$

where $\otimes$ denotes the Kronecker product, $\boldsymbol{\xi} = \left[\boldsymbol{\xi}_1^{\mathrm{T}}, \ldots, \boldsymbol{\xi}_N^{\mathrm{T}}\right]^{\mathrm{T}}$, $\bar{\mathbb{L}}_v = \mathbb{L}_v + \mathrm{diag}\left(\lambda_1^v, \ldots, \lambda_N^v\right)$. Since the Laplacian matrix $\mathbb{L}_v = \left[l_{ij}^v\right] \in \mathbb{R}^{N \times N}$ associated with $\mathbb{W}_v = \left[w_{ij}^v\right] \in \mathbb{R}^{N \times N}$ is symmetric positive semi-definite, it follows that $\bar{\mathbb{L}}_v$ is symmetric positive definite, and so is $\bar{\mathbb{L}}_v^{-1}$. The time derivative of V can be written as

$$\dot{V} = \sum_{i=1}^{N} \lambda_i^p \boldsymbol{e}_i^{\mathrm{T}} \dot{\boldsymbol{e}}_i + \sum_{i=1}^{N} \dot{\boldsymbol{e}}_i^{\mathrm{T}} \left[-2\,\mathbb{M}_v \dot{\boldsymbol{e}}_i - \lambda_i^p \boldsymbol{e}_i - \sum_{j=1}^{N} w_{ij}^p \left(\boldsymbol{e}_i - \boldsymbol{e}_j\right) - \boldsymbol{\zeta}_i\right] + \frac{1}{2}\sum_{i=1}^{N}\sum_{j=1}^{N} w_{ij}^p \left(\boldsymbol{e}_i - \boldsymbol{e}_j\right)^{\mathrm{T}} \left(\dot{\boldsymbol{e}}_i - \dot{\boldsymbol{e}}_j\right)$$
$$+ \frac{1}{2}\ddot{\boldsymbol{\xi}}^{\mathrm{T}} \left(\bar{\mathbb{L}}_v \otimes \mathbb{I}_3\right)^{-1} \left(\mathbb{I}_N \otimes \boldsymbol{\Gamma}\right) \dot{\boldsymbol{\xi}} + \frac{1}{2}\dot{\boldsymbol{\xi}}^{\mathrm{T}} \left(\bar{\mathbb{L}}_v \otimes \mathbb{I}_3\right)^{-1} \left(\mathbb{I}_N \otimes \boldsymbol{\Gamma}\right) \ddot{\boldsymbol{\xi}} \tag{8.47}$$

After some mathematical manipulations, Eq. (8.47) reduces to

$$\dot{V} = -\frac{1}{2}\dot{\boldsymbol{\xi}}^{\mathrm{T}} \left(\bar{\mathbb{L}}_v^{-1} \otimes \boldsymbol{\Theta}\right) \dot{\boldsymbol{\xi}} \leq 0 \tag{8.48}$$

Since the time derivative $\dot{V}$ given by Eq. (8.48) is a function of $\dot{\boldsymbol{\xi}}_i$ rather than $\dot{\boldsymbol{e}}_i$, Matrosov's theorem cannot be directly applied. Instead, the Barbalat's lemma [17] is here used for convergence analysis. In Eq. (8.46), V is a positive definite function of $\left\{\boldsymbol{e}_i, \dot{\boldsymbol{e}}_i, \boldsymbol{e}_i - \boldsymbol{e}_j, \dot{\boldsymbol{\xi}}\right\}$, and Eq. (8.48) implies $\dot{V} \leq 0$, therefore $\left\{\boldsymbol{e}_i, \dot{\boldsymbol{e}}_i, \boldsymbol{e}_i - \boldsymbol{e}_j, \dot{\boldsymbol{\xi}}\right\}$ are all bounded. Since $\boldsymbol{\Phi}$ is Hurwitz, it can be drawn from Eq. (8.43) that $\ddot{\boldsymbol{\xi}}$ is bounded, from which

$$\left|\ddot{V}\right| = \frac{1}{2}\left|\ddot{\boldsymbol{\xi}}^{\mathrm{T}} \left(\bar{\mathbb{L}}_v^{-1} \otimes \boldsymbol{\Theta}\right) \dot{\boldsymbol{\xi}} + \dot{\boldsymbol{\xi}}^{\mathrm{T}} \left(\bar{\mathbb{L}}_v^{-1} \otimes \boldsymbol{\Theta}\right) \ddot{\boldsymbol{\xi}}\right| \leq \left\|\left(\bar{\mathbb{L}}_v^{-1} \otimes \boldsymbol{\Theta}\right) \dot{\boldsymbol{\xi}}\right\| \left\|\ddot{\boldsymbol{\xi}}\right\| \tag{8.49}$$

Equation (8.49) shows that $\ddot{V}$ is also bounded, and hence $\dot{V}$ is uniformly continuous. According to the Barbalat's lemma, $\dot{V} \rightarrow 0$ as $t \rightarrow \infty$, which further results in $\dot{\boldsymbol{\xi}} \rightarrow \mathbf{0}$. From Eq. (8.43) $\dddot{\boldsymbol{\xi}}$ is bounded, therefore $\ddot{\boldsymbol{\xi}}$ is uniformly continuous. Using the Barbalat's lemma, $\ddot{\boldsymbol{\xi}} \rightarrow \mathbf{0}$ as $t \rightarrow \infty$. For the sake of convenience, Eq. (8.43) may be rearranged into a compact form as

$$\dot{\boldsymbol{\xi}} = \left(\mathbb{I}_N \otimes \boldsymbol{\Phi}\right) \boldsymbol{\xi} + \left(\bar{\mathbb{L}}_v \otimes \mathbb{I}_3\right) \boldsymbol{e} \tag{8.50}$$

where $\boldsymbol{e} = \left[\boldsymbol{e}_1^{\mathrm{T}}, \ldots, \boldsymbol{e}_N^{\mathrm{T}}\right]^{\mathrm{T}}$. Recalling that $\left\{\dot{\boldsymbol{\xi}}, \ddot{\boldsymbol{\xi}}\right\} \to \boldsymbol{0}$, Eq. (8.50) states that $\dot{\boldsymbol{e}} \to \boldsymbol{0}$. Likewise, it can be verified that $\ddot{\boldsymbol{e}} \to \boldsymbol{0}$. Using the control law given by Eq. (8.45), the error equation of the formation system described by Eq. (8.30) becomes

$$\ddot{\boldsymbol{e}} + 2\,(\mathbb{I}_N \otimes \mathbb{M}_v)\,\dot{\boldsymbol{e}} + \left(\bar{\mathbb{L}}_p \otimes \mathbb{I}_3\right)\boldsymbol{e} + (\mathbb{I}_N \otimes \boldsymbol{\Gamma})\,\dot{\boldsymbol{\xi}} = \boldsymbol{0} \tag{8.51}$$

Since $\left\{\dot{\boldsymbol{e}}, \ddot{\boldsymbol{e}}, \dot{\boldsymbol{\xi}}\right\} \to \boldsymbol{0}$, therefore $\left(\bar{\mathbb{L}}_p \otimes \mathbb{I}_3\right)\boldsymbol{e} = \boldsymbol{0}$, which amounts to stating that $\boldsymbol{e} \to \boldsymbol{0}$. This complete the proof. □

8.4 Numerical Simulations

To illustrate the performance of the developed consensus-based control law, a formation mission scenario is now discussed, with a chief spacecraft (S_C) and three surrounding deputies (S_i with $i = \{1, 2, 3\}$) that are coordinated through local communication. The chief covers an Earth-synchronized PFDO (i.e. $e_C = 0.0167$) with semimajor axis $a_C = 0.95$ au, and displacement $H_C = 0.05$ au, while all deputies are driven towards the prescribed trajectory with different phases. In the chief rotating reference frame $\mathcal{T}_{R_C}$, the parametric solution of the i-th deputy trajectory (relative to the chief) is assumed to have the following algebraic form

$$\boldsymbol{\rho}_i^{\star} = \left[100\,\sin\left(n_r\,t + \phi_i\right),\ 200\,\cos\left(n_r\,t + \phi_i\right),\ 100\sqrt{3}\,\sin\left(n_r\,t + \phi_i\right)\right]^{\mathrm{T}}\ \text{km} \tag{8.52}$$

where $n_r = 100\,n_P$ is the angular velocity of the relative orbit, and $\phi_i = 2\,(i-1)\,\pi/3$ is the phase angle.

8.4.1 *Case A: Full Relative State Information is Available*

Consider first the case where a full relative state information is available when a path between two deputies exists. The corresponding communication topology graph of the formation system is shown in Fig. 8.8, whereas the position and velocity weighted adjacency matrices among deputies are given by

$$\mathbb{W}_p = \begin{bmatrix} 0 & 1 & 2 \\ 1 & 0 & 0 \\ 2 & 0 & 0 \end{bmatrix}, \qquad \mathbb{W}_v = \begin{bmatrix} 0 & 5\times10^{-3} & 1\times10^{-2} \\ 5\times10^{-3} & 0 & 0 \\ 1\times10^{-2} & 0 & 0 \end{bmatrix} \tag{8.53}$$

The initial conditions of the three deputies are assumed slightly different from the desired trajectories, and the initial errors $\boldsymbol{e}_0 = \left[e_{x_0},\ e_{y_0},\ e_{z_0}\right]^{\mathrm{T}}$ are reported in Table 8.1. Using a trial and error procedure, a satisfactory control performance (in terms of convergence time and tracking accuracy) is obtained when the dimensionless parameters are selected as $\lambda_i^p = 5\times10^3$ and $\lambda_i^v = 25$.

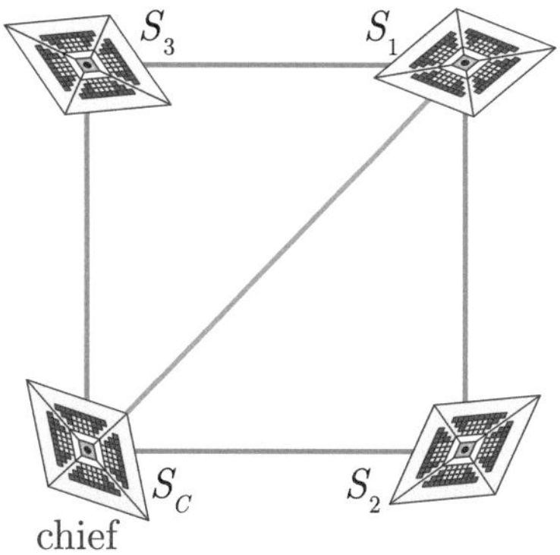

Fig. 8.8 Information exchange topology among the solar sail formation system

Table 8.1 Initial errors of three deputies in the full state feedback case

	e_{x0} [km]	e_{y0} [km]	e_{z0} [km]	$\dot{e}_{x0}$ [m/s]	$\dot{e}_{y0}$ [m/s]	$\dot{e}_{z0}$ [m/s]
S_1	40	−20	20	1×10^{-5}	-2×10^{-5}	3×10^{-5}
S_2	−20	20	−40	-2×10^{-5}	1×10^{-5}	-1×10^{-5}
S_3	−40	−20	40	2×10^{-5}	-1×10^{-5}	-3×10^{-5}

Accordingly, the position and velocity errors of the three deputies are drawn in Figs. 8.9 and 8.10 respectively, while the variations of control inputs $\Delta\varphi_i$, $\Delta\theta_i$, and Δu_i are shown in Fig. 8.11. Figure 8.12 illustrates the actual position of the three deputies (denoted with a circle, a square and an asterisk) when compared with the desired position (denoted with a dot). These positions are calculated at $t = \{0, 5, 10, 15\}$ days and are projected on the $(\hat{\boldsymbol{x}}_{R_C}, \hat{\boldsymbol{y}}_{R_C})$ plane. Note that Fig. 8.12 shows how the control law given by Eq. (8.35) allows the three deputies to be perfectly synchronized with negligible final errors.

8.4.2 Case B: The Relative Velocity Information is Unavailable

In this case, the initial errors are assumed to take the same value as those listed in Table 8.1, and the position and velocity weighted adjacency matrices are given by

$$\mathbb{W}_p = \begin{bmatrix} 0 & 1 & 2 \\ 1 & 0 & 0 \\ 2 & 0 & 0 \end{bmatrix}, \quad \mathbb{W}_v = \begin{bmatrix} 0 & 500 & 1000 \\ 500 & 0 & 0 \\ 1000 & 0 & 0 \end{bmatrix} \tag{8.54}$$

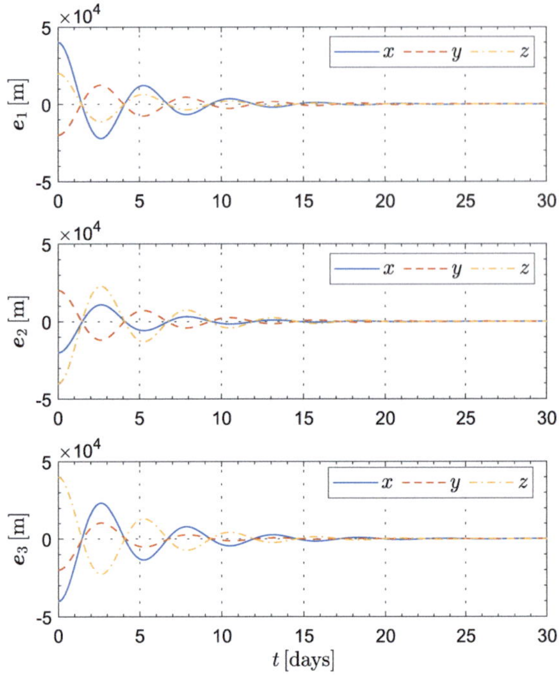

Fig. 8.9 Components of the position error in the full state feedback case

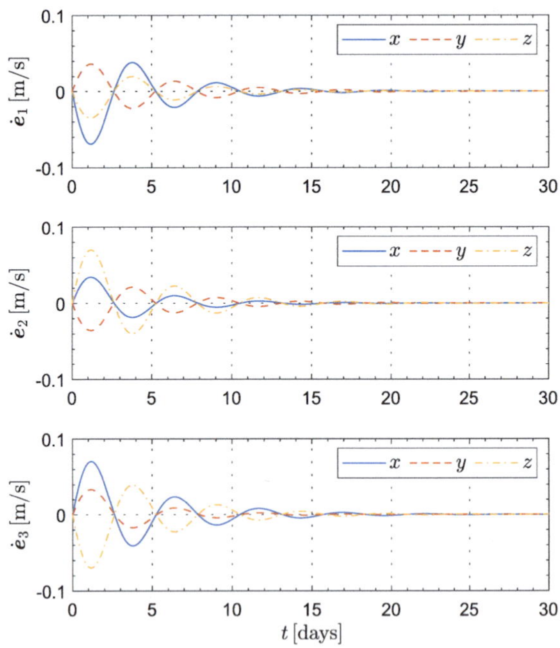

Fig. 8.10 Components of the velocity error in the full state feedback case

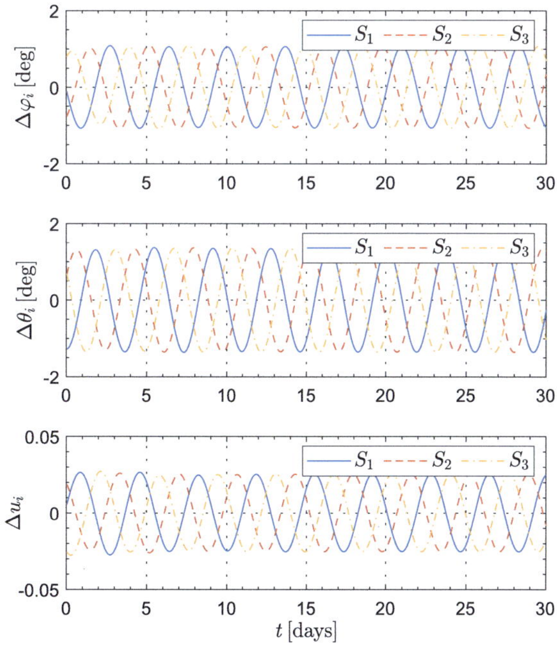

Fig. 8.11 Control input with control law given by Eq. (8.35)

Because the relative velocity information is not incorporated into the feedback control system, $\mathbb{W}_v$ can be freely adjusted. Using a trial and error procedure, a satisfactory control performance is obtained when the parameters in Eq. (8.45) are chosen to be $\lambda_i^p = \lambda_i^v = 100$, $\mathbf{\Phi} = -18\,\mathbb{I}_3$, and $\mathbf{\Gamma} = 10\,\mathbb{I}_3$. Finally, since the initial value of intermediate variables $\boldsymbol{\xi}_i$ can be randomly chosen, here it is assumed that $\boldsymbol{\xi}_{i0} = \mathbf{0}$.

To illustrate the performance of the control law given by Eq. (8.45), the time histories of position and velocity errors are reported in Figs. 8.13 and 8.14, for a time interval of 30 days. Figures 8.15 and 8.16 illustrate the control input and the intermediate variables $\boldsymbol{\xi}_i$, while the instantaneous positions of the three deputies are shown in Fig. 8.17. Note that, even in absence of relative velocity measurements, the steady state errors of the three deputies all converge to zero, while consensus is guaranteed during the whole transition phase. When compared with the results of the full state feedback case, the control law given by Eq. (8.45) produces a more oscillating behavior and requires higher control inputs.

8.5 Conclusions

The problem of solar sail formation flying around a heliocentric displaced orbit has been investigated in this chapter. Distributed architectures of the formation control system have been developed for both a full state feedback case and a relative velocity-free case, the latter being achieved with the introduction of a passive filter

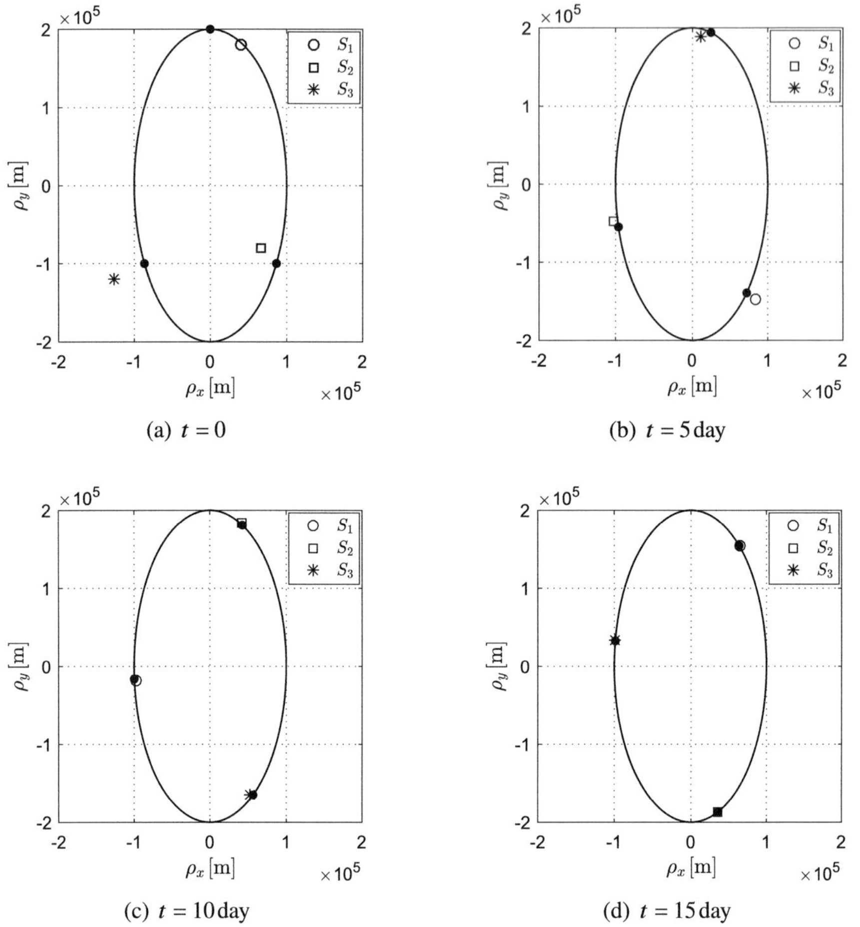

Fig. 8.12 Relative trajectories of the three deputies in the full state feedback case

that removes the relative velocity requirement. In particular, the discussed consensus algorithms rely on a protocol that enables multiple solar sails to maintain a desired formation geometry, while every available neighbor-to-neighbor information coupling is used to obtain an asymptotic coordination.

Compared with a traditional chief-deputy framework, the inclusion of the shared data flow among neighbors into cooperative control system improves the robustness and redundancy of the group, thus preventing an unfavorable situation in which a chief malfunction would imply a failure of the whole formation system.

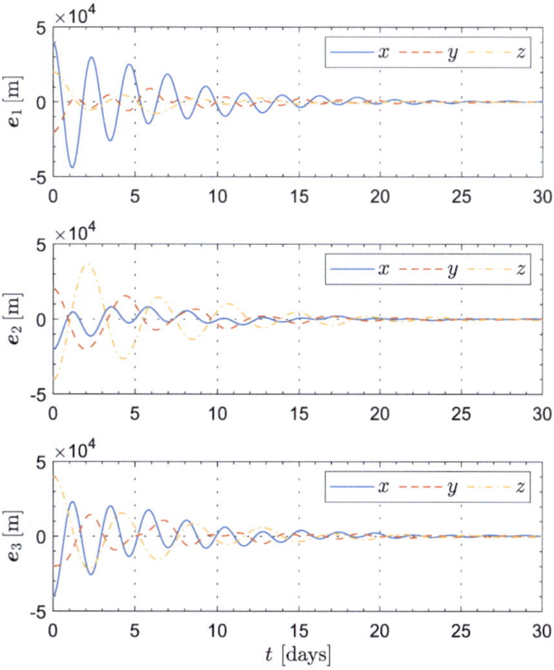

Fig. 8.13 Components of the position errors in the relative velocity-free case

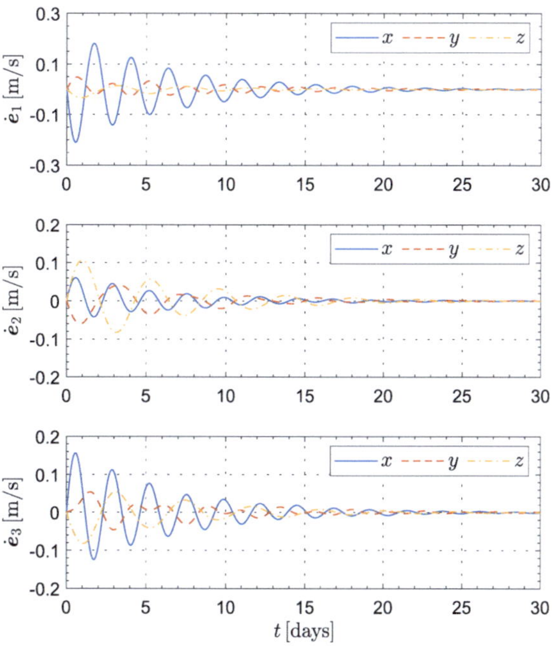

Fig. 8.14 Components of the velocity errors in the relative velocity-free case

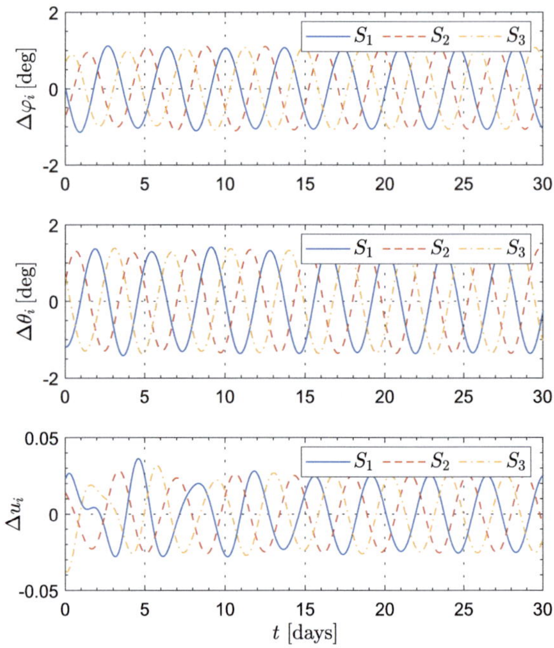

Fig. 8.15 Control input with control law given by Eq. (8.45)

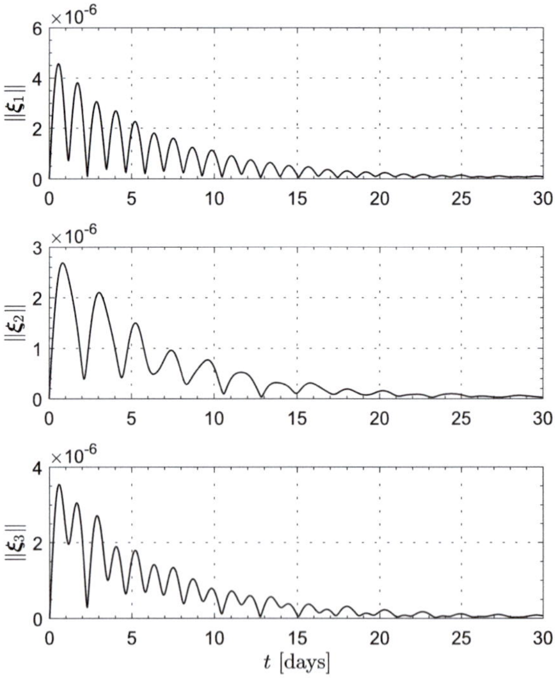

Fig. 8.16 Variations of $\|\boldsymbol{\xi}_i\|$ in Eq. (8.43)

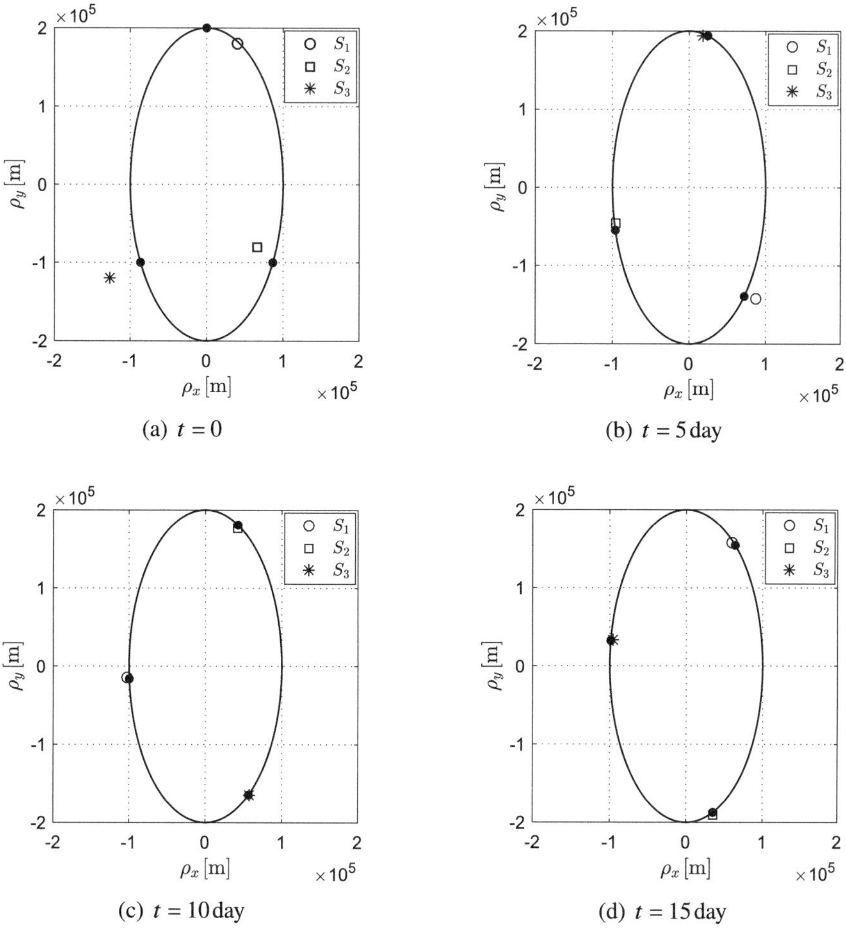

Fig. 8.17 Relative trajectories of the three deputies in the relative velocity-free case

References

1. Parsay W, Schaub H, Designing solar sail formations in Sun-synchronous orbits for geomagnetic tail exploration [J]. Acta Astronautica, 2015, 107: 218–233.
2. Macdonald M, McInnes C R, Alexander D, et al, Geosail: Exploring the magnetosphere using a low-cost solar sail [J], Acta Astronautica, 2006, 59 (8–11): 757–767.
3. Gong S, Baoyin H, Li J. Solar sail formation flying around displaced solar orbits [J]. Journal of Guidance, Control, and Dynamics, 2007, 30 (4): 1147–1151.
4. Wang W, Yuan J, Mengali G, Quarta A A. Invariant manifold and bounds of relative motion between heliocentric displaced orbits [J]. Journal of Guidance, Control, and Dynamics, 2016, 39 (8): 1764–1776.
5. Ren W. Formation keeping and attitude alignment for multiple spacecraft through local interactions [J]. Journal of Guidance, Control, and Dynamics, 2007, 30 (2): 633–638.

6. Niccolai L, Quarta A A, Mengali G, Electric sail elliptic displaced orbits with advanced thrust model [J]. Acta Astronautica, 2017, 138: 503–511.
7. Wang W, Mengali G, Quarta A A, et al. Formation flying for electric sails in displaced orbits. Part I: geometrical analysis [J]. Advances in Space Research, 2017, 60 (6): 1115–1129.
8. Janhunen P. Electric sail for spacecraft propulsion [J]. Journal of Propulsion and Power, 2004, 20 (4): 763–764.
9. Janhunen P. Status report of the electric sail in 2009 [J]. Acta Astronautica, 2011, 68 (5–6): 567–570.
10. Aliasi G, Mengali G, Quarta A A. Artificial Lagrange points for solar sail with electrochromic material panels [J]. Journal of Guidance, Control, and Dynamics, 2013, 36 (5): 1544–1550.
11. Mengali G, Quarta A A. Heliocentric trajectory analysis of Sun-pointing smart dust with electrochromic control [J]. Advances in Space Research, 2016, 57 (4): 991–1001.
12. Mu J, Gong S, Li J. Reflectivity-controlled solar sail formation flying for magnetosphere mission [J]. Aerospace Science and Technology, 2013, 30 (1): 339–348.
13. Dachwald B, Mengali G, Quarta A A, et al. Parametric model and optimal control of solar sails with optical degradation [J]. Journal of Guidance, Control, and Dynamics, 2006, 29 (5): 1170–1178.
14. Dachwald B, Macdonald M, McInnes C R, et al. Impact of optical degradation on solar sail mission performance [J]. Journal of Spacecraft and Rockets, 2007, 44 (4): 740–749.
15. Ren W, Atkins E. Distributed multi-vehicle coordinated control via local information exchange [J]. International Journal of Robust and Nonlinear Control, 2007, 17 (11): 1002–1033.
16. Ren W. Distributed leaderless consensus algorithms for networked Euler-Lagrange systems [J]. International Journal of Control, 2009, 82 (11): 2137–2149.
17. Slotine J E, Li W. Applied Nonlinear Control [M]. Prentice-Hall, New York, 1991.

Chapter 9
Invariant Manifold and Bounds of Relative Motion Between Heliocentric Circular Displaced Orbits

This chapter presents a methodology for modeling the relative motion between heliocentric circular displaced orbits using the Cartesian state variables in combination with a set of displaced orbital elements. Similar to classical Keplerian orbital elements, the newly defined set of displaced orbital elements has a clear physical meaning and provides an alternative approach to obtain a closed-form solution to the relative motion problem between circular displaced orbits, without linearizing or solving nonlinear equations. The invariant manifold of relative motion between two arbitrary circular displaced orbits is determined by coordinate transformations, thus obtaining a straightforward interpretation of the bounds, namely, maximumand minimum relative distances of three directional components. The extreme values of these bounds are then calculated from an analytical viewpoint, both for quasi-periodic orbits in the incommensurable case and periodic orbits in the 1 : 1 commensurable case. Moreover, in some degenerate cases, the extreme values of relative distance bounds can also be solved analytically. For each case, simulation examples are discussed to validate the correctness of the proposed method.

9.1 Introduction

In recent years, non-Keplerian displaced orbits have attracted considerable attention because of their unique advantages in astronomical missions [1–3]: for example, in situ observation for Saturn's rings, solar wind monitoring, and real-time stereographic investigations of a planetary surface [4–6]. A displaced orbit is an orbit for which the orbital plane does not contain the primary body (the Sun in this case). Displaced orbits can be generated by suitably orienting the spacecraft thrust direction in such a way to balance the centrifugal and gravitational components of the acceleration [1]. Spacecraft flying along displaced orbits can be promoted by new kinds of continuous thrust propulsion systems such as solar sails [2, 7] and electric solar wind sails [3],

© The Author(s), under exclusive license to Springer Nature Singapore Pte Ltd. 2026

C. Gao and W. Wang, *Solar Sailing for Deep-Space Explorations*,
https://doi.org/10.1007/978-981-95-0109-0_9

whereas they are very difficult to reach with conventional (either chemical or electric) propulsion systems.

However, using solar sails as an example, somemissions require an extremely large reflector area that cannot be carried out from an engineering standpoint, leaving it impractical to compromise between technical requirements and industrial manufacturing in the near future. To that end, it is necessary to introduce the concept of formation flying, distributing multiple sails in a relatively proximate region and enabling them to construct a larger virtual sensor to achieve greater resolution than a single one. Nevertheless, up to now, studies of relative motion between displaced orbits obtained using photonic solar radiation pressure are scarce in the literature. Further, the linearized dynamic model was adopted among several existing ones [8, 9], which were inherently limited to small-distance and short-term missions. Therefore, it is useful to gain a nonlinear insight into the relative motion problem, especially for large baseline formation flying. Moreover, there have been few published works to explore the relative motion bounds and topology in the general case of two sails flying along displaced orbits, which are of great importance in space mission design and analysis. Hence, they deserve a detailed study.

Recognizing these open issues, the chapter first manages to model relative motion between circular displaced orbits with Cartesian relative state variables in conjunction with a newly established set of displaced orbital elements. Analogous to Keplerian orbital elements, the new set of parameters defines the orientation of a displaced orbit with respect to an inertial reference frame, as well as its real-time motion in a rotating frame. Besides using coordinate transformations that incorporate circular displaced orbital elements, a closed-form solution expressed in a rotating frame can be derived without linearizing or solving the corresponding nonlinear equations. For the relative motion between circular displaced orbits, the solution is time explicit, which is useful to analyze the nature of relative motion.

This chapter also determines the relative motion invariant manifold, which is constituted by the parameterized general solution, and all possible orbits move along the manifold as they evolve. The invariant manifold presents a well-defined region in the configuration space with certain clear bounds, the orbits moving on the manifold manifest periodicity for the commensurable case, and quasi-periodicity (ergodicity) for the incommensurable case. The referred bounds in this chapter are indeed extreme distances of radial, along-track, and cross-track motions, as well as the relative distance between two spacecraft flying along displaced orbits. The directional bounds are determined from an analytical viewpoint, both for the incommensurable case and the $1:1$ commensurable case. Moreover, the relative distance bounds can be calculated analytically in some degenerate cases. An analytical calculation of these bounds is essential for onboard on-off control when the relative distance arrives at the preset boundary, whereas the prediction of relative motion bounds by pure numerical integration is time consuming and may lead to time delays in inner communication. Thus, the designers may select proper elements to maintain a long-term loose formation around circular displaced orbits.

9.2 Set of Displaced Orbital Elements

In the classical two-body problem, the orbital motion in three dimensional space can be described either by the state vector $\boldsymbol{S}_{\text{Keplerian}} = \{\boldsymbol{r},\ \boldsymbol{v}\}^{\text{T}} \in \mathbb{R}^3 \times \mathbb{R}^3$, where $\boldsymbol{r}$ and $\boldsymbol{v}$ are, respectively, the position and velocity vector, or a set of classical orbital elements $\boldsymbol{œ}_{\text{Keplerian}} = \{a,\ e,\ i,\ \Omega,\ \omega,\ f\}^{\text{T}} \in \mathbb{R}^2 \times \mathbb{S}^4$, where a is the semimajor axis, e is the eccentricity, i is the inclination, Ω is the right ascension of the ascending node, ω is the argument of periapsis, and f is the true anomaly. However, the Eulerian angles Ω and ω are both undefined for circular displaced orbits [10–12]. To this end, a new set of (nonsingular) displaced orbital elements $\boldsymbol{œ}_{\text{Displaced}}$ will be defined to describe the motion along a heliocentric circular orbit (of displacement H from the Sun). To formulate the problem conveniently, introduce the following related coordinate systems, see Fig. 9.1:

- 1 A heliocentric inertial frame $\mathcal{T}_I\left(O;\hat{\boldsymbol{x}}_I,\hat{\boldsymbol{y}}_I,\hat{\boldsymbol{z}}_I\right)$ centered at the Sun. The fundamental plane $\left(\hat{\boldsymbol{x}}_I,\hat{\boldsymbol{y}}_I\right)$ coincides with the ecliptic plane, the $\hat{\boldsymbol{x}}_I$ axis points to the vernal equinox, the $\hat{\boldsymbol{z}}_I$ axis points to the north of celestial pole, and $\hat{\boldsymbol{y}}_I = \hat{\boldsymbol{z}}_I \times \hat{\boldsymbol{x}}_I$.
- 2 A rotating reference frame $\mathcal{T}_R\left(S;\hat{\boldsymbol{x}}_R,\hat{\boldsymbol{y}}_R,\hat{\boldsymbol{z}}_R\right)$ centered at the spacecraft center-of-mass S. The fundamental plane $\left(\hat{\boldsymbol{x}}_R,\hat{\boldsymbol{y}}_R\right)$ is the orbital plane with its origin o, the $\hat{\boldsymbol{x}}_R$ axis points from o to the spacecraft, $\hat{\boldsymbol{z}}_R$ axis is positive in the direction of the spacecraft angular velocity vector $\boldsymbol{\omega}$, and the $\hat{\boldsymbol{y}}_R$ axis completes the setup.

First, locate the intersection of the orbital plane with $\mathcal{P}$ which is parallel to the ecliptic plane and passes through the center o of the displaced orbit. The intersection is the line of nodes, and the point on that line where the orbit passes above the plane $\mathcal{P}$ from below is called the ascending node. The unit vector directed from the center o to the ascending node is denoted by $\hat{\boldsymbol{N}}$, as shown in Fig. 9.1. The circular displaced orbital elements are defined as follows: a is the displaced orbital radius, the inclination i is the angle between the orbital plane and the ecliptic plane, measured from the positive $\hat{\boldsymbol{Z}}$ axis to the normal to the orbital plane, the right ascension of the ascending node Ω is measured from the positive $\hat{\boldsymbol{X}}$ axis to the line of nodes,

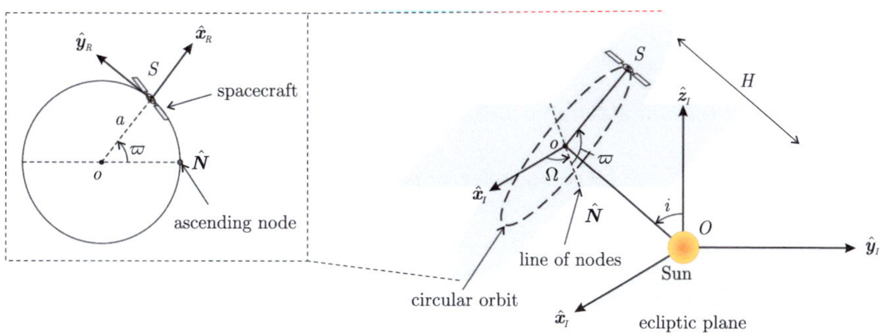

Fig. 9.1 Coordinate systems and related angles

and the mean argument of latitude ϖ is measured from $\hat{\boldsymbol{N}}$ to the position vector $\boldsymbol{r}$. Accordingly, the circular displaced orbit can be characterized by

$$\text{œ}_{\text{Displaced}} \triangleq \{a,\ H,\ \omega,\ i,\ \Omega,\ \varpi\} \in \mathbb{R}^3 \times \mathbb{S}^3 \tag{9.1}$$

Note that the spacecraft's angular velocity ω along the circular displaced orbit is not uniquely, but which can be designed according to mission requirement [13]. Hence, the transformation procedure between $\left[\boldsymbol{S}_{\text{Displaced}}\right]_{\mathcal{T}_I} = [\boldsymbol{r}^{\text{T}},\ \boldsymbol{v}^{\text{T}}]^{\text{T}}_{\mathcal{T}_I}$ and $\text{œ}_{\text{Displaced}}$ is different from that in the Keplerian case, given as follows.

- $\text{œ}_{\text{Displaced}} \mapsto \boldsymbol{S}_{\text{Displaced}}$

  (a) Calculate the spacecraft's state vector $\boldsymbol{S}_{\text{Displaced}}$ in rotating frame $\mathcal{T}_R$:

$$[\boldsymbol{r}]_{\mathcal{T}_R} = [a,\ 0,\ H]^{\text{T}},\quad [\boldsymbol{v}]_{\mathcal{T}_R} = [0,\ \omega a,\ 0]^{\text{T}} \tag{9.2}$$

  (b) Determine the direction matrix $\mathbb{T}_{RI}$ (from $\mathcal{T}_R$ to $\mathcal{T}_I$):

$$\mathbb{T}_{RI}\,(\Omega, i, \varpi) = \mathbb{T}_z\,(-\Omega)\ \mathbb{T}_x\,(-i)\,\mathbb{T}_z\,(-\varpi) \tag{9.3}$$

where $\mathbb{T}_x\ (\cdot)$ and $\mathbb{T}_z\ (\cdot)$ are, respectively, the rotation matrix about the $\hat{\boldsymbol{x}}$ axis and $\hat{\boldsymbol{z}}$ axis, given by

$$\mathbb{T}_x\ (\cdot) = \begin{bmatrix} 1 & 0 & 0 \\ 0 & \cos(\cdot) & \sin(\cdot) \\ 0 & -\sin(\cdot) & \cos(\cdot) \end{bmatrix},\quad \mathbb{T}_z\ (\cdot) = \begin{bmatrix} \cos(\cdot) & \sin(\cdot) & 0 \\ -\sin(\cdot) & \cos(\cdot) & 0 \\ 0 & 0 & 1 \end{bmatrix} \tag{9.4}$$

  (c) Calculate the spacecraft's state vector $\boldsymbol{S}_{\text{Displaced}}$ in rotating frame $\mathcal{T}_I$:

$$[\boldsymbol{r}]_{\mathcal{T}_I} = \mathbb{T}_{RI}\ (\Omega, i, \varpi)\ [\boldsymbol{r}]_{\mathcal{T}_R},\quad [\boldsymbol{v}]_{\mathcal{T}_R} = \mathbb{T}_{RI}\ (\Omega, i, \varpi)\ [\boldsymbol{v}]_{\mathcal{T}_R} \tag{9.5}$$

- $\boldsymbol{S}_{\text{Displaced}} \mapsto \text{œ}_{\text{Displaced}}$

  (a) Calculate the modulus of the state vector $r = \|\boldsymbol{r}\|$ and $v = \|\boldsymbol{v}\|$:

$$\|\boldsymbol{r}\| = \sqrt{\boldsymbol{r}\cdot\boldsymbol{r}},\quad \|\boldsymbol{v}\| = \sqrt{\boldsymbol{v}\cdot\boldsymbol{v}} \tag{9.6}$$

  (b) Calculate the orbital radius a and the angular velocity vector $\boldsymbol{\omega}$ of the displaced orbit:

$$\boldsymbol{\omega}\cdot\boldsymbol{v} = 0,\quad \boldsymbol{\omega}\cdot\boldsymbol{r} = \omega a,\quad \|\boldsymbol{\omega}\| = \omega \tag{9.7}$$

(c) Calculate the displacement H:

$$H = \sqrt{r^2 - a^2} \tag{9.8}$$

(d) Calculate the inclination i:

$$i = \arccos\left(\hat{\boldsymbol{\omega}} \cdot \hat{\boldsymbol{z}}_I\right) \tag{9.9}$$

(e) Calculate the line of nodes $\hat{\boldsymbol{N}}$:

$$\hat{\boldsymbol{N}} = \frac{\hat{\boldsymbol{z}}_I \times \hat{\boldsymbol{\omega}}}{\left\|\hat{\boldsymbol{z}}_I \times \hat{\boldsymbol{\omega}}\right\|} \tag{9.10}$$

(f) Calculate the right ascension of the ascending node Ω from Eq. (9.10):

$$\Omega = \arccos\left(\hat{\boldsymbol{N}} \cdot \hat{\boldsymbol{x}}_I\right) \tag{9.11}$$

(g) Calculate the mean argument of latitude ϖ from Eq. (9.10):

$$\varpi = \arccos\left(\hat{\boldsymbol{N}} \cdot \hat{\boldsymbol{r}}\right) \tag{9.12}$$

Note that the physical significance of $\text{œ}_{\text{Displaced}}$ provids a straightforward approach for describing the spacecraft motion along a circular displaced orbit. In the following study, the geometric topology of spacecraft relative motion will be analyzed based on $\text{œ}_{\text{Displaced}}$.

9.3 Relative Motion Topology

As is shown in Fig. 9.2, assume the chief S_C and deputy S_D to move along two (different) heliocentric circular displaced orbits, and let $\boldsymbol{\rho} \triangleq \boldsymbol{r}_D - \boldsymbol{r}_C$ denote the relative position vector between the chief and deputy. Since the relative motion is usually described in the chief's rotating frame $\mathcal{T}_{R_C}$, the initial frame $\mathcal{T}_I$ can be taken as an intermediate bridge [14]. Accordingly, the relative position vector $\boldsymbol{\rho}$ in frame $\mathcal{T}_{R_C}$ can be written as

$$[\boldsymbol{\rho}]_{\mathcal{T}_{R_C}} = \mathbb{T}_{R_C I}^{\mathrm{T}}\left(\Omega_C, i_C, \varpi_C\right) \mathbb{T}_{R_D I}\left(\Omega_D, i_D, \varpi_D\right) [\boldsymbol{r}_D]_{\mathcal{T}_{R_D}} - [\boldsymbol{r}_C]_{\mathcal{T}_{R_C}} \tag{9.13}$$

where $\mathbb{T}_{R_D I}$ represents the transformation matrix from the deputy's rotating frame $\mathcal{T}_{R_D}$ to the heliocentric initial frame $\mathcal{T}_I$, see Eqs. (9.3) and (9.4). $\mathbb{T}_{R_C I}^{\mathrm{T}} = \mathbb{T}_{R_C I}^{-1} = \mathbb{T}_{I R_C}$

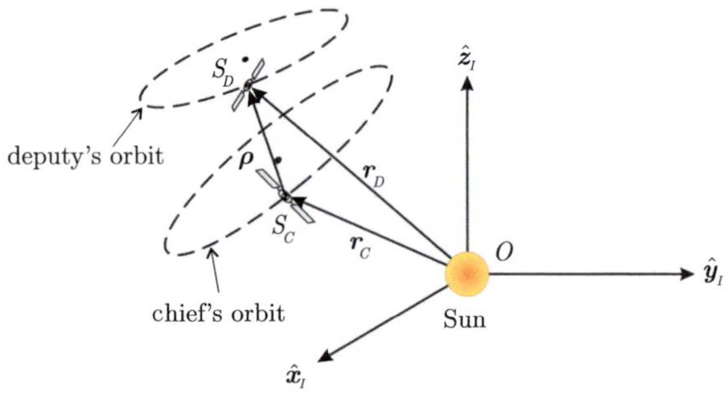

Fig. 9.2 Circular displaced orbits of chief and deputy spacecraft

is the transformation matrix from the heliocentric initial frame $\mathcal{T}_I$ to the chief's rotating frame $\mathcal{T}_{R_C}$. After some mathematical operations, Eq. (9.13) becomes

$$\begin{aligned}\rho_x &= a_D\,\xi_1\,\cos\varpi_C\,\cos\varpi_D - a_D\,\xi_2\,\cos\varpi_C\,\sin\varpi_D + a_D\,\xi_4\,\sin\varpi_C\,\cos\varpi_D\\ &\quad + a_D\,\xi_3\,\sin\varpi_C\,\sin\varpi_D - a_D\,\xi_5\,\cos\varpi_C + a_D\,\xi_6\,\sin\varpi_C - a_C\end{aligned} \tag{9.14}$$

$$\begin{aligned}\rho_y &= a_D\,\xi_4\,\cos\varpi_C\,\cos\varpi_D + a_D\,\xi_3\,\cos\varpi_C\,\sin\varpi_D - a_D\,\xi_1\,\sin\varpi_C\,\cos\varpi_D\\ &\quad + a_D\,\xi_2\,\sin\varpi_C\,\sin\varpi_D + a_D\,\xi_5\,\sin\varpi_C + a_D\,\xi_6\,\cos\varpi_C\end{aligned} \tag{9.15}$$

$$\rho_z = -a_D\,\xi_7\,\cos\varpi_D + a_D\,\xi_8\,\sin\varpi_D + \xi_9 - \frac{H_C}{a_D} \tag{9.16}$$

where the coefficients $\xi_1, \ldots, \xi_9$ are

$$\begin{aligned}&\xi_1 \triangleq \cos(\Omega_D - \Omega_C),\ \xi_2 \triangleq \cos i_D\,\sin(\Omega_D - \Omega_C),\ \xi_3 \triangleq \cos i_C\,\cos i_D\,\cos(\Omega_D - \Omega_C) + \sin i_C\,\sin i_D,\\ &\xi_4 \triangleq \cos i_C\,\sin(\Omega_D - \Omega_C),\\ &\xi_5 \triangleq -\frac{H_D}{a_D}\,\sin i_D\,\sin(\Omega_D - \Omega_C),\ \xi_6 \triangleq \frac{H_D}{a_D}\,[\sin i_C\,\cos i_D - \cos i_C\,\sin i_D\,\cos(\Omega_D - \Omega_C)],\\ &\xi_7 \triangleq \sin i_C\,\sin(\Omega_D - \Omega_C),\ \xi_8 \triangleq \cos i_C\,\sin i_D - \cos i_D\,\sin i_C\,\cos(\Omega_D - \Omega_C),\\ &\xi_9 \triangleq \frac{H_D}{a_D}\,[\cos i_C\,\cos i_D + \sin i_C\,\sin i_D\,\cos(\Omega_D - \Omega_C)]\end{aligned} \tag{9.17}$$

The closed-form solution in the three-dimensional configuration space constitutes the invariant manifold $\mathfrak{R}$ on which the relative motion will evolve all the time. It is evident fram Eqs. (9.14)–(9.16) that $\boldsymbol{\rho}$ is a function of two angular coordinates $\{\varpi_C, \varpi_D\} \in \mathbb{S}^1 \times \mathbb{S}^1$, and the invariant manifold $\mathfrak{R}$ is, most likely, a typical space with the continuous surfaces of a cylinder, torus (toroid), sphere, truncated cone, etc. [15]. Such a manifold has the appearance of a smooth surface and can be treated as a two-dimensional Euclidean space.

There are two qualitatively different cases, depending on whether the frequencies are commensurable or not. For the commensurable case, $\exists p, q \in \mathbb{N}$, so that the ratio of the frequencies is rational, satisfying $\omega_C/\omega_D = p/q$, where $\mathbb{N}$ is the denotes the set of natural numbers. All possible orbits in this case are closed curves on $\mathfrak{R}$. In practice, the 1 : 1 commensurability is more preferable in the engineering sense because most of the formation-flying missions require periodic relative motion on short timescales. For the incommensurable case where the ratio of the frequencies is irrational, i.e., $\forall p, q \in \mathbb{N}$, $\omega_C/\omega_D \neq p/q$, the flow on $\mathfrak{R}$ is referred to as two-periodic or quasi-periodic. This type of orbit unfolds along a helix on the surface of $\mathfrak{R}$, meandering over the whole region of $\mathfrak{R}$. However, every orbit winds around endlessly, never intersecting itself and yet never quite closing. For this case, the orbit is termed dense on $\mathfrak{R}$. In other words, each orbit comes arbitrarily close to any given point on $\mathfrak{R}$. This is not to say that the orbit passes through each point, it just comes arbitrarily close [16].

Consider now a chief heliocentric displaced orbit, paralleled to the ecliptic plane. Note that Ω_C is undefined when $i_C = 0\,\text{deg}$. However, for the sake of simplicity in form and consistency in equations, Ω_C is treated as a meaningless variable that can be set freely, without the need to introduce the mean longitude $l_C = \varpi_C + \Omega_C$ to eliminate the singularity, as is usually done [10]. Notably, this is not a restrictive assumption, as the results are independent of Ω_C.

The normalized variables are used by setting $r_C = \mu_\odot = 1$, where $\mu_\odot$ is the gravitational parameter of the Sun. The displaced orbital elements $\boldsymbol{\text{œ}}_C$ and $\boldsymbol{\text{œ}}_D$ are assumed to be

$$\boldsymbol{\text{œ}}_C = \left\{0.8,\ 0.6,\ 0.5,\ 0\,\text{deg},\ 0\,\text{deg},\ \varpi_C\right\}, \qquad \boldsymbol{\text{œ}}_D = \left\{\frac{2}{\sqrt{7}},\ \frac{\sqrt{3}}{\sqrt{7}},\ \sqrt{0.5},\ 5\,\text{deg},\ 90\,\text{deg},\ \varpi_D\right\} \tag{9.18}$$

According to Eqs. (9.14)–(9.16), the closed-form solution of relative position becomes

$$[\boldsymbol{\rho}]_{\mathcal{T}_{R_C}} = \begin{bmatrix} \rho_x \\ \rho_y \\ \rho_z \end{bmatrix} = \begin{bmatrix} a_D\left(\sin\varpi_C\cos\varpi_D - \cos\varpi_C\sin\varpi_D\cos i_D\right) + H_D\cos\varpi_C\sin i_D - a_C \\ a_D\left(\cos\varpi_C\cos\varpi_D + \sin\varpi_C\sin\varpi_D\cos i_D\right) - H_D\sin\varpi_C\sin i_D \\ a_D\sin\varpi_D\sin i_D + H_D\cos i_D - H_C \end{bmatrix} \tag{9.19}$$

Since the orbit is assumed circular, the mean arguments of latitude of the chief and deputy are explicit function of time t, given by

$$\varpi_C = \varpi_{C_0} + \omega_C t \mod 2\pi, \qquad \varpi_D = \varpi_{D_0} + \omega_D t \mod 2\pi \tag{9.20}$$

where ϖ_{C_0} and ϖ_{D_0} are the initial mean arguments of latitude of the chief and deputy, respectively. It indicates that $\boldsymbol{\rho}$ can be expressed as an explicit function of time. This is of great importance and advantage in understanding the dynamic nature of relative motion between two circular displaced orbits. For example, it can be used to find the bounds of relative motion, as will be illustrated in Sect. 10.4.

Based on the conditions in Eq. (9.18), Fig. 9.3 shows the three-dimensional relative orbit (solid line) lying on the invariant manifold $\mathfrak{R}$. Since the ratio of the angular velocities $\omega_C/\omega_D = \sqrt{0.5}$ is irrational, the motion along $\mathfrak{R}$ is therefore

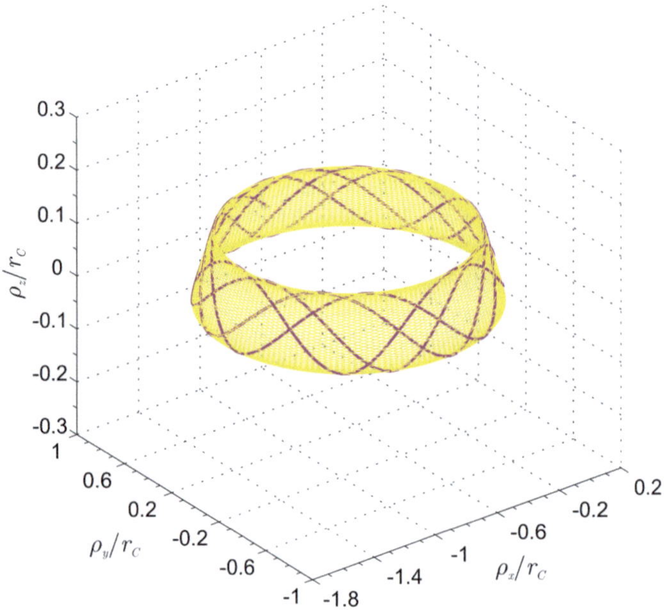

Fig. 9.3 Three-dimensional quasi-periodic relative motion and the invariant manifold upon which it lies

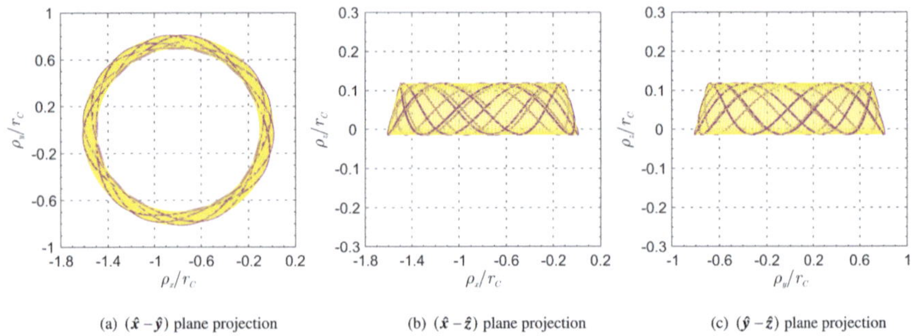

Fig. 9.4 Planar projections of quasi-periodic (incommensurable) relative motion

quasi-periodic. Figure 9.4 depicts the relative motion projected in $(\hat{x}-\hat{y})$, $(\hat{x}-\hat{z})$, and $(\hat{y}-\hat{z})$ planes, respectively. It is evident that, with the given conditions, the invariant manifold resembles the frustum of a cone with a curved rotation surface.

On the other hand, if the angular velocity of deputy ω_D in Eq. (9.18) is modulated so that $\omega_D = \omega_C = 0.5$, satisfying $1:1$ commensurability (resonance) condition, then the relative motion turns to be periodic. Figure 9.5 illustrates ten $1:1$ commensurable periodic relative orbits that are evenly distributed on the invariant manifold $\mathfrak{R}$ with equal initial phase (mean argument of latitude) differences. Figure 9.6 provides the projections in $(\hat{x}-\hat{y})$, $(\hat{x}-\hat{z})$, and $(\hat{y}-\hat{z})$ planes, respectively.

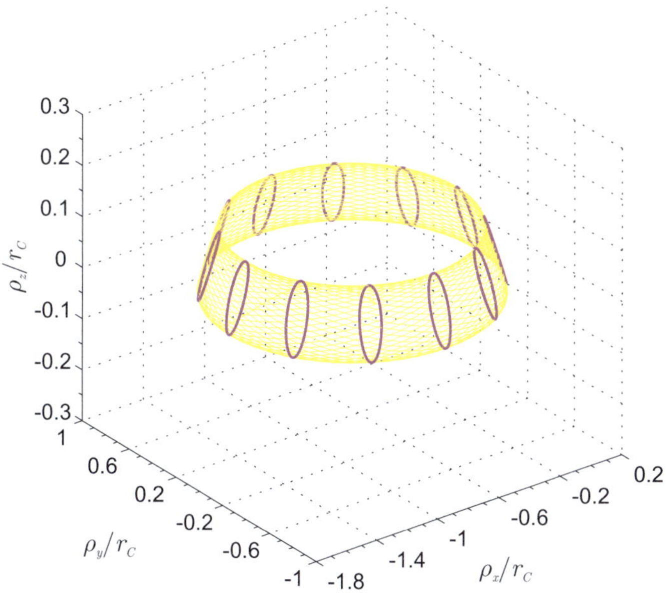

Fig. 9.5 Three-dimensional quasi-periodic relative motion and the invariant manifold upon which it lies

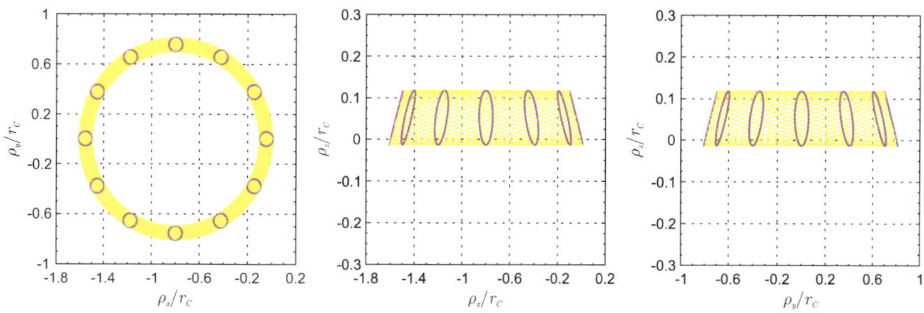

(a) $(\hat{x}-\hat{y})$ plane projection (b) $(\hat{x}-\hat{z})$ plane projection (c) $(\hat{y}-\hat{z})$ plane projection

Fig. 9.6 Planar projections of periodic (1 : 1 commensurable) relative motion

Changing the ratio of the frequencies ω_C/ω_D, more types of commensurable periodic relative orbits are obtained. As is shown in Fig. 9.7, although the periodicity can be guananteed, the scale of relative motion is much larger than that in the 1 : 1 commensurability case. In the following analysis, the upper and lower bounds of spacecraft relative motion for the incommensurable case and 1 : 1 commensurable case will be studied from an analytical point of view.

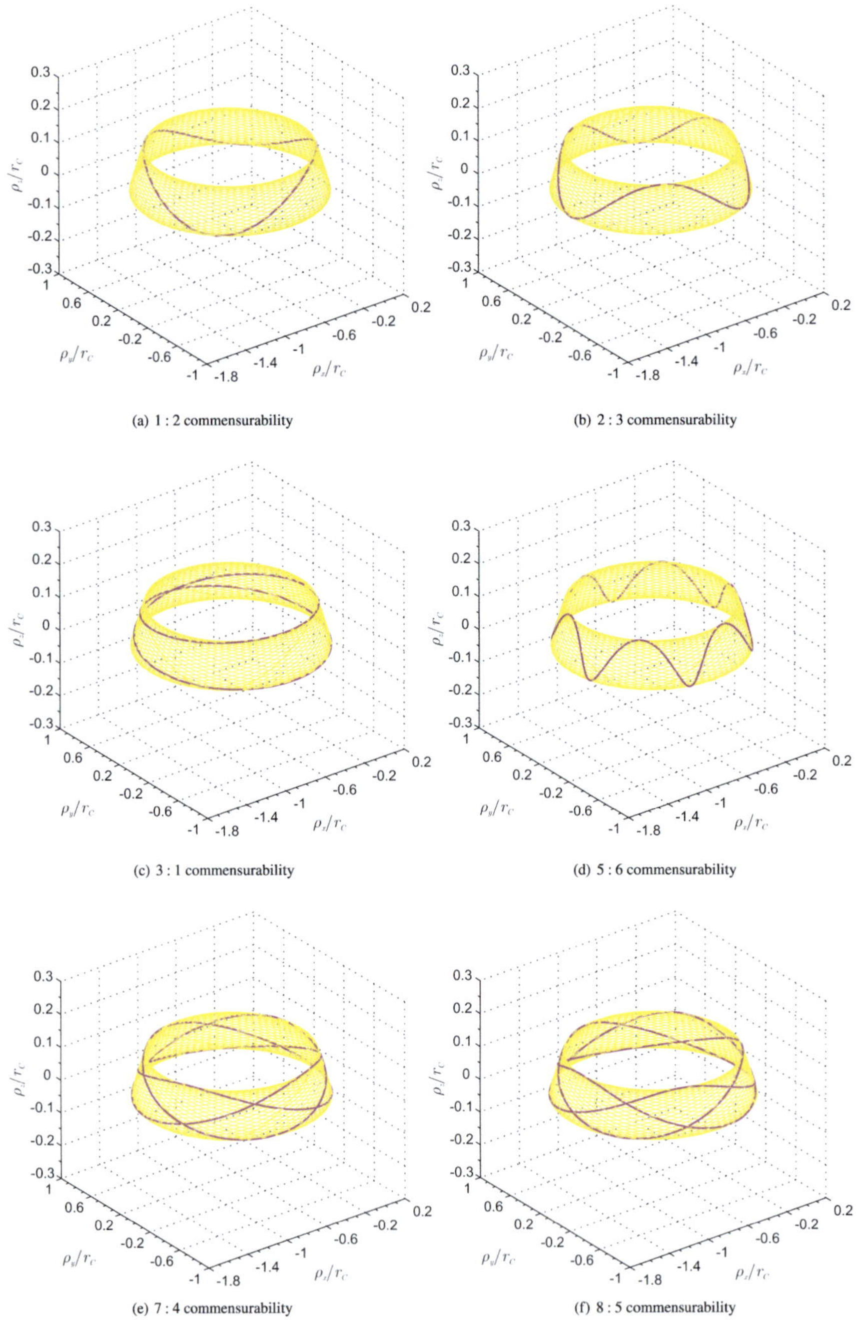

(a) 1 : 2 commensurability

(b) 2 : 3 commensurability

(c) 3 : 1 commensurability

(d) 5 : 6 commensurability

(e) 7 : 4 commensurability

(f) 8 : 5 commensurability

Fig. 9.7 Other types of periodic relative motion

9.4 Bounds of Incommensurable Relative Motion

Equation (9.13) provides a parametric presentation of the relative motion invariant manifold between two displaced orbits. For the incommensurable case, the relative orbit is quasi-periodic and ergodic, thus ϖ_C and ϖ_D should be treated as two independent variables when seeking for the extreme values of each single directional component and relative distance.

9.4.1 *Radial Bounds*

To find the radial bounds of relative motion, it is necessary to enforce

$$\frac{\partial \rho_x}{\partial \varpi_C} = 0 \quad \cap \quad \frac{\partial \rho_x}{\partial \varpi_D} = 0 \tag{9.21}$$

Substituting Eq. (9.21) into Eq. (9.14) results in

$$\begin{aligned} &\xi_1 \sin \varpi_C \cos \varpi_D - \xi_2 \sin \varpi_C \sin \varpi_D - \xi_3 \cos \varpi_C \sin \varpi_D \\ &- \xi_4 \cos \varpi_C \cos \varpi_D - \xi_5 \sin \varpi_C - \xi_6 \cos \varpi_C = 0 \end{aligned} \tag{9.22}$$

$$\xi_3 \sin \varpi_C \cos \varpi_D - \xi_4 \sin \varpi_C \sin \varpi_D - \xi_1 \cos \varpi_C \sin \varpi_D - \xi_2 \cos \varpi_C \cos \varpi_D = 0 \tag{9.23}$$

For convenience, using the notation

$$\lambda_x = \tan \frac{\varpi_C}{2}, \quad \eta_x = \tan \frac{\varpi_D}{2} \tag{9.24}$$

and Eqs. (9.22) and (9.23) are transformed into a set of quartic equations with two unknown variables λ_x and η_x:

$$\begin{aligned} &(\xi_4 - \xi_6)\, \lambda_x^2 \eta_x^2 + 2\,(\xi_1 + \xi_5)\, \lambda_x \eta_x^2 - 2\,\xi_3 \lambda_x^2 \eta_x - (\xi_4 + \xi_6)\, \lambda_x^2 \\ &- (\xi_4 - \xi_6)\, \eta_x^2 + 4\,\xi_2 \lambda_x \eta_x + 2\,(\xi_5 - \xi_1)\, \lambda_x + 2\,\xi_3 \eta_x + (\xi_4 + \xi_6) = 0 \end{aligned} \tag{9.25}$$

$$\xi_2 \lambda_x^2 \eta_x^2 + 2\,\xi_3 \lambda_x \eta_x^2 - 2\,\xi_1 \lambda_x^2 \eta_x - \xi_2 \lambda_x^2 - \xi_2 \eta_x^2 + 4\,\xi_4 \lambda_x \eta_x - 2\,\xi_3 \lambda_x + 2\,\xi_1 \eta_x + \xi_2 = 0 \tag{9.26}$$

Given a favorable initial guess, Eqs. (9.25) and (9.26) can be easily solved in MATLAB® to obtain the critical values $\lambda_x^\star$ and $\eta_x^\star$, and then $\varpi_C^\star$ and $\varpi_D^\star$. Note that the mapping in Eq. (9.24) is discontinuous at point $\varpi_C = (2\,k + 1)\,\pi$ and/or $\varpi_D = (2\,k + 1)\,\pi$ $(k \in \mathbb{Z})$, leading to $\lambda_x = \pm\infty$ and/or $\eta_x = \pm\infty$. Thus, it is necessary to determine whether $\lambda_x = \pm\infty$ and/or $\eta_x = \pm\infty$ correspond to the extreme value points by examining Eqs. (9.22) and (9.23). The critical value $\rho_x^\star$ can be obtained by substituting $\varpi_C^\star$ and $\varpi_D^\star$ into Eq. (9.14).

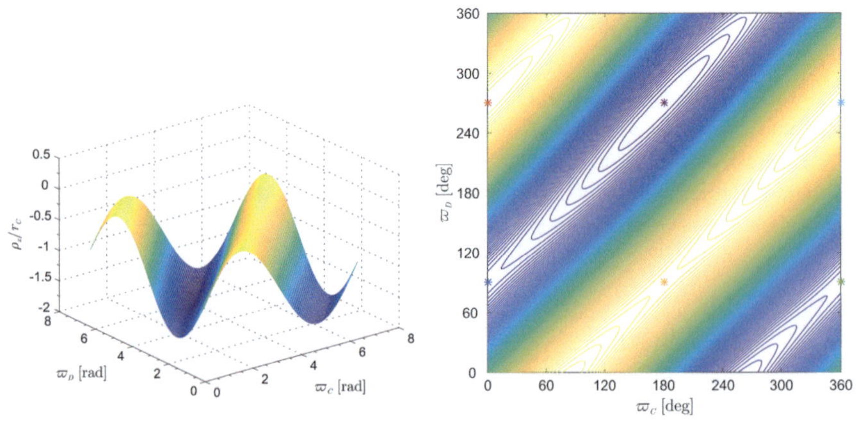

(a) Three-dimensional view (b) Contour plot

Fig. 9.8 Normalized radial relative distance $\tilde{\rho}_x$ as a function of ϖ_C and ϖ_D

For example, assume the displaced orbital elements of the two spacecraft are given by Eq. (9.18). Accordingly, four possible solutions are obtained

$$\left\{\lambda_x^\star, \eta_x^\star\right\} = \left\{\{0, 1\};\ \{0, -1\};\ \{\pm\infty, 1\};\ \{\pm\infty, -1\}\right\} \tag{9.27}$$

The critical mean arguments of latitude $\varpi_C^\star$ and $\varpi_D^\star$ corresponding to these solutions are

$$\varpi_C^\star = k\,\pi, \quad \varpi_D^\star = \frac{2\,k+1}{2}\,\pi \tag{9.28}$$

and the four extreme values of radial relative distance (normalized by r_C, denoted by a tilde) are given by

$$\tilde{\rho}_x^\star = \{-1.496;\ 0.01;\ -0.104;\ -1.61\} \tag{9.29}$$

Figure 9.8 shows the contour lines of normalized radial relative distance $\tilde{\rho}_x \triangleq \rho_x / r_C$ as a function of mean arguments of latitude of two spacecraft ϖ_C and ϖ_D, where the extreme value points are denoted by an asterisk $*$. Due to the periodicity of ϖ_C and ϖ_D, only the result with $\{\varpi_C, \varpi_D\} \in [0, 2\,\pi]$ is provided.

9.4.2 *Along-Track Bounds*

Likewise, the along-track bounds of relative motion entail the following condition:

$$\frac{\partial\,\rho_y}{\partial\,\varpi_C} = 0 \quad \cap \quad \frac{\partial\,\rho_y}{\partial\,\varpi_D} = 0 \tag{9.30}$$

which leads to

$$\xi_4 \sin \varpi_C \cos \varpi_D + \xi_3 \sin \varpi_C \sin \varpi_D - \xi_2 \cos \varpi_C \sin \varpi_D + \xi_1 \cos \varpi_C \cos \varpi_D + \xi_6 \sin \varpi_C - \xi_5 \cos \varpi_C = 0 \tag{9.31}$$

$$\xi_2 \sin \varpi_C \cos \varpi_D + \xi_1 \sin \varpi_C \sin \varpi_D - \xi_4 \cos \varpi_C \sin \varpi_D + \xi_3 \cos \varpi_C \cos \varpi_D = 0 \tag{9.32}$$

Again, using the notation

$$\lambda_y = \tan \frac{\varpi_C}{2}, \quad \eta_y = \tan \frac{\varpi_D}{2} \tag{9.33}$$

and Eqs. (9.31) and (9.32) are therefore transformed into a set of quartic equations with two unknown variables λ_y and η_y:

$$(\xi_1 + \xi_5)\, \lambda_y^2\, \eta_y^2 + 2\,(\xi_6 - \xi_4)\, \lambda_y\, \eta_y^2 + 2\,\xi_2\, \lambda_y^2\, \eta_y + (\xi_5 - \xi_1)\, \lambda_y^2 - (\xi_1 + \xi_5)\, \eta_y^2 + 4\,\xi_3\, \lambda_y\, \eta_y + 2\,(\xi_4 + \xi_6)\, \lambda_y - 2\,\xi_2\, \eta_y + (\xi_1 - \xi_5) = 0 \tag{9.34}$$

$$\xi_3\, \lambda_y^2\, \eta_y^2 - 2\,\xi_2\, \lambda_y\, \eta_y^2 + 2\,\xi_4\, \lambda_y^2\, \eta_y - \xi_3\, \lambda_y^2 - \xi_3\, \eta_y^2 + 4\,\xi_1\, \lambda_y\, \eta_y + 2\,\xi_2\, \lambda_y - 2\,\xi_4\, \eta_y + \xi_3 = 0 \tag{9.35}$$

Similar to Eqs. (9.25) and (9.26), Eqs. (9.34) and (9.35) can also be solved to obtain $\lambda_y^\star$ and $\eta_y^\star$, and then $\varpi_C^\star$ and $\varpi_D^\star$. Also, it is necessary to further examine whether $\varpi_C = (2k+1)\,\pi$ and/or $\varpi_D = (2k+1)\,\pi$ correspond to the extreme values of along-track relative motion $\rho_y^\star$.

Using the same displaced orbital elements by Eq. (9.18) and solving Eqs. (9.34) and (9.35), four solutions are obtained:

$$\left\{\lambda_y^\star, \eta_y^\star\right\} = \left\{\{1, 1\};\ \{1, -1\};\ \{-1, 1\};\ \{-1, -1\}\right\} \tag{9.36}$$

Upon substitution into Eq. (9.33), the critical mean arguments of latitude $\varpi_C^\star$ and $\varpi_D^\star$ are

$$\varpi_C^\star = \frac{2k+1}{2}\,\pi, \quad \varpi_D^\star = \frac{2k+1}{2}\,\pi \tag{9.37}$$

and the four normalized extreme values of along-track relative distance are

$$\tilde{\rho}_y^\star = \{0.696;\ -0.81;\ -0.696;\ 0.81\} \tag{9.38}$$

Figure 9.9 shows the contour lines of normalized along-track relative distance $\tilde{\rho}_y \triangleq \rho_y / r_C$ as a function of mean arguments of latitude of two spacecraft ϖ_C and ϖ_D with $\{\varpi_C, \varpi_D\} \in [0, 2\,\pi]$. Clearly, the extreme value points given by Eq. (9.38) are exactly the along-track bounds exhibited in Fig. 9.9.

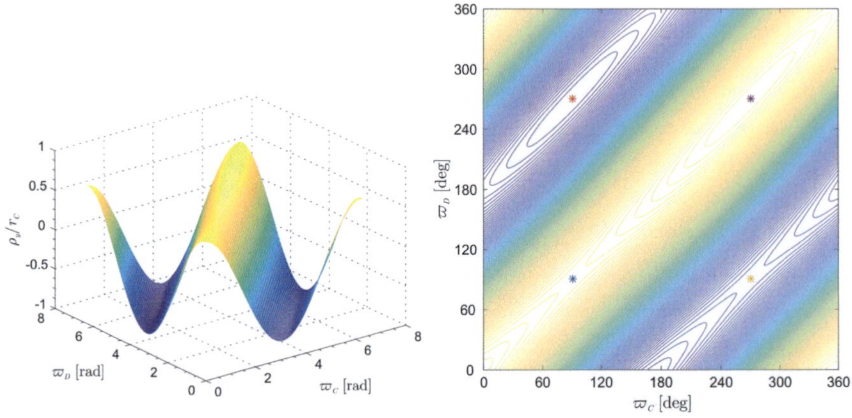

(a) Three-dimensional view

(b) Contour plot

Fig. 9.9 Normalized along-track relative distance $\tilde{\rho}_y$ as a function of ϖ_C and ϖ_D

9.4.3 Cross-Track Bounds

Equation (9.19) implies that the cross-track relative motion is the function of ϖ_D only, therefore the following condition should be imposed to acquire the cross-track bounds of relative motion:

$$\frac{\partial \rho_z}{\partial \varpi_D} = 0 \tag{9.39}$$

which, when substituting into Eq. (9.19), results in

$$\xi_7 \sin \varpi_D + \xi_8 \cos \varpi_D = 0 \tag{9.40}$$

Hence, the critical mean argument of latitude $\varpi_D^\star$ is given by

$$\varpi_D^\star = \begin{cases} k\pi + \arctan \dfrac{\xi_7}{\xi_8} & \xi_8 \neq 0 \\ \dfrac{2k+1}{2}\pi & \xi_8 = 0 \end{cases} \tag{9.41}$$

and two extreme values of cross-track relative distance are obtained using the displaced orbital elements given by Eq. (9.18):

$$\tilde{\rho}_z^\star = \{0.118;\ -0.014\} \tag{9.42}$$

Figure 9.10 illustrates the variation of normalized cross-track relative distance $\tilde{\rho}_z \triangleq \rho_z/r_C$ as a function of mean arguments of latitude of the deputy $\varpi_D \in [0, 2\pi]$. The extreme value points given by Eq. (9.42) are in accordance with the numerical results, verifying the correctness of the proposed method.

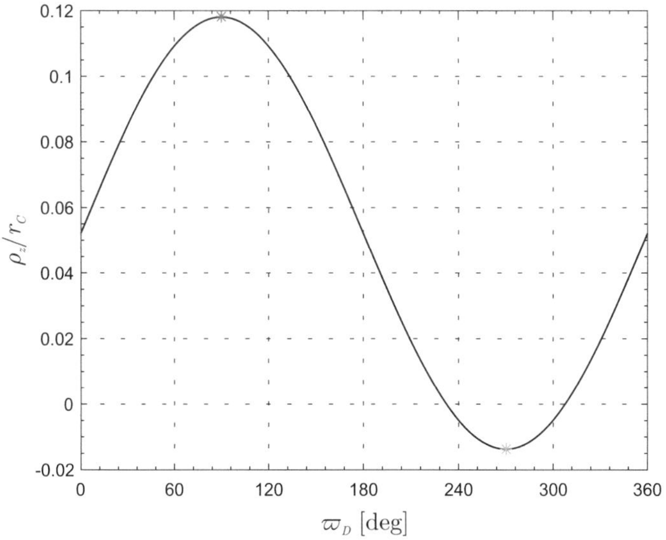

Fig. 9.10 Normalized cross-track relative distance $\tilde{\rho}_z$ as a function of ϖ_D

9.4.4 Relative Distance Bounds

The necessary conditions to seek for the extreme values of relative distance ρ are

$$\frac{\partial^2 \rho}{\partial \varpi_C} = 2\left(\rho_x \frac{\partial \rho_x}{\partial \varpi_C} + \rho_y \frac{\partial \rho_y}{\partial \varpi_C} + \rho_z \frac{\partial \rho_z}{\partial \varpi_C}\right) = 0 \quad \cap \quad \frac{\partial^2 \rho}{\partial \varpi_D} = 2\left(\rho_x \frac{\partial \rho_x}{\partial \varpi_D} + \rho_y \frac{\partial \rho_y}{\partial \varpi_D} + \rho_z \frac{\partial \rho_z}{\partial \varpi_D}\right) = 0 \tag{9.43}$$

Unlike the procedure that calculates the extreme value points of radial, along-track, and cross-track relative motions, Eq. (9.43) will generally result in an equation of eighth degree that is difficult to further simplify into a concise form. However, under some special conditions, e.g., i_C or i_D is either 0 deg or 90 deg, or $\Omega_C = \Omega_D$, the problem degenerates, and the search for extreme values of relative distance bounds can be transformed into seeking for the solution of two simple equations. Using the same displaced orbital elements in Eq. (9.18), Eq. (9.43) leads to

$$\iota_1 \sin\varpi_C \sin\varpi_D - \iota_2 \cos\varpi_C \sin\varpi_D + \iota_3 \sin\varpi_C = 0 \tag{9.44}$$

$$\iota_1 \cos\varpi_C \cos\varpi_D - \iota_2 \sin\varpi_C \sin\varpi_D + \iota_4 \cos\varpi_C = 0 \tag{9.45}$$

where the coefficients $\iota_1, \ldots, \iota_4$ are given by

$$\iota_1 \triangleq 2\, a_C\, a_D \cos i_D,\ \iota_2 \triangleq -2\, a_C\, a_D,\ \iota_3 \triangleq -2\, a_C\, H_D \sin i_D,\ \iota_4 \triangleq -2\, H_C\, a_D \sin i_D \tag{9.46}$$

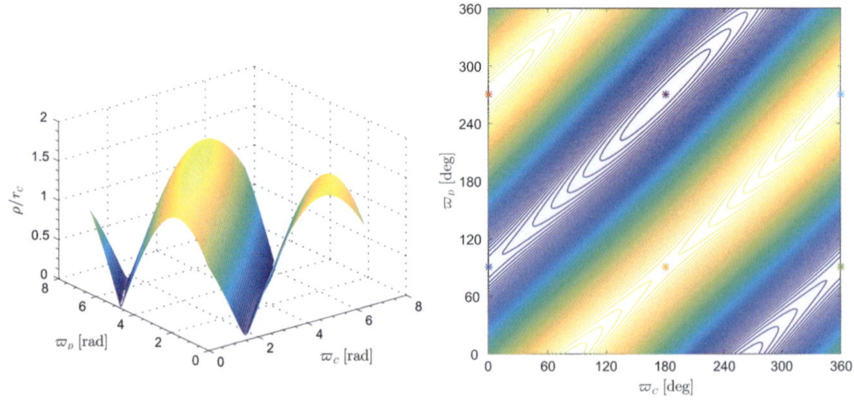

(a) Three-dimensional view (b) Contour plot

Fig. 9.11 Normalized relative distance $\tilde{\rho}$ as a function of ϖ_C and ϖ_D

Solving Eqs. (9.44) and (9.45) yields

$$\varpi_C^\star = k\,\pi, \quad \varpi_D^\star = \frac{2\,k+1}{2}\,\pi \tag{9.47}$$

The relative distance bounds are

$$\rho^\star = \left\{\sqrt{\iota_0+\iota_1+\iota_3+\iota_4};\ \sqrt{\iota_0-\iota_1+\iota_3-\iota_4};\ \sqrt{\iota_0-\iota_1-\iota_3+\iota_4};\ \sqrt{\iota_0+\iota_1-\iota_3-\iota_4}\right\} \tag{9.48}$$

where

$$\iota_0 \triangleq a_C^2 + a_D^2 + H_C^2 + H_D^2 - 2\,H_C\,H_D\,\cos i_D \tag{9.49}$$

Substituting Eq. (9.18) into Eq. (9.48), four normalized extreme values of relative distance are

$$\tilde{\rho}^\star = \{1.5006;\ 0.017;\ 0.1573;\ 1.6102\} \tag{9.50}$$

Figure 9.11 illustrates the variation of normalized relative distance $\tilde{\rho} \triangleq \rho/r_C$ as a function of mean arguments of latitude of two spacecraft $\{\varpi_C, \varpi_D\} \in [0, 2\,\pi]$. Note that the extreme value points given by Eq. (9.50) are exactly in accordance with the numerical results in Fig. 9.11.

9.5 Bounds of 1 : 1 Commensurable Relative Motion

To find the bounds of incommensurable relative orbit, the mean arguments of latitude of the two spacecraft are treated as variables that are independent of each other, due to their ergodic nature. However, for the commensurable case, time is the only variable.

To determine the bounds of directional components, the following condition should be enforced:

$$\begin{bmatrix} \dfrac{\partial \rho_x}{\partial t_x} \\ \dfrac{\partial \rho_y}{\partial t_y} \\ \dfrac{\partial \rho_z}{\partial t_z} \end{bmatrix} = \begin{bmatrix} \dfrac{\partial \rho_x}{\partial \varpi_C}\dfrac{\partial \varpi_C}{\partial t_x} + \dfrac{\partial \rho_x}{\partial \varpi_D}\dfrac{\partial \varpi_D}{\partial t_x} \\ \dfrac{\partial \rho_y}{\partial \varpi_C}\dfrac{\partial \varpi_C}{\partial t_y} + \dfrac{\partial \rho_y}{\partial \varpi_D}\dfrac{\partial \varpi_D}{\partial t_y} \\ \dfrac{\partial \rho_z}{\partial \varpi_C}\dfrac{\partial \varpi_C}{\partial t_z} + \dfrac{\partial \rho_y}{\partial \varpi_D}\dfrac{\partial \varpi_D}{\partial t_z} \end{bmatrix} = \mathbf{0}_3 \tag{9.51}$$

As aforementioned, the 1 : 1 commensurable relative motion is more meaningful for spacecaft formation design than other commensurable cases, the latter exhibiting large-scale relative distances. Thus, only the $\partial \varpi_C/\partial t = \partial \varpi_D/\partial t = \omega$ case is analyzed here. Denoting

$$\begin{cases} \gamma_C \triangleq \tan \dfrac{\varpi_{C_0}}{2}, \quad \gamma_D \triangleq \tan \dfrac{\varpi_{D_0}}{2} \\ \chi_x \triangleq \tan \dfrac{\omega\, t_x}{2}, \quad \chi_y \triangleq \tan \dfrac{\omega\, t_y}{2}, \quad \chi_z \triangleq \tan \dfrac{\omega\, t_z}{2} \end{cases} \tag{9.52}$$

and substituting Eqs. (9.19) and (9.20) and (9.52) into Eq. (9.51), one obtains

$$\sigma_4\, \chi_x^4 + \sigma_3\, \chi_x^3 + \sigma_2\, \chi_x^2 + \sigma_1\, \chi_x + \sigma_0 = 0 \tag{9.53}$$

$$\tau_4\, \chi_y^4 + \tau_3\, \chi_y^3 + \tau_2\, \chi_y^2 + \tau_1\, \chi_y + \tau_0 = 0 \tag{9.54}$$

$$\zeta_2\, \chi_z^2 + \zeta_1\, \chi_z + \zeta_0 = 0 \tag{9.55}$$

where $\sigma_0, \ldots, \sigma_4$, $\tau_0, \ldots, \tau_4$, and $\zeta_0, \ldots, \zeta_2$ are given in the Appendix. Accordingly, the radial, along-track, and cross-track bounds of relative motion can be calculated from Eqs. (9.19) and (9.20).

Likewise, to calculate the relative distance bounds, the following condition should be satisfied

$$\frac{\partial \rho^2}{\partial t_\rho} = \frac{\partial \rho^2}{\partial \omega_C}\frac{\partial \varpi_C}{\partial t_\rho} + \frac{\partial \rho^2}{\partial \varpi_D}\frac{\partial \varpi_D}{\partial t_\rho} = 0 \tag{9.56}$$

Bearing in mind $\partial \varpi_C/\partial t_\rho = \partial \varpi_D \partial/t_\rho = \omega$, hence

$$\frac{\partial \rho^2}{\partial \varpi_C} + \frac{\partial \rho^2}{\partial \varpi_D} = 0 \tag{9.57}$$

Substituting the displaced orbital elements of Eq. (9.18) into Eqs. (9.53)–(9.55) yields

$$\chi_x = \frac{H_D \sin i_D + \sqrt{H_D^2 \sin^2 i_D + 8 a_D^2 (1-\cos i_D)^2}}{2 a_D (1-\cos i_D)} \pm \sqrt{1 + \frac{H_D^2 \sin^2 i_D + H_D \sin i_D \sqrt{H_D^2 \sin^2 i_D + 8 a_D^2 (1-\cos i_D)^2}}{2 a_D^2 (1-\cos i_D)^2}} \tag{9.58}$$

$$\chi_y = \pm 1 \tag{9.59}$$

$$\chi_z = \pm 1 \tag{9.60}$$

and the critical (normalized) times when the directional components of relative motion arrive at these extreme values are

$$\tilde{t}^\star_{x_1} = \frac{1}{\tilde{\omega}} \left\{ (2k+1)\pi - \arcsin\left[\frac{-H_D \sin i_D + \sqrt{H_D^2 \sin^2 i_D + 8 a_D^2 (1-\cos i_D)^2}}{4 a_D (1-\cos i_D)} \right] \right\};$$

$$\tilde{t}^\star_{x_2} = \frac{1}{\tilde{\omega}} \left\{ 2k\pi + \arcsin\left[\frac{-H_D \sin i_D + \sqrt{H_D^2 \sin^2 i_D + 8 a_D^2 (1-\cos i_D)^2}}{4 a_D (1-\cos i_D)} \right] \right\} \tag{9.61}$$

$$\tilde{t}^\star_y = \left\{ \frac{(4k+3)\pi}{2\tilde{\omega}};\ \frac{(4k+1)\pi}{2\tilde{\omega}} \right\} \tag{9.62}$$

$$\tilde{t}^\star_z = \left\{ \frac{(4k+1)\pi}{2\tilde{\omega}};\ \frac{(4k+3)\pi}{2\tilde{\omega}} \right\} \tag{9.63}$$

Substituting Eqs. (9.61)–(9.63) into Eq. (9.19) yields the radial, along-track, and cross-track bounds $\tilde{\rho}^\star_x$, $\tilde{\rho}^\star_y$ and $\tilde{\rho}^\star_z$.

Figures 9.12, 9.13 and 9.14 illustrate the variations of directional relative distance within one period, i.e., $\tilde{t} \in [0,\ T]$, where $T = 2\pi/\omega = 4\pi$. According to Eqs. (9.61)–(9.63), the critical times are calculated to be $\tilde{t}^\star_x = \{0.1004;\ 6.1828\}$, $\tilde{t}^\star_y = \{9.4248;\ 3.1416\}$, $\tilde{t}^\star_z = \{3.1416;\ 9.4248\}$, and the correspongding distance bounds are $\tilde{\rho}^\star_x = \{-0.7429;\ -0.8571\}$, $\tilde{\rho}^\star_y = \{0.8101;\ 0.696\}$, $\tilde{\rho}^\star_z = \{0.118;\ -0.0137\}$.

To find the relative distance bounds between two circular displaced orbits, substituting Eq. (9.57) into Eqs. (9.19) and (9.20) to obtain

$$a_C H_D \sin i_D\, \varsigma - a_D H_C \sin i_D \sqrt{1-\varsigma^2} + a_C a_D (1-\cos i_D) \left(2\varsigma^2 - 1\right) = 0 \tag{9.64}$$

where $\varsigma \triangleq \sin \tilde{\omega} \tilde{t}^\star_\rho$. Theoretically, Eq. (9.64) has four roots, but only two are real and meaningful, with

$$\varsigma_1 = -0.6585; \quad \varsigma_2 = -0.6502 \tag{9.65}$$

and the normalized times corresponding to extreme value points of relative distance are

$$\tilde{t}^\star_{\rho_1} = \frac{2k\pi + \arcsin \varsigma_1}{\tilde{\omega}}; \quad \tilde{t}^\star_{\rho_2} = \frac{(2k+1)\pi + \arcsin \varsigma_2}{\tilde{\omega}} \tag{9.66}$$

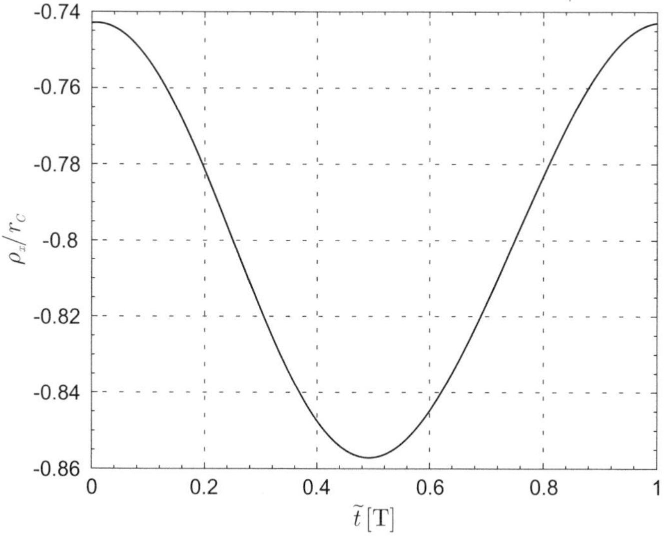

Fig. 9.12 Normalized radial relative distance $\tilde{\rho}_x$ as a function of $\tilde{t}$

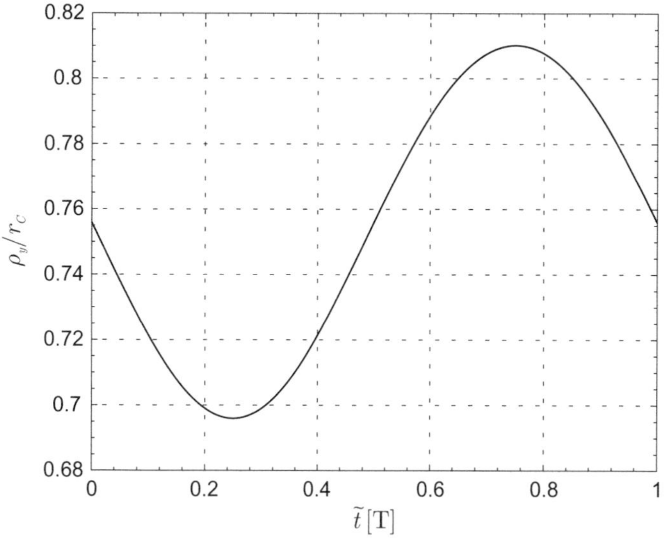

Fig. 9.13 Normalized along-track relative distance $\tilde{\rho}_y$ as a function of $\tilde{t}$

Figure 9.15 shows the variations of relative distance within one period. From Eq. (9.66), the critical times are analytically calculated to be $\tilde{t}_\rho^\star = \{1.4376;\ 7.6989\}$, and the relative distance bounds are $\tilde{\rho}^\star = \{1.1558;\ 1.0461\}$, in full accordance with the numerical results in Fig. 9.15.

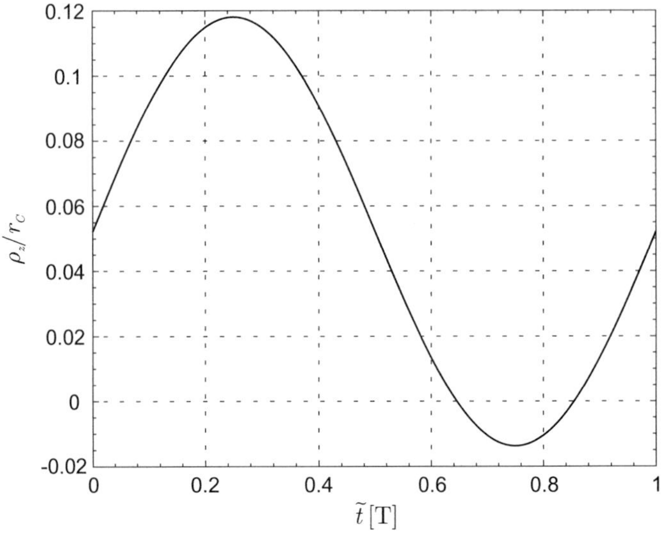

Fig. 9.14 Normalized cross-track relative distance $\tilde{\rho}_z$ as a function of $\tilde{t}$

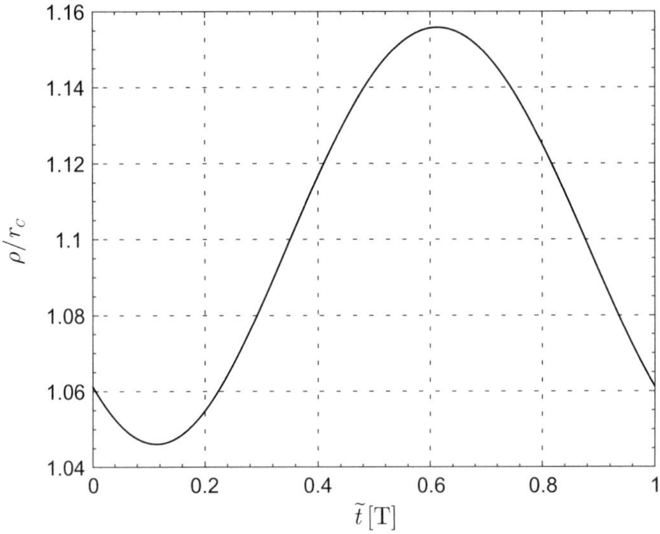

Fig. 9.15 Normalized relative distance $\tilde{\rho}$ as a function of $\tilde{t}$

9.6 Conclusions

This chapter has established a new methodology for modeling the relative motion of two spacecraft that track (different) circular displaced orbits. In particular, a new set of circular displaced orbital elements has been proposed to derive the closed-form solution to the relative motion problem. The orbits evolving on the invariant manifold manifest periodicity for the commensurable case and quasi-periodicity (ergodicity) for the incommensurable case. The bounds of relative motion between circular displaced orbits can be calculated in an analytic form for both the incommensurable and the 1 : 1 commensurable case. The derived expressions of quantitative relative distance bounds are of fundamental importance for safe mission progress because the knowledge of maximum distance is crucial information for communication design, whereas the minimum distance is essential for collision-avoidance purposes.

Appendix

The coefficients in Eqs. (9.53)–(9.55) are

$$\begin{aligned}\sigma_4 = {}&(\xi_4 - \xi_6 - \xi_2) + 2\,(\xi_3 - \xi_1 - \xi_5)\,\gamma_C + 2\,(\xi_3 - \xi_1)\,\gamma_D + 4\,(\xi_2 - \xi_4)\,\gamma_C\,\gamma_D + (\xi_2 + \xi_6 - \xi_4)\,\gamma_C^2 + (\xi_2 - \xi_4 - \xi_6)\,\gamma_D^2\\ &+ 2\,(\xi_1 - \xi_3 - \xi_5)\,\gamma_C\,\gamma_D^2 + 2\,(\xi_1 - \xi_3)\,\gamma_C^2\,\gamma_D + (\xi_4 - \xi_2 + \xi_6)\,\gamma_C^2\,\gamma_D^2\end{aligned} \tag{A-1}$$

$$\begin{aligned}\sigma_3 = {}&2\,(2\xi_1 - 2\xi_3 + \xi_5) + 4\,(\xi_4 - \xi_2)\,\gamma_C + 8\,(\xi_4 - \xi_2)\,\gamma_D + 16\,(\xi_3 - \xi_1)\,\gamma_C\,\gamma_D + 2\,(2\xi_3 - 2\xi_1 - \xi_5)\,\gamma_C^2\\ &+ 2\,(2\xi_3 - 2\xi_1 + \xi_5)\,\gamma_D^2 + 4\,(2\xi_2 - 2\xi_4 - \xi_6)\,\gamma_C\,\gamma_D^2 + 8\,(\xi_2 - \xi_4)\,\gamma_C^2\,\gamma_D + 2\,(2\xi_1 - 2\xi_3 - \xi_5)\,\gamma_C^2\,\gamma_D^2\end{aligned} \tag{A-2}$$

$$\begin{aligned}\sigma_2 = {}&6\,(\xi_2 - \xi_4) + 12\,(\xi_1 - \xi_3)\,\gamma_C + 12\,(\xi_1 - \xi_3)\,\gamma_D + 24\,(\xi_4 - \xi_2)\,\gamma_C\,\gamma_D + 6\,(\xi_4 - \xi_2)\,\gamma_C^2 + 6\,(\xi_4 - \xi_2)\,\gamma_D^2\\ &+ 12\,(\xi_3 - \xi_1)\,\gamma_C\,\gamma_D^2 + 12\,(\xi_3 - \xi_1)\,\gamma_C^2\,\gamma_D + 6\,(\xi_2 - \xi_4)\,\gamma_C^2\,\gamma_D^2\end{aligned} \tag{A-3}$$

$$\begin{aligned}\sigma_1 = {}&2\,(2\xi_3 - 2\xi_1 + \xi_5) + 4\,(2\xi_2 - 2\xi_4 - \xi_6)\,\gamma_C + 8\,(\xi_2 - \xi_4)\,\gamma_D + 16\,(\xi_1 - \xi_3)\,\gamma_C\,\gamma_D + 2\,(2\xi_1 - 2\xi_3 - \xi_5)\,\gamma_C^2\\ &+ 2\,(2\xi_1 - 2\xi_3 + \xi_5)\,\gamma_D^2 + 4\,(2\xi_4 - 2\xi_2 - \xi_6)\,\gamma_C\,\gamma_D^2 + 8\,(\xi_4 - \xi_2)\,\gamma_C^2\,\gamma_D + 2\,(2\xi_3 - 2\xi_1 - \xi_5)\,\gamma_C^2\,\gamma_D^2\end{aligned} \tag{A-4}$$

$$\begin{aligned}\sigma_0 = {}&(\xi_4 + \xi_6 - \xi_2) + 2\,(\xi_3 - \xi_1 + \xi_5)\,\gamma_C + 2\,(\xi_3 - \xi_1)\,\gamma_D + 4\,(\xi_2 - \xi_4)\,\gamma_C\,\gamma_D + 2\,(\xi_2 - \xi_4 - \xi_6)\,\gamma_C^2\\ &+ 2\,(\xi_2 - \xi_4 + \xi_6)\,\gamma_D^2 + 2\,(\xi_1 - \xi_3 + \xi_5)\,\gamma_C\,\gamma_D^2 + 2\,(\xi_1 - \xi_3)\,\gamma_C^2\,\gamma_D + 2\,(\xi_4 - \xi_2 - \xi_6)\,\gamma_C^2\,\gamma_D^2\end{aligned} \tag{A-5}$$

$$\begin{aligned}\tau_4 = {}&(\xi_1 + \xi_5 - \xi_3) + 2\,(\xi_4 - \xi_2 - \xi_6)\,\gamma_C + 2\,(\xi_4 - \xi_2)\,\gamma_D + 4\,(\xi_3 - \xi_1)\,\gamma_C\,\gamma_D + (\xi_3 - \xi_1 - \xi_5)\,\gamma_C^2\\ &+ 2\,(\xi_3 + \xi_5 - \xi_1)\,\gamma_D^2 + 2\,(\xi_2 - \xi_4 - \xi_6)\,\gamma_C\,\gamma_D^2 + 2\,(\xi_2 - \xi_4)\,\gamma_C^2\,\gamma_D + (\xi_1 - \xi_3 - \xi_5)\,\gamma_C^2\,\gamma_D^2\end{aligned} \tag{A-6}$$

$$\begin{aligned}\tau_3 = {}&2\,(2\xi_2 - 2\xi_4 + \xi_6) + 4\,(2\xi_1 - 2\xi_3 + \xi_5)\,\gamma_C + 8\,(\xi_1 - \xi_3)\,\gamma_D + 16\,(\xi_4 - \xi_2)\,\gamma_C\,\gamma_D + 2\,(2\xi_4 - 2\xi_2 - \xi_6)\,\gamma_C^2\\ &+ 2\,(2\xi_4 - 2\xi_2 + \xi_6)\,\gamma_D^2 + 4\,(2\xi_3 - 2\xi_1 + \xi_5)\,\gamma_C\,\gamma_D^2 + 8\,(\xi_3 - \xi_1)\,\gamma_C^2\,\gamma_D + 2\,(2\xi_2 - 2\xi_4 - \xi_6)\,\gamma_C^2\,\gamma_D^2\end{aligned} \tag{A-7}$$

$$\begin{aligned}\tau_2 = {}&6\,(\xi_3 - \xi_1) + 12\,(\xi_2 - \xi_4)\,\gamma_C + 12\,(\xi_2 - \xi_4)\,\gamma_D + 24\,(\xi_1 - \xi_3)\,\gamma_C\,\gamma_D + 6\,(\xi_1 - \xi_3)\,\gamma_C^2 + 6\,(\xi_1 - \xi_3)\,\gamma_D^2\\ &+ 12\,(\xi_4 - \xi_2)\,\gamma_C\,\gamma_D^2 + 12\,(\xi_4 - \xi_2)\,\gamma_C^2\,\gamma_D + 6\,(\xi_3 - \xi_1)\,\gamma_C^2\,\gamma_D^2\end{aligned} \tag{A-8}$$

$$\begin{aligned}\tau_1 = {}&2\,(2\xi_4 - 2\xi_2 + \xi_6) + 4\,(2\xi_3 - 2\xi_1 + \xi_5)\,\gamma_C + 8\,(\xi_3 - \xi_1)\,\gamma_D + 16\,(\xi_2 - \xi_4)\,\gamma_C\,\gamma_D + 2\,(2\xi_2 - 2\xi_4 - \xi_6)\,\gamma_C^2\\ &+ 2\,(2\xi_2 - 2\xi_4 + \xi_6)\,\gamma_D^2 + 4\,(2\xi_1 - 2\xi_3 + \xi_5)\,\gamma_C\,\gamma_D^2 + 8\,(\xi_1 - \xi_3)\,\gamma_C^2\,\gamma_D + 2\,(2\xi_4 - 2\xi_2 - \xi_6)\,\gamma_C^2\,\gamma_D^2\end{aligned} \tag{A-9}$$

$$\begin{aligned}\tau_0 = {}&(\xi_1 - \xi_3 - \xi_5) + 2\,(\xi_4 + \xi_6 - \xi_2)\,\gamma_C + 2\,(\xi_4 - \xi_2)\,\gamma_D + 4\,(\xi_3 - \xi_1)\,\gamma_C\,\gamma_D + 2\,(\xi_3 - \xi_1 + \xi_5)\,\gamma_C^2\\ &+ 2\,(\xi_3 - \xi_1 - \xi_5)\,\gamma_D^2 + 2\,(\xi_2 - \xi_4 + \xi_6)\,\gamma_C\,\gamma_D^2 + 2\,(\xi_2 - \xi_4)\,\gamma_C^2\,\gamma_D + 2\,(\xi_1 - \xi_3 + \xi_5)\,\gamma_C^2\,\gamma_D^2\end{aligned} \tag{A-10}$$

$$\zeta_2 = \xi_8 + 2\xi_7\,\gamma_D - \xi_8\,\gamma_D^2 \tag{A-11}$$

$$\zeta_1 = -2\xi_7 + 4\xi_8\,\gamma_D + 2\xi_7\,\gamma_D^2 \tag{A-12}$$

$$\zeta_0 = -\xi_8 - 2\xi_7\,\gamma_D + \xi_8\,\gamma_D^2 \tag{A-13}$$

References

1. McInnes C R. Solar Sailing: Technology, Dynamics and Mission Applications [M]. Springer-Praxis Series in Space Science and Technology, Springer-Verlag, London, Berlin, 1999.
2. McKay R J, Macdonald M, Biggs J, et al. Survey of highly-non-Keplerian orbits with low-thrust propulsion [J]. Journal of Guidance, Control, and Dynamics, 2011, 34 (3): 645–666.
3. Mengali G, Quarta A A. Non-Keplerian orbits for electric sails [J]. Celestial Mechanics and Dynamical Astronomy, 2009, 15 (1–3): 179–195.
4. McInnes C R. Solar sail mission applications for non-Keplerian orbits [J]. Acta Astronautica, 1999, 45 (4–9): 567–575.
5. Macdonald M, Mckay R J, Vasile M, et al. Low-thrust-enabled highly-non-Keplerian orbits in support of future Mars exploration [J]. Journal of Guidance, Control, and Dynamics, 2011, 34 (5): 1396–1411.
6. Ceriotti M, Heiligers J, McInnes C R. Trajectory and spacecraft design for a pole-sitter mission [J]. Journal of Spacecraft and Rockets, 2014, 51 (1): 311–326.
7. McInnes C R. The existence and stability of families of displaced two-body orbits [J]. Celestial Mechanics and Dynamical Astronomy, 1997, 67 (2): 167–180.
8. Gong S, Baoyin H, Li J. Solar sail formation flying around displaced solar orbits [J]. Journal of Guidance, Control, and Dynamics, 2007, 30 (4): 1147–1151.
9. Gong S, Baoyin H, Li J. Relative orbit design and control of formation around displaced solar orbits [J]. Aerospace Science and Technology, 2008, 12 (2): 195–201.
10. Broucke R A, Cefola P J. On the Equinoctial Orbit Elements [J]. Celestial Mechanics, 1972, 5 (3): 303–310.
11. Gurfil P. Euler Parameters as Nonsingular Orbital Elements in Near-Equatorial Orbits [J]. Journal of Guidance, Control, and Dynamics, 2005, 28 (5): 1079–1083.
12. Cohen C J, Hubbard E C. A nonsingular set of orbit elements [J]. The Astronomical Journal, 1962, 67 (1): 10–15.
13. McInnes C R. Passive control of displaced solar sail orbits [J]. Journal of Guidance, Control, and Dynamics, 1998, 21 (6): 975–982.
14. Gurfil P, Kholshevnikov K V. Manifolds and metrics in the relative spacecraft motion problem [J]. Journal of Guidance, Control, and Dynamics, 2006, 29 (4): 1004–1010.
15. Nayfeh A H, Balachandran B. Applied Nonlinear Dynamics [M]. Wiley-VCH Verlag, Weinheim, Germany, 2004.
16. Strogatz S H. Nonlinear Dynamics and Chaos: With Applications to Physics, Biology, Chemistry, and Engineering [M]. Westview Press-Perseus Books, Cambridge, MA, 1994.

Chapter 10
Extreme Values of Relative Distances for Spacecraft in Elliptic Displaced Orbits

This chapter provides a framework to obtain a semi-analytical approximation of extreme values of the relative distance between two spacecraft that cover elliptic displaced orbits. The relative motion is described in the rotating reference frame of the chief spacecraft and is parameterized with a new set of displaced orbital elements. The extreme values of the radial, along-track and cross-track motion are analytically evaluated (as roots of suitable algebraic equations) both for quasi-periodic orbits in the incommensurable case, and for periodic orbits in the 1 : 1 commensurable case. In particular, in the 1 : 1 commensurable case a Fourier series expansion is used to obtain a time-explicit expression of the relative motion. Finally, some illustrative examples are presented to validate the correctness of the proposed method.

10.1 Introduction

In recent years the development of advanced materials, such as the graphene [1], and the practical use of innovative propulsion system concepts like the photonic solar sail [2–5], have contributed to a growing interest towards non-Keplerian orbits [6, 7], due to their potential benefits offered to astronomical missions. In particular, artificial Lagrange orbits in a three-body dynamical system [8–11] are capable of monitoring solar plasma storms, while displaced orbits in two-body problems [12–14] can be used as planet or star pole sitters [15]. In principle, a displaced orbit, that is, an orbit whose orbital plane does not contain the primary's center-of-mass, can be generated either by means of photonic solar sails [16, 17], or by the more recent electric solar wind sails [7, 18], whereas more conventional propulsion systems are known to be inadequate for these applications [19].

In this context, most of the available literature is dedicated to the study of circular displaced orbits [7, 12–17] or to artificial Lagrange points [8–11] in the restricted three-body problem, whereas elliptic displaced orbits have attracted smaller inter-

© The Author(s), under exclusive license to Springer Nature Singapore Pte Ltd. 2026

C. Gao and W. Wang, *Solar Sailing for Deep-Space Explorations*,
https://doi.org/10.1007/978-981-95-0109-0_10

est. However, some celestial bodies such as Mercury (or some near-Earth asteroids) track orbits with considerable eccentricity and, in that case, a full-time observation of their pole region could not be obtained using circular displaced orbits.An interesting option for those problems is the use of elliptic displaced orbits. In particular, Gong and Li [20] point out that a photonic solar sail with reflection control devices (electrochromic panels) could be able to achieve and maintain such kind of non-Keplerian orbit. This is possible by exploiting the fact that the mean optical properties of the sail's film can be adjusted, within some limits, by switching on or off the state of each electrochromic panel [21, 22].

Moreover, some scientific explorations of celestial bodies that cover eccentric orbits, requiring multi-aspect observations that can only be accomplished using a formation of spacecraft. For instance, the interaction of solar wind with the magnetic field of Mercury is still, to some extent, an unknown phenomenon. An exhaustive comprehension of the magnetotail structure and dynamics of Mercury is possible only if a correlation between these two acting causes may be found, which requires the simultaneous observation and measurement of both the magnetotail and the solar wind from different positions using multiple spacecraft. Nevertheless, the only existing papers on solar sail relative motion around displaced orbits use a linearized mathematical model [17, 23, 24], and thus can only be applied to small-distance and short-term missions. Especially for a large baseline formation flying, a nonlinear insight is indispensable for looking at some inherent properties of the formation structure, such as the bounds and the geometric topology of relative motion [25], which might be useful for collision avoidance and communication analysis.

The aim of this chapter is to give a comprehensive analysis of closed-form solutions and bounds of relative motion between elliptic displaced orbits in a heliocentric scenario, thus generalizing the recently developed theory [26] wherein the relative motion between circular displaced orbits has been investigated. A new set of displaced orbital elements is first introduced to describe an elliptic displaced orbit and to define its orientation with respect to an inertial reference frame. Based on the newly defined parameters, a closed-form solution expressed in the chief sail's rotating frame is derived without the need of solving or linearizing the relative dynamics. The relative motion emanated from the parametric solution is quasi-periodic for the incommensurable case and periodic for the commensurable case. In both cases, the relative motion evolves along its invariant manifold that presents a well-defined accessible domain [27, 28].

The general solution is then used to look for the bounds along each coordinate axis. As far as the incommensurable case is concerned, the relative motion exhibits ergodicity, therefore the true anomaly of the chief and the eccentric anomaly of the deputy can be treated as independent of each other. By taking the partial derivatives of the directional components with respect to the two independent variables, the extreme values of these bounds are found to be the roots of algebraic equations. For the $1:1$ commensurable case, a Fourier series expansion is used to relate the true anomaly and eccentric anomaly to time. Accordingly, a first-order closed-form approximate solution in terms of small eccentricity is obtained. Similarly, by taking the partial derivatives of the components of the relative position vector with respect to time, the problem of looking for the bounds is transformed into that of calculating the roots of algebraic equations. The derived closed-form solutions are of general applicability and do not suffer from the usual assumption of small relative distance between the two orbits, as in the linearized case. The general solution also provides physical insights to the design of feasible relative orbits, both for the incommensurable and the commensurable case.

10.2 Spacecraft Relative Position Vector

Consider a spacecraft that covers a heliocentric displaced orbit using a continuous-thrust propulsion system and a suitable control law. Such a non-Keplerian orbit is elliptic, with semimajor axis a and eccentricity $e \neq 0$ (both constant) and belongs to a plane $\mathcal{P}$ placed at a distance H from Sun's center-of-mass O. The special case of circular displaced orbit ($e = 0$) is thoroughly discussed in Ref. [26], where the usefulness of a particular set of orbital elements has been pointed out to obtain a comprehensive description of the spacecraft dynamics along its orbit. By analogy with the Keplerian motion, let o be one of the two foci of the displaced orbit, and assume that such a point plays the same role of primary focus as it happens in a classic two-body problem (however, in our problem the primary body is not placed at o). In particular, note that the segment O-o is orthogonal to the plane $\mathcal{P}$ of the displaced orbit. Accordingly, the spacecraft position along its elliptic orbit can be located using a sort of true anomaly f, measured anticlockwise from the line joining the focus o with the orbit's pericenter.

To formulate the problem conveniently, it is useful to introduce three right-handed reference frames, see Fig. 10.1. The first one is the classical heliocentric-ecliptic (inertial) reference frame $\mathcal{T}_I$ $(O; x_I, y_I, z_I)$, where the plane (x_I, y_I) coincides with the ecliptic plane, the x_I axis points to the vernal equinox, and the z_I axis points to the north of celestial pole. The second coordinate system $\mathcal{T}_R$ $(S; x_R, y_R, z_R)$ is a rotating reference frame centered at the spacecraft center-of-mass S, whose plane (x_R, y_R) coincides with plane $\mathcal{P}$, the x_R axis points from o to the spacecraft, and z_R axis is positive in the direction of the spacecraft angular velocity vector. The unit vectors of $\mathcal{T}_R$ are $\hat{\boldsymbol{i}}_R$, $\hat{\boldsymbol{j}}_R$, and $\hat{\boldsymbol{k}}_R$. Finally $\mathcal{T}_P$ $(o; x_P, y_P, z_P)$ is a classical perifocal reference frame centered at o, where the plane (x_P, y_P) coincides with plane $\mathcal{P}$, and x_P axis is directed from o to the periapsis of the elliptic displaced orbit.

Paralleling the classical approach of Keplerian motion, the orientation of the elliptic displaced orbit with respect to the inertial frame $\mathcal{T}_I$ is described by three Euler angles: the inclination $i \in [0, \pi]$, i.e., the angle between z_I-axis and the $O - o$ line; the right ascension of the ascending node $\Omega \in [0, 2\pi]$, i.e., the angle between the line of nodes and the direction of x_I-axis; and the argument of periapsis $\omega \in [0, 2\pi]$, i.e., the angle between the line of nodes and the direction of the periapsis, see Fig. 10.1. In this context, the line of nodes is the intersection of plane $\mathcal{P}$ with plane $\mathcal{E}$, which is parallel to the ecliptic plane (x_I, y_I) and passes through the focus o of the displaced

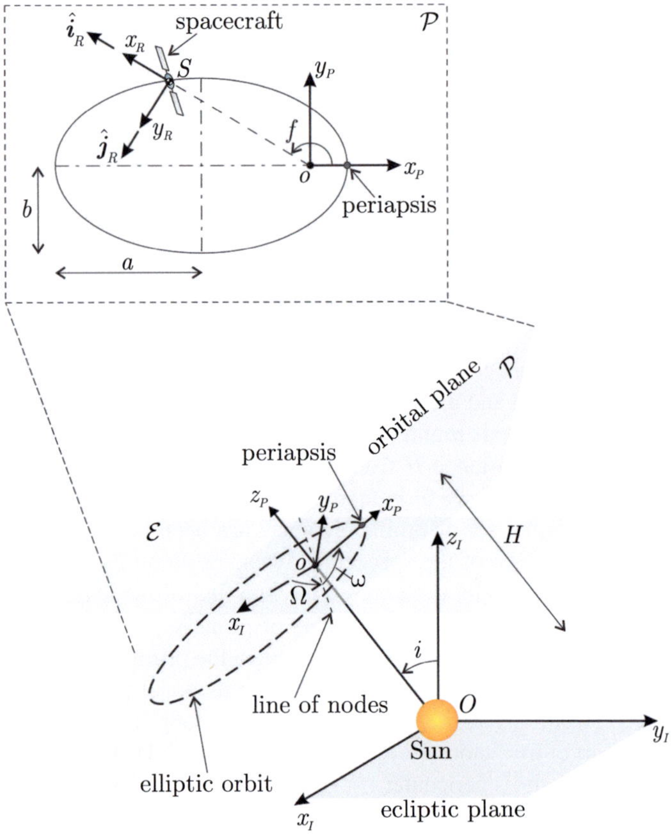

Fig. 10.1 Reference frames

orbit. In the special case in which the plane of the displaced orbit is parallel to the ecliptic plane, the line of nodes is assumed to coincide with the line parallel to x_I and passing through the focus o.

Note that the absolute position vector $\boldsymbol{r}$ of the spacecraft (i.e. its vector position calculated in the inertial frame $\mathcal{T}_I$) can be written as a function of its true anomaly f as

$$\boldsymbol{r} = H\,\hat{\boldsymbol{k}}_R + \frac{a\,(1-e^2)}{1+e\,\cos f}\,\hat{\boldsymbol{i}}_R \tag{10.1}$$

where $H\,\hat{\boldsymbol{k}}_R$ is the position vector (in the inertial frame) of the focus o of the displaced orbit. Accordingly, the components of the position vector $\boldsymbol{r}$ in the rotating reference frame $\mathcal{T}_R$ are expressed as

$$[\boldsymbol{r}]_{\mathcal{T}_R} = \left[\frac{a\,(1-e^2)}{1+e\,\cos f},\; 0,\; H \right]^{\mathrm{T}} \tag{10.2}$$

Likewise, the components of $\boldsymbol{r}$ in the perifocal reference frame $\mathcal{T}_P$ (recall that the displaced orbit's plane does not contain the Sun's center-of-mass) are

$$[\boldsymbol{r}]_{\mathcal{T}_P} = [a\,(\cos E - e)\,,\; b\,\sin E,\; H]^{\mathrm{T}} \tag{10.3}$$

where $b = a\sqrt{1-e^2}$ is the semiminor axis of the elliptic displaced orbit, and E is the spacecraft eccentric anomaly whose value, as a function of the true anomaly, is given by the well known equation

$$\tan\frac{E}{2} = \sqrt{\frac{1-e}{1+e}}\,\tan\frac{f}{2} \tag{10.4}$$

Consider now the general scenario of two spacecraft flying along two different elliptic displaced orbits. The first spacecraft is referred to as the chief (subscript C), and the other as the deputy (subscript D), see Fig. 10.2. The spacecraft relative position vector is given by

$$\boldsymbol{\rho} = \boldsymbol{r}_D - \boldsymbol{r}_C \tag{10.5}$$

where $\boldsymbol{r}_C$ (or $\boldsymbol{r}_D$) is the inertial position vector of chief (or deputy). The components of $\boldsymbol{\rho}$ are usually expressed in the Cartesian space. However, such a choice provides little physical insight within a three-dimensional scenario, especially for emphasizing the extreme values of the spacecraft relative distances, i.e., the extreme values of each component of $\boldsymbol{\rho}$. For this reason, it is better to write the components of $\boldsymbol{\rho}$ as a function of both the true anomaly f and the geometrical characteristics (given by the set $\{a, e, \Omega, \omega, i, H\}$) of the two elliptic displaced orbit.

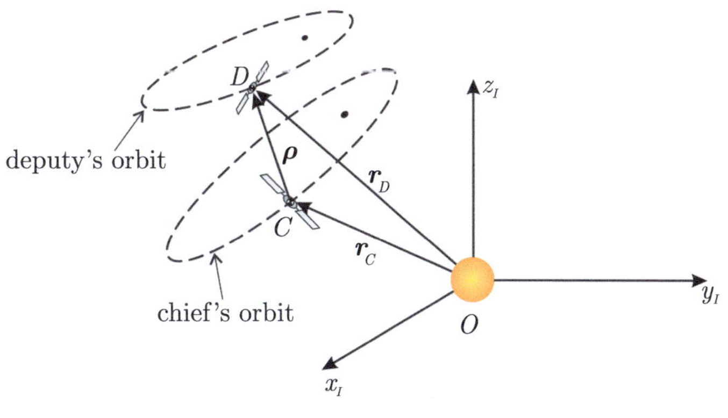

Fig. 10.2 Elliptic displaced orbits of chief and deputy spacecraft

The components of $\boldsymbol{\rho}$ are now expressed in the chief's rotating frame $\mathcal{T}_{R_C}$, and three coordinate transformations are carried out to find the components of $\boldsymbol{r}_D$ in $\mathcal{T}_{R_C}$. To that end, the inertial frame $\mathcal{T}_I$ is used as an intermediate bridge [25]. The first transformation is from the deputy's perifocal frame $\mathcal{T}_{P_D}$ to the inertial frame $\mathcal{T}_I$, the second is from $\mathcal{T}_I$ to the chief's periodical frame $\mathcal{T}_{P_C}$, and the last one is from $\mathcal{T}_{P_C}$ to $\mathcal{T}_{R_C}$. Taking into account that the transformation matrix $\mathbb{T}_{PI}$ between $\mathcal{T}_P$ and $\mathcal{T}_I$ is a function of the triplet $\{\Omega, i, \omega\}$ via the equation

$$\mathbb{T}_{PI}\left(\Omega, i, \omega\right) = \begin{bmatrix} \cos\omega\,\cos\Omega - \sin\omega\,\cos i\,\sin\Omega & -\sin\omega\,\cos\Omega - \cos\omega\,\cos i\,\sin\Omega & \sin i\,\sin\Omega \\ \cos\omega\,\sin\Omega + \sin\omega\,\cos i\,\cos\Omega & \cos\omega\,\cos i\,\cos\Omega - \sin\omega\,\sin\Omega & -\sin i\,\cos\Omega \\ \sin\omega\,\sin i & \cos\omega\,\sin i & \cos i \end{bmatrix} \tag{10.6}$$

and that the transformation matrix $\mathbb{T}_{PR}$ between the perifocal frame and the rotating frame is

$$\mathbb{T}_{PR}\left(f\right) = \begin{bmatrix} \cos f & \sin f & 0 \\ -\sin f & \cos f & 0 \\ 0 & 0 & 1 \end{bmatrix} \tag{10.7}$$

Equation (10.5) provides

$$\left[\boldsymbol{\rho}\right]_{\mathcal{T}_{R_C}} = \mathbb{T}_{PR}\left(f_C\right)\,\mathbb{T}^{\mathrm{T}}_{P_C I}\left(\Omega_C, i_C, \omega_C\right)\,\mathbb{T}_{P_D I}\left(\Omega_D, i_D, \omega_D\right)\,\left[\boldsymbol{r}_D\right]_{\mathcal{T}_{P_D}} - \left[\boldsymbol{r}_C\right]_{\mathcal{T}_{R_C}} \tag{10.8}$$

where $[\boldsymbol{r}_D]_{\mathcal{T}_{P_D}}$ is given by Eqs. (10.3) and (10.4) as a function of $\{a_D, e_D, f_D, H_D\}$, and $[\boldsymbol{r}_C]_{\mathcal{T}_{R_C}}$ is given by Eq. (10.2) as a function of $\{a_C, e_C, f_D, H_C\}$. For the sake of conciseness, the eccentric anomaly E_D is used in place of the true anomaly f_D for the deputy, and the following shorthand notation is introduced

$$\mathbb{T}_{P_D P_C} = \mathbb{T}^{\mathrm{T}}_{P_C I}\left(\Omega_C, i_C, \omega_C\right)\,\mathbb{T}_{P_D I}\left(\Omega_D, i_D, \omega_D\right) \tag{10.9}$$

where $\mathbb{T}_{P_D P_C}$ represents the transformation matrix between the chief's perifocal frame and the deputy's perifocal frame.

Substituting Eqs. (10.2) and (10.3), (10.6) and (10.7) into Eq. (10.8), the components ρ_x, ρ_y, and ρ_z of relative position vector $\boldsymbol{\rho}$ in $\mathcal{T}_{R_C}$ can be written in terms of elliptic displaced orbit characteristics of both the chief and the deputy spacecraft. The result is

$$\begin{aligned} \rho_x = {} & a_D\,T_{11}\,\cos f_C\,\cos E_D + a_D\,T_{21}\,\sin f_C\,\cos E_D + b_D\,T_{12}\,\cos f_C\,\sin E_D \\ & + b_D\,T_{22}\,\sin f_C\,\sin E_D + \left(H_D\,T_{13} - a_D\,e_D\,T_{11}\right)\cos f_C \\ & + \left(H_D\,T_{23} - a_D\,e_D\,T_{21}\right)\sin f_C - \frac{a_C\left(1 - e_C^2\right)}{1 + e_C\,\cos f_C} \end{aligned} \tag{10.10}$$

$$\begin{aligned} \rho_y = {} & a_D\,T_{21}\,\cos f_C\,\cos E_D - a_D\,T_{11}\,\sin f_C\,\cos E_D + b_D\,T_{22}\,\cos f_C\,\sin E_D - b_D\,T_{12}\,\sin f_C\,\sin E_D \\ & + \left(H_D\,T_{23} - a_D\,e_D\,T_{21}\right)\cos f_C + \left(a_D\,e_D\,T_{11} - H_D\,T_{13}\right)\sin f_C \end{aligned} \tag{10.11}$$

$$\rho_z = a_D\,T_{31}\,\cos E_D + b_D\,T_{32}\,\sin E_D + \left(H_D\,T_{33} - H_C - a_D\,e_D\,T_{31}\right) \tag{10.12}$$

where T_{ij} denotes the entry of row i and column j of matrix $\mathbb{T}_{P_D P_C}$, see Appendix A. Note that Eqs. (10.10)–(10.12) provide a parametric representation of the chief-deputy relative distance, with two angular coordinates (f_C, E_D) being the variables, thus comprising the invariant manifold $\mathfrak{R}$.

There are two qualitatively different cases, depending on whether the mean motion of the two orbits, n_C and n_D respectively, are commensurable or not. Recall that in the commensurable case, the ratio n_C/n_D is rational, whereas for the incommensurable case the ratio n_C/n_D is irrational. In the following sections, the bounds of each directional coordinate will be calculated analytically in both cases.

10.3 Bounds of Incommensurable Relative Motion

As far as the incommensurable case is concerned, i.e., when the ratio of mean motions is irrational, the relative motion evolving on the invariant manifold is open, but bounded, and is usually referred to as quasi-periodic. Due to its ergodic nature, the two angular variables in Eqs. (10.10)–(10.12), f_C and E_D, can be considered as independent of each other when differentiating the radial, along-track, and cross-track components with respect to them. In other terms, the necessary conditions to find the bounds of each component ρ_i are

$$\frac{\partial \rho_i}{\partial f_C} = 0 \quad \cap \quad \frac{\partial \rho_i}{\partial E_D} = 0 \qquad \text{with} \qquad i = \{x, y, z\} \tag{10.13}$$

whose corresponding solutions in terms of the chief's true anomaly and deputy's eccentric anomaly are referred to as $f_C = f_C^\star$ and $E_D = E_D^\star$, respectively. The bounds of the radial distance components are then calculated from Eqs. (10.10)–(10.12).

10.3.1 Radial Bounds

According to Eq. (10.13), the extreme values of radial relative distance (i.e. the component of vector $\boldsymbol{\rho}$ along the x_R axis), are given by enforcing the necessary condition

$$\frac{\partial \rho_x}{\partial f_C} = 0 \quad \cap \quad \frac{\partial \rho_x}{\partial E_D} = 0 \tag{10.14}$$

In general, it is not possible to obtain a closed-form solution of $f_C^\star$ using the first condition of Eq. (10.14). However, assuming that $e_C < 1$, a first-order approximation of the chief's focus-spacecraft distance can be obtained using a Taylor series expansion of the last term in Eq. (10.10), viz.,

$$\frac{a_C\,\left(1-e_C^2\right)}{1+e_C\,\cos f_C} \simeq a_C\,\left(1-e_C^2\right)\,(1-e_C\,\cos f_C) \tag{10.15}$$

Substituting Eq. (10.15) into Eq. (10.10), and then into Eq. (10.14), the result is

$$\begin{aligned} &\varsigma_2\,\sin f_C\,\sin E_D - \varsigma_1\,\sin f_C\,\cos E_D + \varsigma_3\,\cos f_C\,\sin E_D \\ &+ \varsigma_4\,\cos f_C\,\cos E_D + \varsigma_5\,\sin f_C + \varsigma_6\,\cos f_C = 0 \end{aligned} \tag{10.16}$$

$$\varsigma_3\,\sin f_C\,\cos E_D - \varsigma_4\,\sin f_C\,\sin E_D - \varsigma_1\,\cos f_C\,\sin E_D - \varsigma_2\,\cos f_C\,\cos E_D = 0 \tag{10.17}$$

where the constant coefficients $\{\varsigma_1, \ldots, \varsigma_6\}$ depending on the characteristics of two elliptic displaced orbits are given by

$$\begin{aligned} &\varsigma_1 = a_D\,T_{11},\ \varsigma_2 = -b_D\,T_{12},\ \varsigma_3 = b_D\,T_{22},\ \varsigma_4 = a_D\,T_{21},\ \varsigma_5 = a_D\,e_D\,T_{11} - H_D\,T_{13} - a_C\,e_C, \\ &\varsigma_6 = H_D\,T_{23} - a_D\,e_D\,T_{21} \end{aligned} \tag{10.18}$$

Using the notation

$$\iota = \tan\frac{f_C}{2}, \qquad \upsilon = \tan\frac{E_D}{2} \tag{10.19}$$

Equations (10.16)–(10.17) are transformed into a new set of two nonlinear equations in two unknowns ι and υ, viz.,

$$\begin{aligned} &(\varsigma_4-\varsigma_6)\,\iota^2\,\upsilon^2 + 2\,(\varsigma_1+\varsigma_5)\,\iota\,\upsilon^2 - 2\,\varsigma_3\,\iota^2\,\upsilon - (\varsigma_4+\varsigma_6)\,\iota^2 - (\varsigma_4-\varsigma_6)\,\upsilon^2 \\ &+4\,\varsigma_2\,\iota\,\upsilon + 2\,(\varsigma_5-\varsigma_1)\,\iota + 2\,\varsigma_3\,\upsilon + (\varsigma_4+\varsigma_6) = 0 \end{aligned} \tag{10.20}$$

$$\varsigma_2\,\iota^2\,\upsilon^2 + 2\,\varsigma_3\,\iota\,\upsilon^2 - 2\,\varsigma_1\,\iota^2\,\upsilon - \varsigma_2\,\iota^2 - \varsigma_2\,\upsilon^2 + 4\,\varsigma_4\,\iota\,\upsilon - 2\,\varsigma_3\,\iota + 2\,\varsigma_1\,\upsilon + \varsigma_2 = 0 \tag{10.21}$$

Given an initial guess, Eqs. (10.20) and (10.21) can be easily solved as a function of ι and υ, and then the critical value of f_C and E_D can be obtained from Eq. (10.19). Due to the inherent singularities of the mapping from (f_C, E_D) to (ι, υ) in Eq. (10.19), it must be further checked whether $\iota \to \pm\infty$ and $\upsilon \to \pm\infty$ correspond to the extreme values. In this case, since $\iota \to \pm\infty$ when $f_C \to k\,\pi$ and $\upsilon \to \pm\infty$ when $E_D \to k\,\pi$ (where $k \in \mathbb{N}$), it is sufficient to investigate whether $\cos f_C = \pm 1$, $\sin f_C = 0$, $\cos E_D = \pm 1$, and $\sin E_D = 0$ correspond to extreme values by substitutings them into Eqs. (10.16) and (10.17).

10.3.2 Along-Track Bounds

Bearing in mind Eq. (10.11), the necessary condition of Eq. (10.13) for the along-track motion leads to

$$\varsigma_4 \sin f_C \cos E_D + \varsigma_3 \sin f_C \sin E_D - \varsigma_2 \cos f_C \sin E_D + \varsigma_1 \cos f_C \cos E_D + \varsigma_6 \sin f_C - \varsigma_7 \cos f_C = 0 \tag{10.22}$$

$$\varsigma_2 \sin f_C \cos E_D + \varsigma_1 \sin f_C \sin E_D - \varsigma_4 \cos f_C \sin E_D + \varsigma_3 \cos f_C \cos E_D = 0 \tag{10.23}$$

where $\varsigma_1, \ldots, \varsigma_6$ are given by Eq. (10.18), and

$$\varsigma_7 = a_D\, e_D\, T_{11} - H_D\, T_{13} \tag{10.24}$$

Taking into account Eq. (10.19), Eqs. (10.22) and (10.23) provide

$$\begin{aligned}&(\varsigma_1 + \varsigma_7)\, \iota^2\, \upsilon^2 + 2\,(\varsigma_6 - \varsigma_4)\, \iota\, \upsilon^2 + 2\, \varsigma_2\, \iota^2\, \upsilon + (\varsigma_7 - \varsigma_1)\, \iota^2 - (\varsigma_1 + \varsigma_7)\, \upsilon^2 \\ &+ 4\, \varsigma_3\, \iota\, \upsilon + 2\,(\varsigma_4 + \varsigma_6)\, \iota - 2\, \varsigma_2\, \upsilon + (\varsigma_1 - \varsigma_7) = 0\end{aligned} \tag{10.25}$$

$$\varsigma_3\, \iota^2\, \upsilon^2 - 2\, \varsigma_2\, \iota\, \upsilon^2 + 2\, \varsigma_4\, \iota^2\, \upsilon - \varsigma_3\, \iota^2 - \varsigma_3\, \upsilon^2 + 4\, \varsigma_1\, \iota\, \upsilon + 2\, \varsigma_2\, \iota - 2\, \varsigma_4\, \upsilon + \varsigma_3 = 0 \tag{10.26}$$

Also, it must be further checked whether $\iota \to \pm\infty$ and $\upsilon \to \pm\infty$ correspond to extreme values using the same approach of determining the radial bounds.

10.3.3 Cross-Track Bounds

To find the bounds of cross-track motion, the necessary condition of Eq. (10.13) results in

$$\varsigma_8 \sin E_D + \varsigma_9 \cos E_D = 0 \tag{10.27}$$

where

$$\varsigma_8 = a_D\, T_{31}, \qquad \varsigma_9 = b_D\, T_{32} \tag{10.28}$$

In this case, the critical values of the deputy's eccentric anomaly corresponding to the cross-track bounds can be directly written in an explicit form as

$$E_D^\star = \begin{cases} \dfrac{(2\,k + 1)\,\pi}{2} & \text{if} \quad \varsigma_8 = 0 \\[2ex] \dfrac{(2\,k + 1)\,\pi}{2} - \arctan \dfrac{\varsigma_8}{\varsigma_9} & \text{if} \quad \varsigma_8 \neq 0 \end{cases} \tag{10.29}$$

with $k \in \mathbb{N}$, whereas the extreme values of z component are obtained by substituting Eq. (10.29) into Eq. (10.12).

10.4 Bounds of 1 : 1 Commensurable Relative Motion

In the commensurable case, i.e., when the relative motion is periodic, an important scenario is the 1 : 1 commensurable relative motion. Indeed, in this case the period of the relative motion is the least common multiple of the two orbital periods, which often turns out to be unacceptably long if the condition of 1 : 1 is not satisfied. Accordingly, the following discussion is confined to the 1 : 1 commensurable case only, and the condition $n_C = n_D = n$ is enforced.

Unlike the incommensurable relative motion where the critical time approaches infinity, the relative motion in the commensurable case exhibits periodicity and, therefore time t should be treated as the only independent variable when looking for corresponding extreme values of the bounds. However, from Eqs. (10.10)–(10.12), the spacecraft relative distance is time-implicit. In this case, to obtain a time-explicit solution, the true anomaly and eccentric anomaly must be written as an explicit function of time. To this end, first perform a nonlinear mapping from true anomaly f to mean anomaly M using a Fourier series expansions [29]:

$$\sin f = 2\sqrt{1-e^2}\sum_{h=1}^{\infty}\frac{\mathrm{d}\,J_h\,(h\,e)}{\mathrm{d}\,e}\,\frac{\sin\,(h\,M)}{h} \tag{10.30}$$

$$\cos f = \frac{2\left(1-e^2\right)}{e}\sum_{h=1}^{\infty}J_h\,(h\,e)\,\cos\,(h\,M) - e \tag{10.31}$$

where $J_h\,(\cdot)$ denotes the Bessel function of the first kind of order h defined as

$$J_h\,(\nu) = \sum_{j=0}^{\infty}\frac{(-1)^j\;\nu^{h+2j}}{2^{h+2j}\;j!\;(h+j)!} \tag{10.32}$$

Likewise, the eccentric anomaly can be related to the mean anomaly, as

$$\sin E = \sum_{h=1}^{\infty}\left[J_{h+1}\,(h\,e) + J_{h-1}\,(h\,e)\right]\frac{\sin\,(h\,M)}{h} = \frac{2}{e}\sum_{h=1}^{\infty}\frac{\sin\,(h\,M)}{h}\,J_h\,(h\,e) \tag{10.33}$$

$$\cos E = \sum_{h=1}^{\infty}\frac{1}{h}\left[J_{h-1}\,(h\,e) - J_{h+1}\,(h\,e)\right]\cos\,(h\,M) - \frac{e}{2} = \sum_{h=1}^{\infty}\frac{2\,\cos\,(h\,M)}{h^2}\,\frac{\mathrm{d}\,J_h\,(h\,e)}{\mathrm{d}\,e} - \frac{e}{2} \tag{10.34}$$

The generalized expansion theorem is now used to get

$$\frac{1-e^2}{1+e\,\cos f} = 1 + \frac{e^2}{2} - 2\,e\sum_{h=1}^{\infty}\frac{\mathrm{d}\,J_h\,(h\,e)}{\mathrm{d}\,e}\,\frac{\cos\,(h\,M)}{h^2} \tag{10.35}$$

The first order term is sufficient for elliptic displaced orbits with a small eccentricity. Accordingly, Eqs. (10.30) and (10.31), (10.33)–(10.35) are rewritten as

$$\sin f \simeq \sin M + e \sin(2M) + \mathcal{O}\left(e^2\right) \tag{10.36}$$

$$\cos f \simeq -e + \cos M + e \cos(2M) + \mathcal{O}\left(e^2\right) \tag{10.37}$$

$$\sin E \simeq \sin M + \frac{e}{2} \sin(2M) + \mathcal{O}\left(e^2\right) \tag{10.38}$$

$$\cos E \simeq -\frac{e}{2} + \cos M + \frac{e}{2} \cos(2M) + \mathcal{O}\left(e^2\right) \tag{10.39}$$

$$\frac{a\left(1-e^2\right)}{1+e\cos f} \simeq a\left(1 - e\cos M\right) + \mathcal{O}\left(e^2\right) \tag{10.40}$$

Substituting Eqs. (10.36)–(10.40) into Eqs. (10.10)–(10.12) and neglecting second-order terms, an approximate solution in terms of the two spacecraft mean anomalies is obtained as

$$\begin{aligned}
\rho_x \simeq & -\left(H_D T_{13} e_C + a_C\right) + \left(-\frac{3}{2} a_D T_{11} e_D + H_D T_{13} + a_C e_C\right) \cos M_C - a_D T_{11} e_C \cos M_D \\
& + \left(-\frac{3}{2} a_D T_{21} e_D + H_D T_{23}\right) \sin M_C - b_D T_{12} e_C \sin M_D + H_D T_{13} e_C \cos 2M_C + H_D T_{23} e_C \sin 2M_C \\
& + a_D T_{11} \cos M_C \cos M_D + b_D T_{12} \cos M_C \sin M_D + a_D T_{21} \sin M_C \cos M_D + b_D T_{22} \sin M_C \sin M_D \\
& + \frac{1}{2} a_D T_{11} e_D \cos M_C \cos 2M_D + \frac{1}{2} b_D T_{12} e_D \cos M_C \sin 2M_D + \frac{1}{2} a_D T_{21} e_D \sin M_C \cos 2M_D \\
& + \frac{1}{2} b_D T_{22} e_D \sin M_C \sin 2M_D + b_D T_{12} e_C \cos 2M_C \sin M_D + a_D T_{11} e_C \cos 2M_C \cos M_D \\
& + a_D T_{21} e_C \sin 2M_C \cos M_D + b_D T_{22} e_C \sin 2M_C \sin M_D
\end{aligned} \tag{10.41}$$

$$\begin{aligned}
\rho_y \simeq & -H_D T_{23} e_C + \left(-\frac{3}{2} a_D T_{21} e_D + H_D T_{23}\right) \cos M_C - a_D T_{21} e_C \cos M_D \\
& + \left(\frac{3}{2} a_D T_{11} e_D - H_D T_{13}\right) \sin M_C - b_D T_{22} e_C \sin M_D + H_D T_{23} e_C \cos 2M_C - H_D T_{13} e_C \sin 2M_C \\
& + a_D T_{21} \cos M_C \cos M_D + b_D T_{22} \cos M_C \sin M_D - a_D T_{11} \sin M_C \cos M_D - b_D T_{12} \sin M_C \sin M_D \\
& + \frac{1}{2} a_D T_{21} e_D \cos M_C \cos 2M_D + \frac{1}{2} b_D T_{22} e_D \cos M_C \sin 2M_D - \frac{1}{2} a_D T_{11} e_D \sin M_C \cos 2M_D \\
& - \frac{1}{2} b_D T_{12} e_D \sin M_C \sin 2M_D + a_D T_{21} e_C \cos 2M_C \cos M_D + b_D T_{22} e_C \cos 2M_C \sin M_D \\
& - a_D T_{11} e_C \sin 2M_C \cos M_D - b_D T_{12} e_C \sin 2M_C \sin M_D
\end{aligned} \tag{10.42}$$

$$\begin{aligned}
\rho_z \simeq & \left(-\frac{3}{2} a_D T_{31} e_D + H_D T_{33} - H_C\right) + a_D T_{31} \cos M_D + b_D T_{32} \sin M_D \\
& + \frac{1}{2} a_D T_{31} e_D \cos 2M_D + \frac{1}{2} b_D T_{32} e_D \sin 2M_D
\end{aligned} \tag{10.43}$$

Assuming an initial mean anomaly $M_C(t_0) = M_D(t_0) = 0$, after some symbolic operations and simplifications, Eqs. (10.41)–(10.43) can be written in a more compact form as

$$\rho_x \simeq \sigma_0 + \sigma_c \cos(n\,t) + \sigma_s \sin(n\,t) + \sigma_{c2} \cos(2\,n\,t) + \sigma_{s2} \sin(2\,n\,t) + \sigma_{c3} \cos(3\,n\,t) + \sigma_{s3} \sin(3\,n\,t) \quad (10.44)$$

$$\rho_y \simeq \tau_0 + \tau_c \cos(n\,t) + \tau_s \sin(n\,t) + \tau_{c2} \cos(2\,n\,t) + \tau_{s2} \sin(2\,n\,t) + \tau_{c3} \cos(3\,n\,t) + \tau_{s3} \sin(3\,n\,t) \quad (10.45)$$

$$\rho_z \simeq \xi_0 + \xi_c \cos(n\,t) + \xi_s \sin(n\,t) + \xi_{c2} \cos(2\,n\,t) + \xi_{s2} \sin(2\,n\,t) \quad (10.46)$$

where the relevant coefficients are given in the Appendix B. Equations (10.44)–(10.46) provide a first-order approximate closed-form solution, which is useful for understanding the nature of the relative motion due to its time-explicit property. Equations (10.44)–(10.46) also serve as a basis for studying the extreme values of the spacecraft distance along each coordinate axis, as is illustrated in the next sections.

To that end, the necessary condition to be met is

$$\frac{\partial\,\rho_i}{\partial\,t} = 0 \qquad \text{with} \qquad i = \{x, y, z\} \quad (10.47)$$

where the components ρ_i are given by Eqs. (10.44)–(10.46). Bearing in mind the structure of the three components of $\boldsymbol{\rho}$, it is convenient to make the change of variable

$$s = \tan\frac{n\,t}{2} \qquad \text{with} \qquad n\,t \neq (2\,k + 1)\,\pi \quad \text{and} \quad k \in \mathbb{N} \quad (10.48)$$

Consequently, the necessary condition of Eq. (10.47) becomes

$$\frac{\partial\,\rho_i}{\partial\,s} = 0 \quad \text{if} \quad n\,t \neq (2\,k + 1)\,\pi \quad \text{with} \quad k \in \mathbb{N} \quad (10.49)$$

Once the positive values of s that satisfy Eq. (10.49) are found, the corresponding time values are obtained from Eq. (10.48). These time values, when substituted into Eqs. (10.44)–(10.46), provide a set of stationary values of ρ_i. The maximum/minimum of these stationary values can now be calculated along with the corresponding quantities $s^\star$ and $t^\star$ at which those extreme values are obtained.

Note that the change of variable in Eq. (10.48), can be performed provided that $n\,t \neq (2\,k + 1)\,\pi$. Therefore, it is necessary to check the generic component ρ_i in the particular case $n\,t = (2\,k + 1)\,\pi$. To that end, it is necessary to substitute the condition $n\,t = (2\,k + 1)\,\pi$ into Eqs. (10.44)–(10.46). The corresponding expressions of the components of the position vector become

$$\rho_x \simeq \sigma_0 - \sigma_c + \sigma_{c2} - \sigma_{c3} \quad (10.50)$$

$$\rho_y \simeq \tau_0 - \tau_c + \tau_{c2} - \tau_{c3} \quad (10.51)$$

$$\rho_z \simeq \xi_0 - \xi_c + \xi_{c2} \quad (10.52)$$

It is now sufficient to compare the values of the generic component ρ_i given by Eqs. (10.50)–(10.52) with those calculated using the necessary condition Eq. (10.49) in order to get (by direct comparison) the extreme values of each component of $\boldsymbol{\rho}$.

10.4.1 Radial Bounds

Substituting Eq. (10.44) into Eq. (10.49) and recalling Eq. (10.48), the following polynomial in the variable s is obtained

$$\lambda_6\, s^6 + \lambda_5\, s^5 + \lambda_4\, s^4 + \lambda_3\, s^3 + \lambda_2\, s^2 + \lambda_1\, s + \lambda_0 = 0 \tag{10.53}$$

where $\lambda_1, \lambda_2, \ldots, \lambda_6$ are constant coefficients given by

$$\lambda_0 = -\sigma_s + 2\,\sigma_{s2} - 3\,\sigma_{s3} \tag{10.54}$$

$$\lambda_1 = 2\,\sigma_c + 8\,\sigma_{c2} + 18\,\sigma_{c3} \tag{10.55}$$

$$\lambda_2 = -\sigma_s - 10\,\sigma_{s2} + 45\,\sigma_{s3} \tag{10.56}$$

$$\lambda_3 = 4\,\sigma_c - 60\,\sigma_{c3} \tag{10.57}$$

$$\lambda_4 = \sigma_s - 10\,\sigma_{s2} - 45\,\sigma_{s3} \tag{10.58}$$

$$\lambda_5 = 2\,\sigma_c - 8\,\sigma_{c2} + 18\,\sigma_{c3} \tag{10.59}$$

$$\lambda_6 = \sigma_s + 2\,\sigma_{s2} + 3\,\sigma_{s3} \tag{10.60}$$

where $\sigma_s, \sigma_{s2}, \sigma_{s3}, \sigma_c, \sigma_{c2}$, and σ_{c3} are given in the Appendix B. Note that Eq. (10.53) is a sixth-order polynomial whose real and positive roots can be easily numerically found. A similar procedure applies for the other two cases of ρ_y and ρ_z.

10.4.2 Along-Track Bounds

In this case, the necessary condition of Eq. (10.47), where ρ_y is taken from Eq. (10.45), results in

$$\eta_6\, s^6 + \eta_5\, s^5 + \eta_4\, s^4 + \eta_3\, s^3 + \eta_2\, s^2 + \eta_1\, s + \eta_0 = 0 \tag{10.61}$$

where the constant coefficients $\eta_0, \eta_1, \ldots, \eta_6$ are given by (see also the Appendix B)

$$\eta_0 = -\tau_s + 2\,\tau_{s2} - 3\,\tau_{s3} \tag{10.62}$$

$$\eta_1 = 2\,\tau_c + 8\,\tau_{c2} + 18\,\tau_{c3} \tag{10.63}$$

$$\eta_2 = -\tau_s - 10\,\tau_{s2} + 45\,\tau_{s3} \tag{10.64}$$

$$\eta_3 = 4\,\tau_c - 60\,\tau_{c3} \tag{10.65}$$

$$\eta_4 = \tau_s - 10\,\tau_{s2} - 45\,\tau_{s3} \tag{10.66}$$

$$\eta_5 = 2\,\tau_c - 8\,\tau_{c2} + 18\,\tau_{c3} \tag{10.67}$$

$$\eta_6 = \tau_s + 2\,\tau_{s2} + 3\,\tau_{s3} \tag{10.68}$$

10.4.3 Cross-Track Bounds

Finally, to get the cross-track bounds, the necessary condition of Eq. (10.47) with $\rho_i = \rho_z$ should be satisfied. Assuming that $n\,t \neq (2\,k + 1)\,\pi$, the result is

$$\kappa_4\,s^4 + \kappa_3\,s^3 + \kappa_2\,s^2 + \kappa_1\,s + \kappa_0 = 0 \tag{10.69}$$

where the constant coefficients $\kappa_0, \ldots, \kappa_4$ are

$$\kappa_0 = -\xi_s - 2\,\xi_{s2} \tag{10.70}$$

$$\kappa_1 = 2\,\xi_c + 8\,\xi_{c2} \tag{10.71}$$

$$\kappa_2 = 12\,\xi_{s2} \tag{10.72}$$

$$\kappa_3 = 2\,\xi_c - 8\,\xi_{c2} \tag{10.73}$$

$$\kappa_4 = \xi_s - 2\,\xi_{s2} \tag{10.74}$$

10.5 Case Study

To verify the consistency of the proposed semi-analytical mathematical model, two illustrative examples are now discussed in which both the incommensurable and commensurable case are considered.

10.5.1 Incommensurable Case

Introduce a distance unit DU $\triangleq a_C$, and a time unit TU $\triangleq 1/n_C$. In order to obtain an incommensurable case, the semimajor axis of the deputy's elliptic displaced orbit is assumed to be $a_D = 1.02$ DU, so that the deputy's mean motion becomes $n_D = \sqrt{1/1.02^3}$ TU^{-1} and the ratio n_C/n_D is irrational. In this scenario, the geometrical characteristics of the two displaced orbits are summarized in Table 10.1.

Table 10.1 Characteristics of the displaced orbits for the incommensurable case

	a [DU]	H [DU]	e [deg]	i [deg]	Ω [deg]	ω [deg]	f [deg]
Chief	1	0.1	0.05	0.001	0	0	0
Deputy	1.02	0.08	0.2	5	5	0	0

Table 10.2 Extreme values of the radial, along-track and cross-track distance for the incommensurable case

Component	$\iota^\star$	$\upsilon^\star$	$f_C^\star$ mod π [rad]	$E_D^\star$ mod π [rad]	Extreme values [rad]	
ρ_x	0.0971	0.0519	0.1935	0.1037	0.1743	−2.1742
ρ_y	−1.4379	0.0445	1.7028	0.0437	0.8158	−1.2241
ρ_z	–	–	–	1.5708	0.0668	−0.1074

When the relative motion arrives at its radial, along-track and cross-track boundaries, the principal value of the corresponding (critical) true anomaly of the chief and eccentric anomaly of the deputy are listed in Table 10.2, along with the extreme values of the generic component ρ_i. In fact, for the radial component of the relative position vector, from Eqs. (10.20) and (10.21) it is found that $\iota^\star = 0.0971$ and $\upsilon^\star = 0.0519$. On the other hand, for the along-track component of the relative position vector, from Eqs. (10.25)–(10.26) one obtains $\iota^\star = -1.4379$ and $\upsilon^\star = 0.0445$.

Figure 10.3 shows the relative motion (dotted line) and its invariant manifold $\mathfrak{R}$ along which it evolves in the chief-fixed rotating frame $\mathcal{T}_{R_C}$. Since the mean motions are incommensurable, the relative motion is quasi-periodic, and the relative orbit unfolds along a helix on the surface of $\mathfrak{R}$, winding round the whole region of the invariant manifold. The bounds values (dotted lines) of the radial, along-track motion, and the cross-track motion are plotted in Fig. 10.4. The extreme values calculated analytically are able to successfully predict the inner and outer boundaries, with maximum errors in the three directions being less than 0.12%, 0.09% and 0.0002% respectively.

10.5.2 1 : 1 Commensurable Case

In this case, the semimajor axis of the deputy's elliptic displaced orbit is assumed to be $a_D = a_C$, in order to obtain $n_D = n_C$ (1 : 1 commensurable case). The relevant data are shown in Table 10.3, and the deputy is assumed to track the Mercury's orbit with a constant displacement.

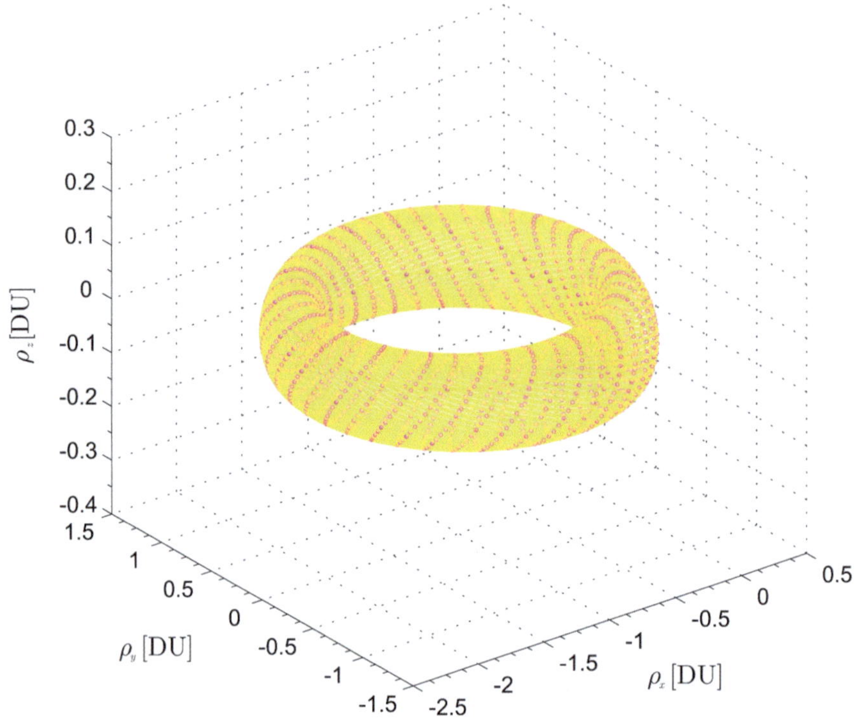

Fig. 10.3 Three-dimensional incommensurable relative motion with its invariant manifold $\mathfrak{R}$

Table 10.4 shows the extreme values of each component of the relative position vector, whereas Fig. 10.5 shows the relative motion generated by the first-order approximation (dashed lines) and the exact solution (solid lines). The relative motion exhibits periodicity since the mean motions are 1 : 1 commensurable and, notably, the first-order solution fits well the numerical solution. In particular, the proposed methodology for determining extreme values via first-order approximate solution is accurate in predicting the distance bounds, with maximum errors in the three directions being less than 0.44%, 0.79% and 0.13% respectively, see also Fig. 10.6.

Since the approximate extreme values of relative distances are calculated under the "small eccentricity" assumption, it is necessary to check the accuracy of the results as a function of the orbital eccentricity of deputy spacecraft. To this end, define the maximum relative distance errors along the three directions, i.e. the maximum differences of the results obtained by using the semi-analytical method and the (exact) numerical method (denoted by subscript " n"):

$$\epsilon_i \triangleq \max \left| \frac{\rho_i^\star - \rho_{i_\text{n}}}{\rho_{i_\text{n}}} \right| \qquad \text{with} \qquad i = \{x, y, z\} \tag{10.75}$$

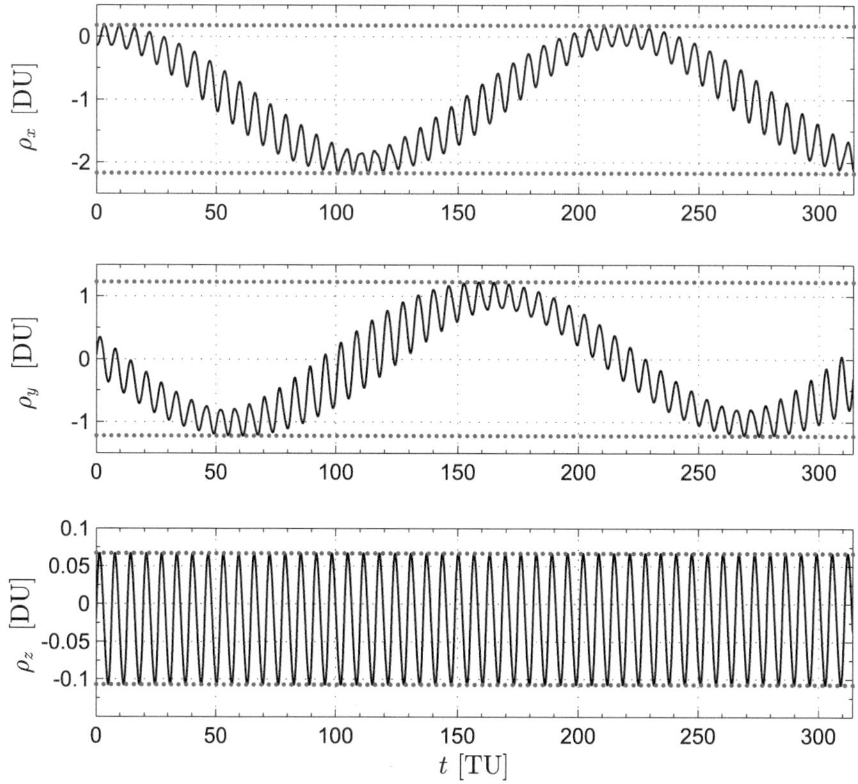

Fig. 10.4 Analytic bounds (dotted lines) and numerical simulation (solid line) of relative motion for the incommensurable case

Table 10.3 Characteristics of the displaced orbits for the 1 : 1 commensurable case

	a [DU]	H [DU]	e [deg]	i [deg]	Ω [deg]	ω [deg]	f [deg]
Chief	1	0.1	0.05	0.001	50	80	0
Deputy	1	0.08	0.2056	7	48.33	77.45	0

Table 10.4 Extreme values of the radial, along-track and cross-track bounds for the 1 : 1 commensurable case

Component	$s^\star$		$t^\star$ mod 2π [TU]		Extreme values [DU]	
ρ_x	21.7824	−0.0968	3.0498	6.0902	0.1549	−0.1740
ρ_y	−1.0057	0.9192	1.4866	4.7067	0.2480	−0.3896
ρ_z	−6.9272	0.0919	0.1833	3.4283	0.0768	−0.1671

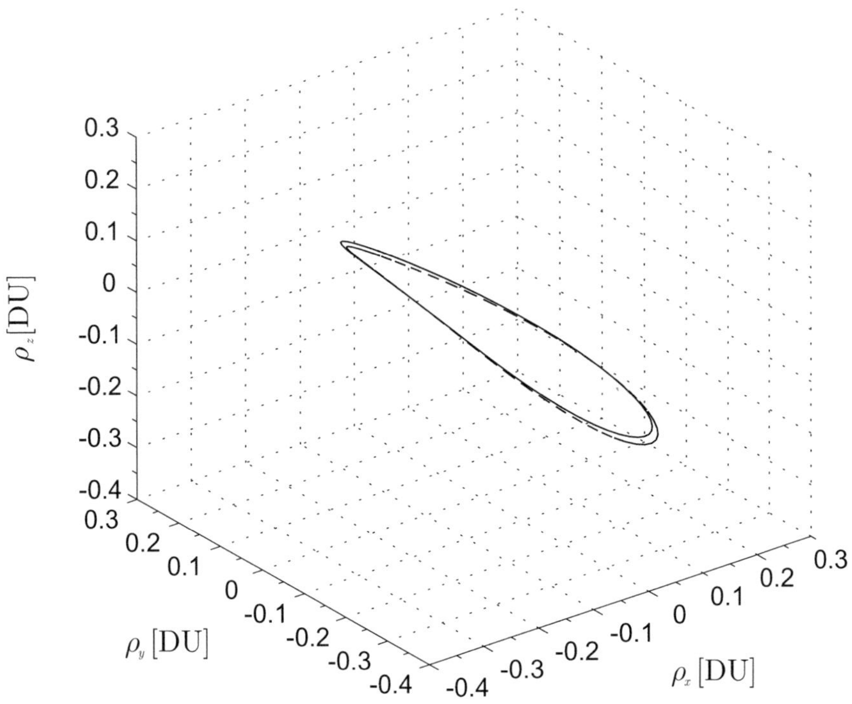

Fig. 10.5 Relative motion: comparison of the first-order approximation and the exact solution

Figure 10.7 illustrates the relative distance errors by varying the chief's eccentricity in a range $e_C \in [0,\ 0.5]$, whereas the deputy's eccentricity is assumed the same as that of Mercury, i.e. $e_D = 0.2056$. Note, however, that even though the approximate solution is able to get the relative distance bounds with a good accuracy, the approximate solution gradually becomes unable to track the relative orbit as e_C approaches 0.1, see Fig. 10.8.

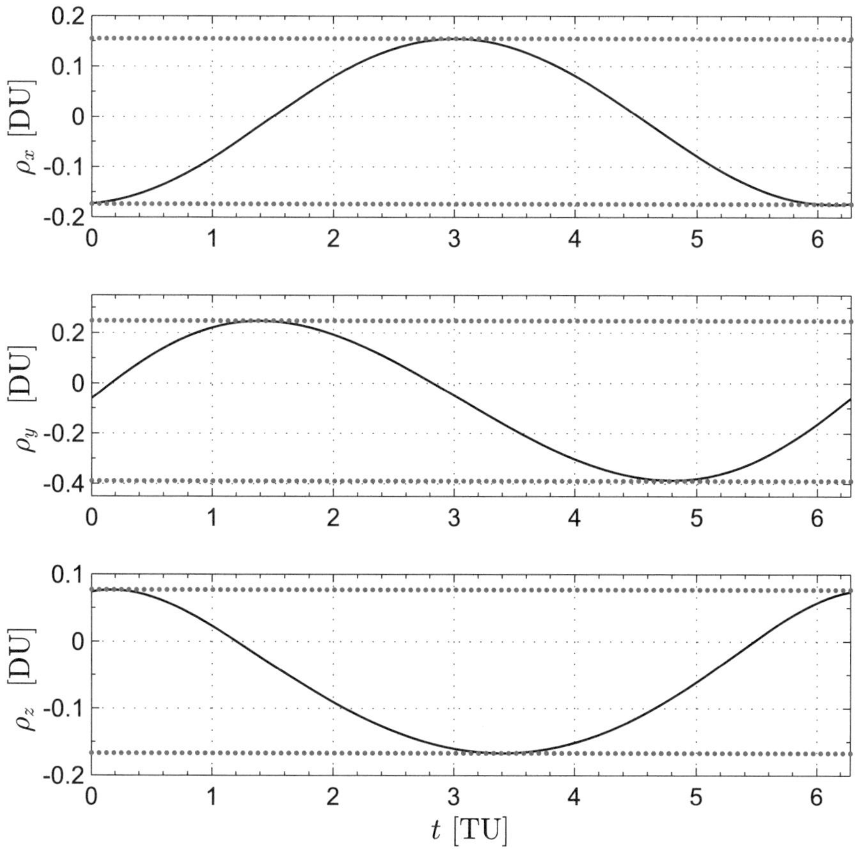

Fig. 10.6 Approximate semi-analytic bounds and simulated directional relative motion for the 1 : 1 commensurable case

10.6 Conclusions

The relative motion between elliptic displaced orbits has been parameterized via a set of newly defined orbital elements, and a semi-analytical approach for evaluating the relative distance has been discussed. In this context, the relative motion evolves along its invariant manifold and exhibits periodicity (quasi-periodicity) for the commensurable (incommensurable) case.

For the incommensurable case, by conducting the partial derivatives of the directional components with respect to both the chief's true anomaly and deputy's eccentric anomaly, the extreme values of relative distance are found as the real roots of a set of algebraic equations. For the 1 : 1 commensurable case, a Fourier series expansion is adopted to obtain a first-order time-explicit approximate solution, which is utilized to seek the bounds by differentiating the directional components of the position vector with respect to time. In this case, the extreme values for the radial and

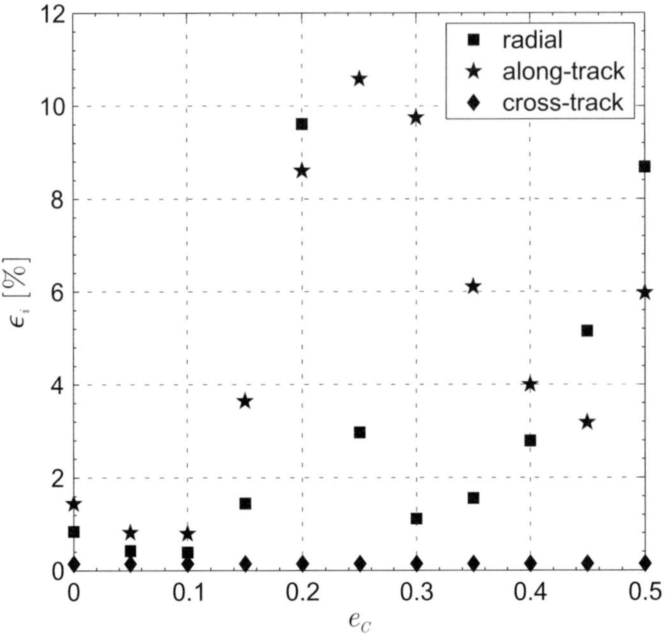

Fig. 10.7 Maximum relative distance errors ϵ_i along the three directions, see Eq. (10.75)

along-track motion are the real roots of a sextic equation, whereas in the cross-track motion, they satisfy a quartic equation.

The illustrative examples have shown that the discussed approach is capable of successfully predicting the inner and outer bounds of relative motion between elliptic displaced orbits and, when compared to the pure numerical integration (which is time consuming and may give rise to time delays in inner communication), this method is more efficient in computation. Therefore, the semi-analytical calculation of these bounds is essential for onboard control when the relative motion arrives at the pre-designed boundary.

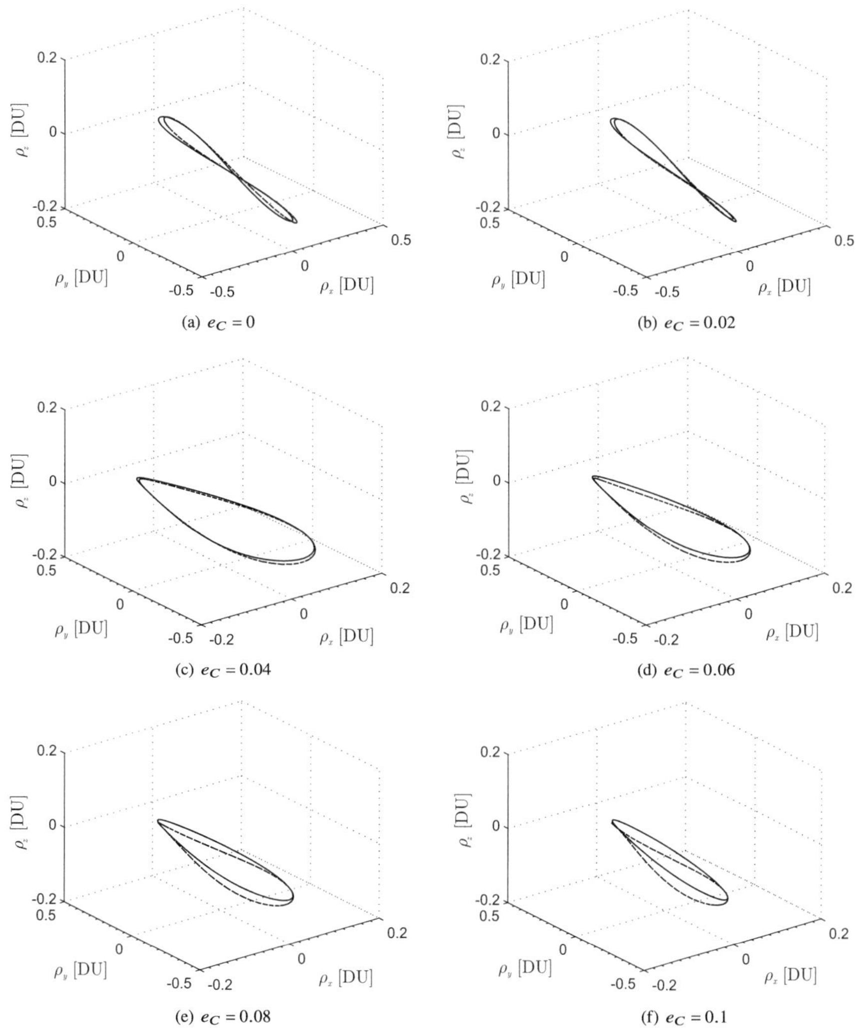

Fig. 10.8 Three dimensional relative motion as a function of chief's orbital eccentricity e_C

Appendix A

The entries T_{ij} of matrix $\mathbb{T}_{P_D P_C}$ in Eqs. (10.10)–(10.12) are

$$\begin{aligned} T_{11} = &\left(\cos\omega_C\, \cos\omega_D + \sin\omega_C\, \sin\omega_D\, \cos i_C\, \cos i_D\right) \cos\left(\Omega_C - \Omega_D\right) \\ &+ \left(\cos\omega_C\, \sin\omega_D\, \cos i_D - \sin\omega_C\, \cos\omega_D\, \cos i_C\right) \sin\left(\Omega_C - \Omega_D\right) + \sin\omega_C\, \sin\omega_D\, \sin i_C\, \sin i_D \end{aligned} \tag{A-1}$$

$$\begin{aligned} T_{12} = &\left(\sin\omega_C\, \cos\omega_D\, \cos i_C\, \cos i_D - \cos\omega_C\, \sin\omega_D\right) \cos\left(\Omega_C - \Omega_D\right) \\ &+ \left(\sin\omega_C\, \sin\omega_D\, \cos i_C + \cos\omega_C\, \cos\omega_D\, \cos i_D\right) \sin\left(\Omega_C - \Omega_D\right) + \sin\omega_C\, \cos\omega_D\, \sin i_C\, \sin i_D \end{aligned} \tag{A-2}$$

$$T_{13} = \cos\omega_C \sin i_D \sin(\Omega_D - \Omega_C) - \sin\omega_C \cos i_C \sin i_D \cos(\Omega_C - \Omega_D) + \sin\omega_C \sin i_C \cos i_D \quad \text{(A-3)}$$

$$T_{21} = (\sin\omega_C \sin\omega_D \cos i_D + \cos\omega_C \cos\omega_D \cos i_C)\sin(\Omega_D - \Omega_C) + (\cos\omega_C \sin\omega_D \cos i_C \cos i_D - \sin\omega_C \cos\omega_D)\cos(\Omega_C - \Omega_D) + \cos\omega_C \sin\omega_D \sin i_C \sin i_D \quad \text{(A-4)}$$

$$T_{22} = (\sin\omega_C \sin\omega_D + \cos\omega_C \cos\omega_D \cos i_C \cos i_D)\cos(\Omega_C - \Omega_D) + (\sin\omega_C \cos\omega_D \cos i_D - \cos\omega_C \sin\omega_D \cos i_C)\sin(\Omega_D - \Omega_C) + \cos\omega_C \cos\omega_D \sin i_C \sin i_D \quad \text{(A-5)}$$

$$T_{23} = \sin\omega_C \sin i_D \sin(\Omega_C - \Omega_D) - \cos\omega_C \cos i_C \sin i_D \cos(\Omega_C - \Omega_D) + \cos\omega_C \sin i_C \cos i_D \quad \text{(A-6)}$$

$$T_{31} = \cos\omega_D \sin i_C \sin(\Omega_C - \Omega_D) + \sin\omega_D \sin i_C \cos i_D \cos(\Omega_C - \Omega_D) + \sin\omega_D \cos i_C \sin i_D \quad \text{(A-7)}$$

$$T_{32} = \sin\omega_D \sin i_C \sin(\Omega_D - \Omega_C) - \cos\omega_D \sin i_C \cos i_D \cos(\Omega_C - \Omega_D) + \cos\omega_D \cos i_C \sin i_D \quad \text{(A-8)}$$

$$T_{33} = \sin i_C \sin i_D \cos(\Omega_C - \Omega_D) + \cos i_C \cos i_D \quad \text{(A-9)}$$

Appendix B

The coefficients in Eqs. (10.44)–(10.46) are

$$\sigma_0 = -H_D T_{13} e_C - a_C + \frac{a_D T_{11}}{2} + \frac{b_D T_{22}}{2} \quad \text{(B-1)}$$

$$\sigma_c = -\frac{5}{4} a_D T_{11} e_D + H_D T_{13} + a_C e_C - \frac{1}{2} a_D T_{11} e_C + \frac{1}{4} b_D T_{22} e_D + \frac{1}{2} b_D T_{22} e_C \quad \text{(B-2)}$$

$$\sigma_s = -\frac{7}{4} a_D T_{21} e_D + H_D T_{23} - \frac{3}{2} b_D T_{12} e_C + \frac{1}{4} b_D T_{12} e_D + \frac{1}{2} a_D T_{21} e_C \quad \text{(B-3)}$$

$$\sigma_{c2} = H_D T_{13} e_C + \frac{1}{2} a_D T_{11} - \frac{1}{2} b_D T_{22} \quad \text{(B-4)}$$

$$\sigma_{s2} = H_D T_{23} e_C + \frac{1}{2} b_D T_{12} + \frac{1}{2} a_D T_{21} \quad \text{(B-5)}$$

$$\sigma_{c3} = \frac{1}{4} a_D T_{11} e_D + \frac{1}{2} a_D T_{11} e_C - \frac{1}{4} b_D T_{22} e_D - \frac{1}{2} b_D T_{22} e_C \quad \text{(B-6)}$$

$$\sigma_{s3} = \frac{1}{4} b_D T_{12} e_D + \frac{1}{2} a_D T_{21} e_C + \frac{1}{4} a_D T_{21} e_D + \frac{1}{2} b_D T_{12} e_C \quad \text{(B-7)}$$

$$\tau_0 = -H_D T_{23} e_C + \frac{1}{2} a_D T_{21} - \frac{1}{2} b_D T_{12} \quad \text{(B-8)}$$

$$\tau_c = -\frac{5}{4} a_D T_{21} e_D + H_D T_{23} - \frac{1}{2} a_D T_{21} e_C - \frac{1}{4} b_D T_{12} e_D - \frac{1}{2} b_D T_{12} e_C \quad \text{(B-9)}$$

$$\tau_s = \frac{7}{4} a_D T_{11} e_D - H_D T_{13} - \frac{3}{2} b_D T_{22} e_C + \frac{1}{4} b_D T_{22} e_D - \frac{1}{2} a_D T_{11} e_C \quad \text{(B-10)}$$

$$\tau_{c2} = H_D T_{23} e_C + \frac{1}{2} a_D T_{21} + \frac{1}{2} b_D T_{12} \quad \text{(B-11)}$$

$$\tau_{s2} = -H_D T_{13} e_C + \frac{1}{2} b_D T_{22} - \frac{1}{2} a_D T_{11} \quad \text{(B-12)}$$

$$\tau_{c3} = \frac{1}{4} a_D T_{21} e_D + \frac{1}{2} a_D T_{21} e_C + \frac{1}{4} b_D T_{12} e_D + \frac{1}{2} b_D T_{12} e_C \quad \text{(B-13)}$$

$$\tau_{s3} = \frac{1}{4} b_D T_{22} e_D - \frac{1}{2} a_D T_{11} e_C - \frac{1}{4} a_D T_{11} e_D + \frac{1}{2} b_D T_{22} e_C \tag{B-14}$$

$$\xi_0 = -\frac{3}{2} a_D T_{31} e_D + H_D T_{33} - H_C \tag{B-15}$$

$$\xi_c = a_D T_{31} \tag{B-16}$$

$$\xi_s = b_D T_{32} \tag{B-17}$$

$$\xi_{c2} = \frac{1}{2} a_D T_{31} e_D \tag{B-18}$$

$$\xi_{s2} = \frac{1}{2} b_D T_{32} e_D \tag{B-19}$$

References

1. Matloff G L. Graphene: the ultimate interstellar solar sail material? [J]. Journal of the British Interplanetary Society, 2012, 65: 378–381.
2. Tsuda Y, Mori O, Funase R, et al. Flight status of IKAROS deep space solar sail demonstrator [J]. Acta Astronautica, 2011, 69 (9–10): 833–840.
3. Tsuda Y, Mori O, Funase R, et al. Achievement of IKAROS–Japanese deep space solar sail demonstration mission [J]. Acta Astronautica, 2013, 82 (2): 183–188.
4. Johnson L, Whorton M, Heaton A, et al. NanoSail-D: A solar sail demonstration mission [J]. Acta Astronautica, 2011, 68 (5–6): 571–575.
5. Johnson L, Young R, Montgomery E, et al. Status of solar sail technology within NASA [J]. Advances in Space Research, 2011, 48 (11): 1687–1694.
6. McKay R, Macdonald M, Biggs J, et al. Survey of highly non-Keplerian orbits with low-thrust propulsion [J]. Journal of Guidance, Control, and Dynamics, 2011, 34 (3): 645–666.
7. Mengali G, Quarta A A. Non-Keplerian orbits for electric sails [J]. Celestial Mechanics and Dynamical Astronomy, 2009, 15 (1–3): 179–195.
8. McInnes C R. Solar sail parking in restricted three-body systems [J]. Journal of Guidance, Control, and Dynamics, 1994, 17 (2): 399–406.
9. Baoyin H, McInnes C R. Solar sail orbits at artificial Sun-Earth libration points [J]. Journal of Guidance, Control, and Dynamics, 2005, 28 (6): 1328–1331.
10. Baoyin H, McInnes C R. Solar sail halo orbits at the Sun-Earth artificial L_1 point [J]. Celestial Mechanics and Dynamical Astronomy, 2006, 94 (2): 155–171.
11. Baoyin H, McInnes C R. Solar sail equilibria in the elliptical restricted three-body problem [J]. Journal of Guidance, Control, and Dynamics, 2006, 29 (3): 538–543.
12. McInnes C R. The existence and stability of families of displaced two-body orbits [J]. Celestial Mechanics and Dynamical Astronomy, 1997, 67 (2): 167–180.
13. McInnes C R, Simmons J F L. Solar sail halo orbits I: heliocentric case [J]. Journal of Spacecraft and Rockets, 1992, 29 (4): 466–471.
14. McInnes C R, Simmons J F L. Solar sail halo orbits II: geocentric case [J]. Journal of Spacecraft and Rockets, 1992, 29 (4): 472–479.
15. Ceriotti M, Heiligers J, McInnes C R. Trajectory and spacecraft design for a pole-sitter mission [J]. Journal of Spacecraft and Rockets, 2014, 51 (1): 311–326.
16. McInnes C R. Passive control of displaced solar sail orbits [J]. Journal of Guidance, Control, and Dynamics, 1998, 21 (6): 975–982.
17. Gong S, Li J, Baoyin H. Analysis of displaced solar sail orbits with passive control [J]. Journal of Guidance, Control, and Dynamics, 2008, 31 (3): 782–785.

18. Janhunen P. Electric sail for spacecraft propulsion [J]. Journal of Propulsion and Power, 2004, 20 (4): 763–764.
19. McInnes C R. Solar Sailing: Technology, Dynamics and Mission Applications [M]. Springer-Praxis Series in Space Science and Technology, Springer-Verlag, London, Berlin, 1999.
20. Gong S, Li J. Solar sail halo orbit control using reflectivity control devices [J]. Transactions of the Japan Society for Aeronautical and Space Sciences, 2014, 57 (5): 279–288.
21. Mu J, Gong S, Li J. Coupled control of reflectivity modulated solar sail for Geosail formation flying [J]. Journal of Guidance, Control, and Dynamics, 2015, 38 (4): 740–751.
22. Mu J, Gong S, Ma P, et al. Dynamics and control of flexible spinning solar sails under reflectivity modulation [J]. Advances in Space Research, 2015, 56 (8): 1737–1751.
23. Gong S, Baoyin H, Li J. Solar sail formation flying around displaced solar orbits [J]. Journal of Guidance, Control, and Dynamics, 2007, 30 (4): 1147–1151.
24. Gong S, Gao Y, Li J. Solar sail formation flying on an inclined Earth orbit [J]. Acta Astronautica, 2011, 68 (1–2): 226–239.
25. Gurfil P, Kholshevnikov K V. Manifolds and metrics in the relative spacecraft motion problem [J]. Journal of Guidance, Control, and Dynamics, 2006, 29 (4): 1004–1010.
26. Wang W, Yuan J, Mengali G, et al. Invariant manifold and bounds of relative motion between heliocentric displaced orbits [J]. Journal of Guidance, Control, and Dynamics, 2016, 39 (8): 1764–1776.
27. Topputo F. Fast numerical approximation of invariant manifolds in the circular restricted three-body problem [J]. Journal of Guidance, Control, and Dynamics, 2016, 32: 89–98.
28. Mingotti G, Topputo F, Bernelli-Zazzera F. Transfers to distant periodic orbits around the Moon via their invariant manifolds [J]. Acta Astronautica, 2012, 79: 20–32.
29. Battin R H. An Introduction to the Mathematics and Methods of Astrodynamics [M]. AIAA Education Series, New York, 1987.

MIX
Papier aus verantwortungsvollen Quellen
Paper from responsible sources
FSC® C105338

If you have any concerns about our products,
you can contact us on
ProductSafety@springernature.com

In case Publisher is established outside the EU,
the EU authorized representative is:
Springer Nature Customer Service Center GmbH
Europaplatz 3, 69115 Heidelberg, Germany

Printed by Libri Plureos GmbH
in Hamburg, Germany